ENVIRONMENTAL BIOLOGY

ENVIRONMENTAL BIOLOGY

Dr. Atul Kumar Tiwari

Delve
PUBLISHING

www.delvepublishing.com

Environmental Biology

Dr. Atul Kumar Tiwari

Delve Publishing
4164 Lakeshore Road
Burlington, ON L7L 1A4
Canada
www.delvepublishing.com
Email: orders@arclereducation.com

ISBN: 978-1-77956-480-1 (Hardcover)

ABOUT THE AUTHOR

Dr. Atul Kumar Tiwari, an eminent academic and researcher, is currently serving as the Associate Professor and Head of the School of Environmental Biology at A.P.S. University, Rewa, Madhya Pradesh. With over two decades of teaching experience, Dr. Tiwari has held esteemed positions at various institutions, including Government Science College, Ambikapur, and Dr. Bhanvar Singh Porte Government College, Pendra. His academic credentials include a Doctor of Philosophy in Zoology and a Master of Science from Awadhesh Pratap Singh University, Rewa. Throughout his illustrious career, he has been actively involved in administrative roles, such as Registrar at Kushabhau Thakre Patrakarita Avam Jansanchar Vishavidyalaya, and Chairman of the Board of Studies in Environmental Biology at A.P.S. University.

Dr. Tiwari has an extensive portfolio of scholarly contributions, including numerous research publications on topics ranging from environmental biology to molecular biomarkers and ecotourism. He is a recipient of several awards, including the Lal Bahadur Shastri Global Award for Biodiversity and the Rajya Puraskar in Scouting. A prolific author and editor, he has contributed to significant works on cancer diagnosis, antimicrobial resistance, and environmental issues, alongside publishing books such as The Catfish: M. seenghala. His professional affiliations span Ph.D. supervision and membership on boards and committees across various universities, underscoring his expertise and influence in the field of environmental biology and allied sciences.

TABLE OF CONTENTS

LIST OF FIGURES

PREFACE

The urgency of addressing environmental issues has never been more pressing. As the global population continues to grow and technological advancements reshape our world, we are faced with unprecedented challenges in sustaining the delicate balance of nature that supports life on Earth.

"*Environmental Biology*" is designed with the intent to explore the interconnectedness of living organisms, their habitats, and the delicate systems that sustain life. The book serves as an essential resource for students, researchers, and anyone interested in understanding the scientific principles underlying environmental issues and their implications for the future.

Our planet is currently experiencing profound changes, many of which are driven by human activities. From the depletion of biodiversity to the escalating crisis of climate change, pollution, and resource depletion, these environmental challenges are at the heart of the biological sciences today. The relationship between humans and the environment is more complex than ever, and understanding these dynamics is essential to shaping policies, behaviors, and scientific innovations that will guide us toward a more sustainable future.

This book explores the concepts of environmental biology through the lens of ecosystems, biodiversity, climate change, energy, and human impact on natural systems.

Chapter 1 provides an overview of ecosystems and biodiversity, their importance to human well-being, and the threats they face from climate change, pollution, and unsustainable development. The chapter also introduces renewable energy sources and environmental policies, highlighting the crucial role they play in mitigating climate change and promoting sustainability.

Chapter 2 deals with the fundamental concepts of matter and energy, focusing on their roles in biological systems. This chapter offers a deeper understanding of the interactions between matter and energy, particularly in ecosystems and cellular processes, and explores the implications of these interactions for the functioning of life on Earth.

Chapter 3 takes a closer look at the various types of ecosystems, including terrestrial, aquatic, and urban environments. It explores how energy flows through ecosystems and how nutrients cycle, as well as the importance of biodiversity in maintaining ecosystem health. This chapter also highlights the human impacts on ecosystems, such as pollution, deforestation, and climate change, and discusses the urgent need for conservation efforts.

In Chapter 4, we explore the dynamics of communities and populations in ecosystems. This chapter focuses on species interactions, community structure, and ecological niches, as well as the factors that influence population growth. It also includes case studies on specific ecosystems, such as coral reefs, forests, and grasslands, to illustrate the principles of community and population ecology in real-world settings.

Chapter 5 addresses the critical issue of biodiversity loss and the strategies that can be employed to conserve species and ecosystems. From the creation of protected areas to the importance of sustainable resource management, this chapter provides a comprehensive overview of the efforts to halt the loss of biodiversity and the role of legislation and international agreements in these efforts.

In Chapter 6, we examine the various environmental hazards that impact human health, such as air pollution, water contamination, and soil pollution. The chapter explores the links between environmental factors and diseases, including respiratory conditions, cardiovascular issues, and the increasing prevalence of vector-borne diseases due to climate change.

Chapter 7 focuses on the critical issue of water scarcity, one of the most pressing global challenges. This chapter discusses the sources of water, the global distribution of freshwater, and the challenges faced by regions suffering from water scarcity. It also highlights the importance of water conservation and sustainable water management practices in agriculture, industry, and domestic use.

Chapter 8 addresses the intersection of environmental biology and food security. This chapter provides an in-depth analysis of the causes of hunger, including economic, political, and environmental factors, and explores strategies to achieve global food security. It also examines the nutritional aspects of food and the role of technological advancements in increasing food production.

In Chapter 9, we explore the interconnectedness of air pollution, climate change, and the depletion of the ozone layer. This chapter highlights the major sources of air pollution, the health effects of polluted air, and the global efforts to mitigate these issues, including international treaties such as the Montreal Protocol and the Paris Agreement.

Finally, Chapter 10 provides a detailed examination of energy systems, comparing the environmental impacts of conventional energy sources such as fossil fuels and nuclear energy with those of sustainable energy sources like solar, wind, and geothermal. The chapter discusses the economic and environmental implications of transitioning to renewable energy and the technological innovations that are paving the way for a more sustainable energy future.

The importance of environmental biology cannot be overstated. As we move further into the 21st century, our ability to adapt to environmental challenges will determine the future of life on Earth. This book aims to provide the knowledge and tools necessary to understand these challenges, as well as the strategies needed to create a more sustainable world.

–Author

CHAPTER 1

ENVIRONMENTAL SCIENCE

Contents

INTRODUCTION TO ENVIRONMENTAL SCIENCE

Environmental Science serves as a multidimensional field that integrates various scientific disciplines to address pressing ecological issues. At its core, this discipline investigates the interaction between human activities and the natural world, emphasizing the complex interrelationships that define ecosystems. By employing methods from biology, chemistry, geology, and social sciences, environmental science provides a comprehensive framework for understanding phenomena such as climate change, biodiversity loss, and pollution. This holistic approach elucidates the mechanisms underlying environmental challenges and fosters an understanding of the societal implications of these issues, ultimately highlighting the necessity of sustainable practices for the well-being of both humanity and the planet. One of the primary goals of Environmental Science is to identify and propose solutions to the detrimental impacts of human behavior on the ecosystem. For example, the study of renewable energy sources aims to mitigate greenhouse gas emissions driven by fossil fuel dependency. Additionally, environmental scientists engage in the assessment of natural resources, advocating for strategies like conservation and resource management to promote long-term sustainability. By emphasizing a balance between ecological preservation and human economic needs, this discipline seeks to develop strategies that benefit both people and ecosystems. Ultimately, the contributions of Environmental Science are crucial in forming responsible policies that guide societal practices toward a more sustainable future. The education and awareness raised by Environmental Science are vital for fostering an informed public and engaged citizenry. Through various outreach programs, academic courses, and community initiatives, knowledge about environmental issues is disseminated widely, encouraging individuals to adopt eco-friendly practices in their daily lives. Additionally, environmental science empowers communities to advocate for legislation that protects their local environments and ensures equitable access to resources. As individuals become more educated about the complexities of ecological systems, they recognize their role in promoting sustainability. By cultivating a sense of responsibility and stewardship, Environmental Science informs policy and inspires collective action necessary for addressing the environmental challenges facing our world today.

1.1. DEFINITION AND SCOPE OF ENVIRONMENTAL SCIENCE

Environmental science encompasses a multidisciplinary framework aimed at understanding the intricate relationships between human activity and the natural environment. This field integrates principles from various scientific disciplines such as biology, chemistry, geology, and meteorology to analyze environmental systems and processes. The scope of environmental science extends beyond mere observation; it engages in critical evaluation and synthesis of environmental data to address pressing challenges such as pollution, climate change, and biodiversity loss. By applying scientific methodologies, environmental scientists seek to comprehend ecological dynamics and to develop sustainable solutions that balance ecological health and human needs. Within the expansive framework of environmental science, the concept of environmental security emerges as a crucial component. Defined variously across disciplines, environmental security emphasizes the importance of safeguarding ecosystems to ensure the continued survival of both humans and wildlife. As noted, public authorities are responsible for ensuring environmental security (Rosiek et al., 2015). This highlights the intersection between governance and ecological stewardship, wherein legal and policy frameworks are essential for mitigating negative environmental impacts. The broad implications of environmental security are manifest, influencing strategic planning in sectors like water management, where sustainable practices are increasingly imperative to balance resource use with ecological preservation. Furthermore, environmental science involves risk assessment as a critical tool for evaluating potential environmental hazards and their impacts on human health and ecosystems. The rigorous analysis of risks associated with various environmental factors, such as chemical exposure and ecological disruptions, provides a foundation for informed decision-making in policy and practice. A reevaluation of existing risk assessments illustrates this point, suggesting that professional frameworks can enhance our understanding of risk beyond traditional models (Brown et al., 1990). By integrating innovative risk assessment methods into environmental science, practitioners can better address complex environmental challenges, ultimately fostering an adaptive and resilient approach to securing a sustainable future for both society and the planet.

1.2. IMPORTANCE OF STUDYING ENVIRONMENTAL SCIENCE

A significant impact of studying environmental science is its ability to illuminate the intricate interactions between human activities and natural ecosystems. By examining these relationships, researchers, and students alike can better understand how urbanization and industrialization affect local and global environments. For instance, as demonstrated in various studies, the evolution of urban waters has been heavily influenced by anthropogenic factors such as drainage systems and sewage management strategies (Kaushal et al., 2015). These insights enable communities to anticipate future environmental challenges and implement more sustainable practices. The field highlights the importance of protecting ecosystems and underscores the need for informed decision-making that balances economic development with ecological preservation. An essential aspect of environmental science is its role in fostering awareness about the consequences of climate change and resource depletion. For example, the Millennium Assessment (2003) discusses how human activities have reshaped the planet, emphasizing the urgency of addressing these challenges. Integrated understanding from diverse disciplines within environmental science equips individuals and policymakers to devise strategies that mitigate adverse effects on ecosystems. This is particularly vital in an age where the threat of climate-change-induced natural disasters looms larger than ever. Studying environmental science invigorates the discourse on sustainability, prompting crucial discussions about consumption patterns, biodiversity loss, and the resilience of ecosystems undergoing unprecedented transformations. Moreover, environmental science serves as a cornerstone for developing innovative solutions to pressing global problems. As we confront issues such as food security, urban planning, and water quality management, the field merges scientific inquiry with practical applications. An illustration of this is the transformation of urban drainage systems, where researchers apply principles of hydrology and engineering to create more efficient and ecologically sensitive designs (Belt et al., 2015). By synthesizing scientific research with community needs, environmental science empowers individuals to advocate for systemic changes that promote sustainability and resilience. The interconnectedness of ecosystems and human well-being stresses the importance of education and collaboration, ultimately shaping a more sustainable future for generations to come.

1.3. INTERDISCIPLINARY NATURE OF ENVIRONMENTAL SCIENCE

The complex issues faced by contemporary environmental science require an integrated approach that draws on various disciplines, ranging from biology and chemistry to sociology and economics. This interdisciplinary perspective is crucial for understanding the multifaceted nature of environmental problems, as each discipline contributes unique insights and methodologies. For instance, the intersection of ecological studies with social sciences illuminates how human behaviors impact natural ecosystems. By employing a mix of qualitative and quantitative methods, researchers can gather a holistic understanding of both environmental dynamics and human interactions with them. This aligns with the findings that many conservation scientists use mixed methods, reinforcing the notion that one-dimensional approaches are inadequate for comprehensive environmental assessment (Kinnebrew et al., 2020). A key aspect of the interdisciplinary nature of environmental science is its emphasis on collaboration among experts from diverse fields. This collaboration is essential for developing effective solutions to pressing issues such as climate change, resource depletion, and habitat destruction. For example, incorporating perspectives from urban planning and public policy can inform better land management strategies that consider local communities' needs while preserving ecosystems. As indicated in the literature, the integration of various research methods—whether through informed nesting of qualitative data within quantitative frameworks or bidirectional synthesis—can foster a richer understanding of environmental phenomena (Kinnebrew et al., 2020). Consequently, interdisciplinary cooperation enhances the effectiveness of scientific research and fosters shared responsibility among stakeholders. Moreover, the educational approaches within environmental science increasingly reflect this interdisciplinary ethos. The rise of STEAM (Science, Technology, Engineering, Arts, and Mathematics) education exemplifies how holistic frameworks can facilitate a broader understanding of environmental sustainability. By encouraging students to engage with environmental issues through multiple lenses, such as engineering solutions and social impacts, educational programs can cultivate critical thinking and problem-solving skills essential for addressing complex environmental challenges. As articulated in recent studies, the interdisciplinary nature of environmental science is not simply academic; it is imperative for fostering innovative solutions to enhance environmental sustainability in a rapidly changing world. This shift suggests that nurturing an interdisciplinary approach in education is a foundational step toward preparing future leaders to tackle these critical global issues effectively.

1.4. ECOSYSTEMS AND BIODIVERSITY

A crucial component of environmental science is understanding ecosystems and their intricate web of interdependencies, which collectively sustain biodiversity. The loss of biodiversity poses significant threats to ecosystem stability and function. When species within an ecosystem diminish, the balance of food chains is disrupted, leading to unforeseen consequences such as invasive species proliferation and habitat degradation. Recent research emphasizes that human-induced stressors, such as climate change and pollution, have led to alarming decreases in biodiversity, affecting plant and animal populations and vital ecosystem services. For instance, effective biodiversity assessment and conservation management are indispensable for navigating these challenges, particularly in the Anthropocene era, where anthropogenic influences are most pronounced (Jin et al., 2021). We can better formulate strategies to preserve both ecosystems and the myriad species that inhabit them by enhancing our understanding of these connections. The implications of poor biodiversity management extend beyond the loss of species; they also encompass the degradation of ecosystems that provide essential services to humanity. Urban ecosystems, for example, are often overlooked in conservation efforts despite their critical role in supporting human populations and mitigating climate impacts. There is a pressing need to bridge the gap between scientific understanding and the practical implementation of conservation strategies, particularly in the context of urban land-use planning. As highlighted in recent assessments, municipalities like eThekwini in South Africa recognize this challenge and are working to incorporate scientific knowledge into local governance. By addressing the science-action gap, urban areas can promote biodiversity while simultaneously enhancing resilience against urbanization impacts (Boon et al., 2016).

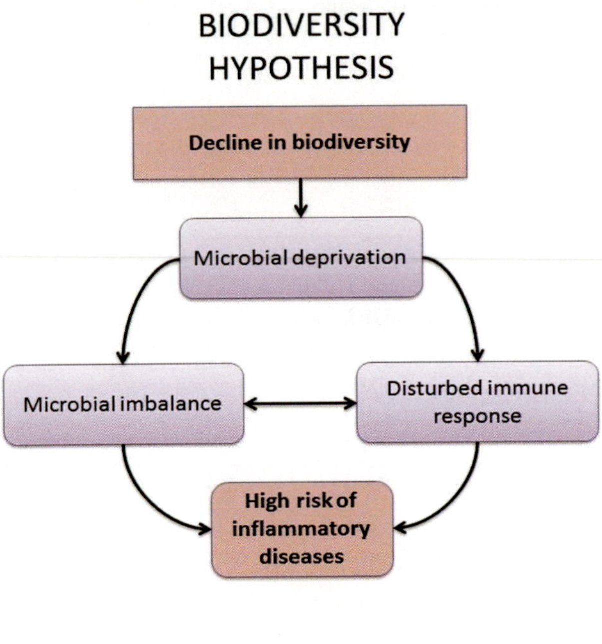

Figure 1.1. The importance of biodiversity.

Source: By Suvi V – Own work, CC BY-SA 4.0, https://commons.wikimedia.org/w/index.php?curid=66937550.

Effective environmentally-focused policies can thus enhance conservation efforts and ensure the ongoing health of urban ecosystems. Moving forward, the integration of environmental DNA (eDNA) technology into biodiversity assessment represents a transformative advancement in understanding ecosystems. This innovative approach allows researchers to detect and monitor species presence through genetic material recovered from various environments, such as soil and water, without the need for physical specimens. eDNA metabarcoding, in particular, has revolutionized biodiversity studies, especially in aquatic ecosystems where traditional surveying methods fall short (Jin et al., 2021). With the ability to identify common, endangered, and invasive species efficiently, eDNA techniques enable more informed decision-making for conservation management. As we strive to maintain biodiversity amidst mounting environmental challenges, embracing these cutting-edge methodologies will be vital in fostering sustainability and resilience across diverse ecosystems. This strategic embrace of technology can profoundly enhance our capacity to protect the environment and the rich tapestry of life it supports (Figure 1.1).

1.4.1. Types of Ecosystems and Their Characteristics

One of the most biodiverse ecosystems, tropical rainforests, is characterized by high humidity, dense flora, and remarkable species diversity. These ecosystems provide myriad services, including carbon sequestration, habitat for countless organisms, and regulation of local climates. In their intricate layers—from the emergent, towering trees to the understory teeming with life—tropical rainforests exemplify complex ecological interactions. Moreover, they play a crucial role in global oxygen production and water cycling, while also serving as reservoirs of medicinal resources. The degradation of these ecosystems due to deforestation and climate change threatens biodiversity and disrupts the vital services they provide, underscoring the need for sustainable management practices to preserve their integrity, as emphasized in the framework for assessments of ecosystem condition outlined in environmental reports (Lomas et al., 2017). In contrast, deserts are another type of ecosystem that illustrates the resilience of life in extreme conditions. Characterized by minimal precipitation, extreme temperature fluctuations, and sparse vegetation, deserts possess unique adaptations that enable survival, such as drought-resistant plants like cacti and animals with specialized behaviors for conserving moisture. These ecosystems are crucial for regulating local climates and maintaining biodiversity, despite their seemingly inhospitable nature. The interactions between organisms and their environment in deserts highlight the importance of ecological balance and adaptation. However, desert ecosystems are increasingly threatened by urban development and climate change, which can disrupt their delicate balance and lead to habitat loss. Understanding the conditions and characteristics of deserts helps inform strategies to mitigate human impact and promote conservation efforts, drawing on previous insights into ecosystem mapping (Lomas et al., 2017). Wetlands represent another vital and unique ecosystem, often described as the earths kidneys due to their ability to filter pollutants and manage water flow.

These ecosystems, which include swamps, marshes, and bogs, are characterized by water-saturated soils and a diverse array of plant and animal life that thrive in such conditions. Wetlands provide essential services, such as flood control, water purification, and habitats for migratory birds and aquatic species. Additionally, they play a pivotal role in carbon storage, which is increasingly recognized in environmental accounting practices. The interplay between nutrient cycling and habitat provision in wetlands demonstrates the interconnectedness of ecosystem services and the need for effective management strategies to combat threats like pollution and habitat destruction. By integrating knowledge of the diverse functions of wetlands, policymakers can better explore nature conservation strategies and the ecological integrity highlighted in research studies (Maes et al., 2016) (Figure 1.2).

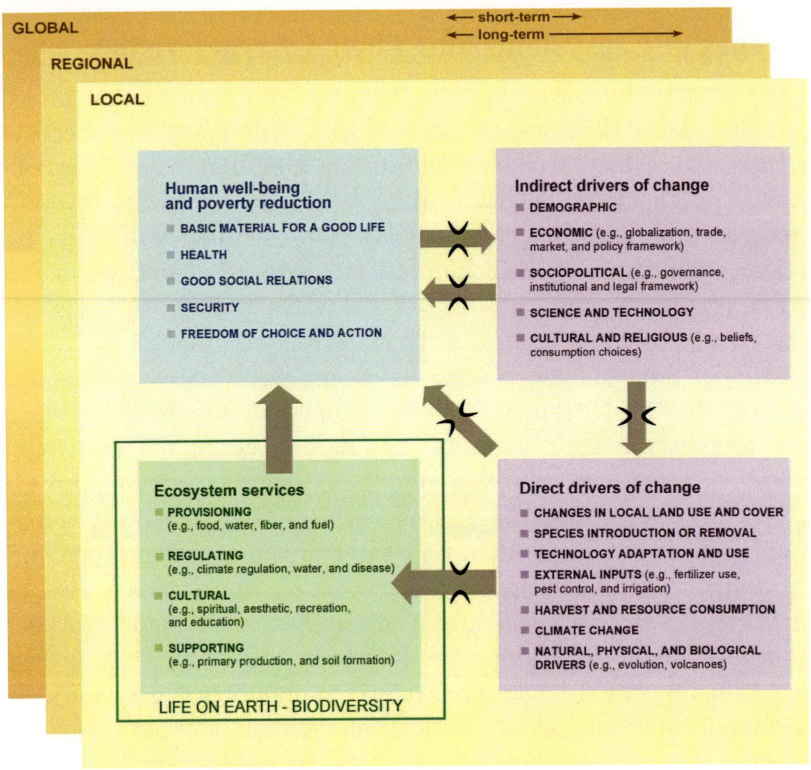

Figure 1.2. *Types of ecosystems.*

Source: By Millennium Ecosystem Assessment, 2005. Ecosystems and Human Well-being: Biodiversity Synthesis. World Resources Institute, Washington, DC. – Millennium Ecosystem Assessment, CC BY 3.0, https://commons.wikimedia.org/w/index.php?curid=5668602.

1.4.2. Importance of Biodiversity for Ecosystem Health

The intricate interplay among species in an ecosystem underscores the necessity of biodiversity for maintaining ecosystem health. Diverse biological communities enhance resilience to environmental changes, as multiple species can fulfill similar ecological roles. This redundancy ensures that if one species declines due to disease or climate change, others can take its place, thereby supporting essential ecosystem functions such as nutrient cycling, pollination, and carbon sequestration. Moreover, the genetic diversity within and among species contributes to their adaptability, enabling populations to respond to shifting environmental pressures. As emphasized in (Thomson et al., 2021), genetic diversity is crucial for evolutionary potential and resilience, factors that are increasingly threatened by human impacts and climate change. Protecting biodiversity is not merely an ethical mandate; it is a critical component for sustaining ecosystem services that humans rely on for survival and quality of life. In addition to ecological stability, biodiversity is a key driver of the socioeconomic benefits derived from natural ecosystems. Healthy ecosystems provide vital resources such as food, clean water, and medicine, all of which depend on the diverse interactions among species. The importance of conserving genetic diversity extends beyond ecological considerations; it directly affects coastal communities and marine industries reliant on healthy fish stocks and thriving ecosystems. By maintaining genetic diversity in marine environments, communities can better ensure food security and foster resilience against economic fluctuations.

According to Frediani et al (2024), the integration of traditional ecological knowledge and empirical science can enhance ecosystem management practices by recognizing the multifaceted relationships between humans and their environment. The sustainable management of biodiversity, therefore, serves to protect ecosystems and to safeguard human health and economic well-being. Furthermore, the decline of biodiversity poses severe threats to ecosystem health, with cascading effects that jeopardize both ecological and human systems. Disturbances such as habitat destruction, pollution, and climate change can precipitate species loss, leading to diminished ecosystem function and stability. A loss of biodiversity can compromise the ability of the ecosystem to support essential services such as water purification, soil fertility, and climate regulation, ultimately resulting in an environment less capable of supporting human populations. Institutions and policymakers must prioritize biodiversity conservation within broader environmental protection strategies to mitigate these impacts effectively; engaging in interdisciplinary collaboration between scientists, environmental managers, and communities is crucial. Such efforts can lead to sustainable practices that protect individual species and the interconnected web of life that supports all forms of existence (Frediani et al., 2024; Thomson et al., 2021) (Figure 1.3).

Figure 1.3. *The importance of biodiversity for human health.*

Source: By Kkmd at the English-language Wikipedia, CC BY-SA 3.0, https://commons.wikimedia.org/w/index.php?curid=3853162.

1.4.3. Threats to Biodiversity and Conservation Efforts

The interplay between human activities and environmental health presents a formidable challenge to the preservation of biodiversity. Increasingly, anthropogenic disturbances such as deforestation, industrialization, and urban expansion reshape ecosystems and contribute to habitat loss. As reported, the vast transformation in Sabahs forest coverage—an 85.25% reduction between 1975 and 1992—illustrates the alarming pace of deforestation driven by agricultural and economic pursuits (Chew et al., 2004). This loss of habitat not only threatens the flora and fauna reliant on these ecosystems but also disrupts essential ecological services. Such changes prompt a decline in plant and animal populations, leading to diminished genetic diversity and increased extinction rates, which further jeopardize the stability of global biodiversity. Conservation efforts are challenged by direct environmental degradation and by the complex socio-economic dynamics that fuel such harm. Competition for land between economic development and conservation creates a significant dilemma for policymakers and conservationists. As people prioritize material advancement, the slightest compromise often results in irreversible damage to vulnerable habitats. Solutions like land-use planning and sustainable practices are vital, yet they often fall short due to insufficient political will or public support. Moreover, the integration of traditional ecological knowledge with scientific research can enhance conservation strategies, allowing for more effective management of natural resources in a way that simultaneously respects local communities and enhances biodiversity protection efforts (Figure 1.4).

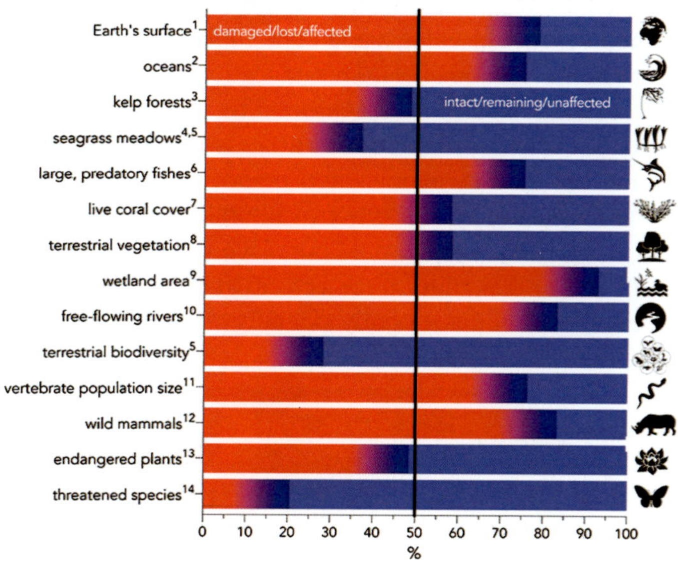

Figure 1.4. *Main threats to biodiversity.*

Source: https://en.wikipedia.org/wiki/File:Summary_of_major_environmental-change_categories_expressed_as_a_percentage_change_(red)_relative_to_base-line_-_fcosc-01-615419-g001.jpg.

The task of effectively addressing the threats to biodiversity necessitates a holistic and collaborative approach. Initiatives in places like the Lower Mekong region underscore the importance of strategic partnerships between various stakeholders, including NGOs, local communities, and international donors. Evaluations of conservation investments, as conducted by the MacArthur Foundation, reveal that thoughtful planning and shared objectives can significantly enhance the impact of conservation programs (Pilgrim et al., 2015). To achieve long-lasting success, it is crucial to engage diverse perspectives and foster cooperation across sectors. By understanding the intricate links between economic activity, habitat protection, and community well-being, we can develop sustainable practices that not only mitigate threats to biodiversity but also ensure ecological resilience for future generations.

1.5. CLIMATE CHANGE

Addressing the multifaceted impacts of climate change requires an interdisciplinary approach that integrates scientific knowledge, indigenous practices, and community action. As climate change influences weather patterns, agricultural productivity, and ecological balance, the need for comprehensive educational frameworks becomes imperative. Notably, climate change education is crucial in empowering individuals and communities to adapt to and mitigate the effects of environmental changes. This education should build upon existing environmental education initiatives, promoting environmental citizenship across all societal levels. By fostering awareness and understanding, climate change education can inform policies and practices that leverage indigenous knowledge systems, ensuring solutions are culturally relevant and contextually appropriate. Consequently, an informed populace is better equipped to engage in sustainable practices that address the current and anticipated environmental challenges facing communities globally (Ogbuigwe et al., 2009). (Figure 1.5).

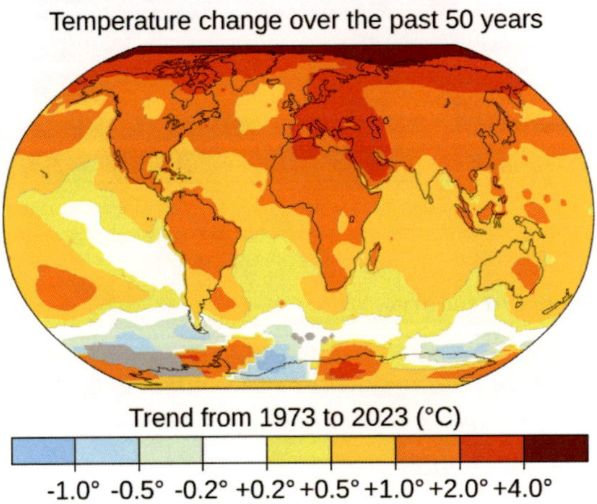

Figure 1.5. *Climate change.*

Source: By NASA's Scientific Visualization Studio, Key, and Title by uploader (Eric Fisk) – https://data.giss.nasa.gov/gistemp/maps/index_v4.html, Public Domain, https://commons.wikimedia.org/w/index.php?curid=86928646.

The interaction between climate change and biodiversity loss is particularly pronounced in coastal regions, where ecosystems are increasingly at risk. Communities in these areas face threats such as rising sea levels, habitat loss, and altered ecosystems, which endanger species and destabilize local economies. Shorebirds, for instance, heavily depend on specific coastal habitats for breeding and migration. As these areas undergo irreversible changes due to climate impacts, effective adaptation strategies become essential for protecting such vulnerable species. Nature-based solutions, including living shorelines and beach nourishment, emerge as key methods to combat these threats. By employing these sustainable ecosystem-based strategies, communities can enhance their resilience while simultaneously conserving biodiversity. There is a need for ongoing research to optimize these solutions and ensure they are socially, economically, and environmentally sustainable, ultimately contributing to the broader field of sustainability science (Gauthier et al., 2021). As the complexities of climate change continue to unfold, the necessity for robust, cross-disciplinary research and collaboration cannot be overstated. Understanding the interactions among different environmental factors is critical for developing effective climate adaptation and mitigation strategies. This requires the involvement of diverse fields, including biology, ecology, social sciences, and public policy. By synthesizing knowledge from various disciplines, researchers can identify innovative solutions that address specific regional challenges while considering the socio-economic dimensions of climate actions. Moreover, this collaborative approach fosters a more comprehensive understanding of the impacts of climate change and underscores the interconnectedness of human and environmental systems. Emphasizing a nuanced and inclusive research agenda will ensure that the responses to climate change are effective, equitable, and sustainable, ultimately supporting the well-being of both people and the planet (Ogbuigwe et al., 2009).

1.5.1. Causes of Climate Change

The interplay between human activities and climate drivers is a fundamental factor contributing to climate change. The extensive burning of fossil fuels, such as coal, oil, and natural gas, releases significant quantities of carbon dioxide and other greenhouse gases into the atmosphere. These emissions trap heat, leading to an increase in global temperatures. This phenomenon is primarily associated with industrial processes, transportation, and electricity generation, all of which heavily rely on fossil fuels. The consequences of these human interventions are dire, as they induce alterations in weather patterns and increase the frequency of extreme weather events. Moreover, activities like deforestation and land-use changes further exacerbate the situation by diminishing the planet's capacity to absorb carbon dioxide, highlighting the critical role of anthropogenic actions in driving climate change. Natural factors also contribute to climate change, although their impact is often overshadowed by human influence. Variations in solar radiation and volcanic activity can lead to significant climatic shifts over geological time scales. For instance, volcanic eruptions release aerosols and gases that can temporarily cool the Earth by reflecting sunlight back into space. However, these effects are generally short-lived compared to the long-term warming driven by human actions. As noted, the greenhouse effect is a foundational cause of global warming, which initiates a cascade of climatic phenomena, including disruptions in weather systems and alterations in ecosystems (Meilinda et al., 2017).

This understanding underscores the complexity of climate change, as it is the product of human endeavors and influenced by natural processes interacting with anthropogenic factors. Awareness of the causes of climate change is crucial for developing effective mitigation strategies. Education plays a pivotal role in fostering understanding and changing attitudes towards environmental sustainability. Research indicates that many misconceptions about climate change stem from media representations rather than factual scientific evidence (Meilinda et al., 2017). Scientists, educators, and policymakers need to collaborate in disseminating accurate information, emphasizing that climate change encompasses both warming and cooling trends. Addressing the issue requires a multifaceted approach that combines scientific knowledge with community engagement and robust policy measures to reduce emissions and enhance resilience. Ultimately, understanding the causes of climate change, both human-induced and natural, paves the way for effective actions and inspires collective responsibility toward preserving the environment for future generations (Figure 1.6).

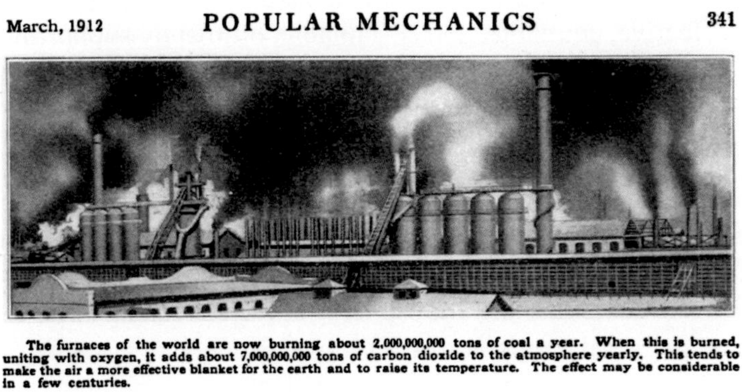
March, 1912 POPULAR MECHANICS 341

The furnaces of the world are now burning about 2,000,000,000 tons of coal a year. When this is burned, uniting with oxygen, it adds about 7,000,000,000 tons of carbon dioxide to the atmosphere yearly. This tends to make the air a more effective blanket for the earth and to raise its temperature. The effect may be considerable in a few centuries.

Figure 1.6. *Cause of climate change.*

Source: By Writers for Popular Mechanics magazine, March 1912. – Popular Mechanics magazine, March 1912, page 341. Image from Google Books., Public Domain, https://commons.wikimedia.org/w/index.php?curid=109852217.

1.5.2. Impacts of Climate Change on the Environment

The effects of climate change are profoundly reshaping marine environments, affecting both the ecological balance and livelihoods dependent on these ecosystems. In regions like the German North Sea, stakeholders from sectors such as fisheries and marine renewable energy are observing significant alterations in their working conditions and resources. As Benkens et al (2020) highlight, participants from various marine sectors noted perceivable impacts from climate change, which have necessitated adaptations in their professional practices. This illustrates that climate change is not only an environmental issue but also a socioeconomic one, pushing communities to reassess their strategies for sustainability. With changing marine conditions, the potential for conflict between sectors increases, as new uses of the ocean space clash with traditional practices, emphasizing the urgent need for collaborative governance and effective policymaking that can accommodate these emerging challenges. Alongside marine ecosystems, terrestrial environments are equally vulnerable to the far-reaching consequences of climate change. Rising temperatures and shifting precipitation patterns disrupt established ecological processes, leading to shifts in species distribution, altered life cycles, and even extinctions.

Figure 1.7. *Climate change impacts.*

Source: https://www.noaa.gov/education/resource-collections/climate/climate-change-impacts.

The degradation of habitats is further compounded by human activities, which exacerbate the stresses on biodiversity. For instance, intensifying agricultural practices and urban development diminish natural spaces, making species more susceptible to the impacts of climate change. The psychological toll on communities aware of these shifts also cannot be overlooked. According to Bostwick et al (2021), students studying environmental science studies report significant anxiety regarding climate impacts, illustrating the deep intertwining of ecological degradation with mental health challenges in affected populations. This showcases that the repercussions of climate change extend beyond physical environments, permeating the social fabric of communities as they navigate this increasingly altered landscape (Figure 1.7).

The urgency to address climate change necessitates immediate action and long-term strategic planning that integrates ecological, social, and economic perspectives. Enhanced understanding of climate impacts, as highlighted by the stakeholders in the North Sea region, fosters awareness and drives initiatives for adaptation and resilience. Policies must account for the complex interplay between climate change and human activity, ensuring that sustainable practices are prioritized across sectors. Moreover, fostering collaborative networks among scientists, policymakers, and community members can lead to more robust solutions that are adept at adjusting to the dynamic challenges posed by climate change. As communities grapple with the emotional effects of environmental degradation, educational institutions play a critical role in equipping the next generation with the knowledge and tools necessary to confront these challenges head-on. Recognizing that climate change is an interdisciplinary concern will be instrumental in shaping a more resilient future for both the environment and society.

1.5.3. Mitigation Strategies to Combat Climate Change

One effective approach to combat climate change involves adopting sustainable land management practices that enhance carbon sequestration while minimizing emissions. Transitioning to methods such as agroforestry, no-till agriculture, and crop rotation can optimize land use for both agricultural production and environmental conservation. These strategies not only encourage biodiversity but also contribute to achieving the goals of global environmental conventions. According to Cowie et al.), land management techniques that reduce greenhouse gas emissions can simultaneously benefit biodiversity and combat desertification. Policymakers and stakeholders need to recognize the intricate relationships between land use, food security, and ecological health. By implementing these synergistic strategies, we can develop stronger resilience against climate change while fostering sustainable development for diverse ecosystems. In addition to sustainable land practices, enhancing energy efficiency is crucial to mitigate climate change. Industries, residential sectors, and transportation systems can adopt advanced technologies and energy-efficient systems that reduce overall energy consumption and reliance on fossil fuels. For instance, retrofitting buildings with improved insulation, energy-efficient appliances, and renewable energy sources like solar or wind power can significantly decrease carbon footprints. The imperative lies in shifting toward a circular economy where waste is minimized and resources are reused, further curbing emissions. Moreover, public awareness and policy support for energy-efficient interventions can drive significant change (Figure 1.8).

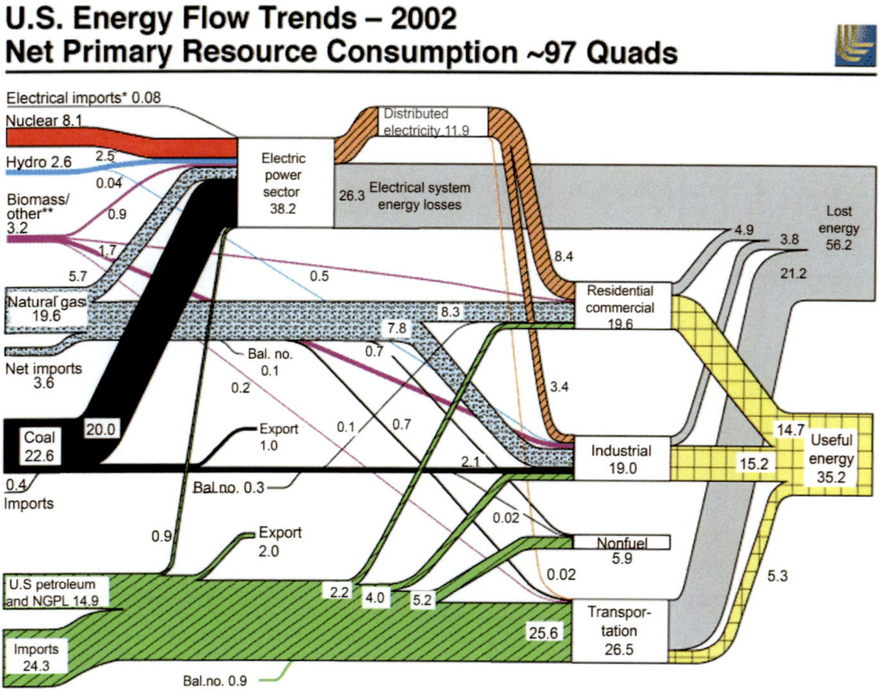

Figure 1.8. *Major radiative forcing geoengineering technologies that aim to alter the earth's radiative energy budget to stabilize or reduce global temperatures. These technologies include stratospheric aerosol injection, marine sky brightening, cirrus cloud thinning, space-based mirrors and surface-based brightening.*

Source: By Energy and Environment Directorate, Lawrence Livermore National Laboratory – http://eed.llnl.gov/flow/02flow.php, Public Domain, https://commons.wikimedia.org/w/index.php?curid=1782286.

Encouraging collaboration between various sectors is vital to foster innovative solutions that align with climate goals, ensuring that energy efficiency becomes a standard practice across all levels of society. Lastly, robust climate adaptation policies must accompany mitigation strategies to address the inevitable impacts of climate change. As highlighted in (Aggarwal et al., 2010), vulnerability assessments that take into account climatic variables and socio-economic factors are essential for developing targeted adaptation measures. By recognizing the multifaceted nature of climate-related risks, policymakers can create frameworks that support communities in enhancing their resilience to unpredictable weather patterns and food security challenges. This dual approach of mitigation and adaptation ensures that we work to reduce greenhouse gas emissions and prepare societies to cope with the effects of climate change. Engaging local communities and fostering public-private partnerships can elevate these strategies, leading to sustainable solutions tailored to the unique characteristics of different regions, ultimately contributing to a holistic response to climate challenges.

1.6. POLLUTION AND WASTE MANAGEMENT

One significant aspect of pollution and waste management is the relationship between environmental consciousness and the role of community resources such as libraries. Libraries, traditionally centers of knowledge and information, have the potential to foster pro-environmental behaviors (PEBs) among their patrons. According to a study conducted in Lagos, respondents from both academic and public libraries expressed a consensus on the inadequacy of information regarding PEBs, highlighting a gap in environmental education (Alabi, 2020). By leveraging their resources to provide workshops, informational materials, and programs focused on sustainability, libraries can substantially influence community attitudes toward waste management and pollution reduction. This integrated approach enriches the library's role within the community and empowers individuals to take proactive steps in addressing environmental challenges.

Furthermore, corporate practices play an equally crucial role in pollution prevention and waste management efforts. The Pollution Prevention Act of 1990 established a framework encouraging manufacturers to innovate in reducing toxic waste output. A study analyzing the Toxics Release Inventory (TRI) data highlighted that organizations that actively involve employees in pollution prevention strategies experience significantly greater reductions in toxic emissions when compared to those that do not adopt such practices (Bunge et al., 1995). This underscores the importance of employee engagement as a driving force in corporate environmental responsibility. By creating a culture that prioritizes sustainability and encourages employee input on waste management processes, manufacturers can enhance their operational efficiency while simultaneously contributing to broader environmental goals. Lastly, addressing pollution and waste management effectively requires a multifaceted approach that includes both educational initiatives and corporate accountability. Environmental awareness campaigns that target individuals and businesses alike can foster a collective commitment to sustainable practices. Public institutions, like libraries, and private sectors, through employee-focused strategies, can collaborate to amplify their impacts. The imperative for action is clear; as the study in Lagos indicates, insufficient information can hinder progress (Alabi, 2020), while the corporate environment reveals that participation enhances efficacy in waste reduction (Bunge et al., 1995). By intertwining community resources with active corporate responsibility, a more informed and engaged society can emerge, significantly mitigating the detrimental effects of pollution and waste on our environment (Figure 1.9).

Figure 1.9. *Sustainable waste management.*

Source: By GST HBK – Own work, Public Domain, https://commons.wikimedia.org/w/index.php?curid=6402254.

1.6.1. Types of Pollution: Air, Water, and Soil

Air pollution poses one of the most significant threats to environmental health, arising primarily from industrial emissions, vehicle exhaust, and the burning of fossil fuels. The introduction of harmful pollutants such as particulate matter, nitrogen oxides, and sulfur dioxide can lead to severe respiratory issues and cardiovascular diseases in human populations. Furthermore, contaminated air detrimentally affects ecosystems by altering weather patterns and disrupting wildlife habitats. According to (Archana et al., 2016), the degradation of the environment, particularly air quality, has become increasingly visible as an issue that must be addressed through stringent policies and public awareness. Efforts like the National Air Quality Monitoring Programme illustrate governmental attempts to monitor and mitigate air pollutions effects, yet major challenges remain due to persistent industrial practices and urbanization. To effectively combat air pollution, a collective commitment to reducing emissions and implementing greener technologies is imperative. Water pollution is another critical concern, dramatically affecting aquatic ecosystems and human health. Sources such as agricultural runoff, industrial discharge, and urban waste often cause the contamination of rivers, lakes, and groundwater, leading to eutrophication and biodiversity loss. Toxic substances found in polluted water can result in serious health problems for those who rely on these water sources for drinking and bathing. The harmful effects of water pollution highlight the need for regulatory frameworks aimed at safeguarding water quality (Figure 1.10).

Figure 1.10. *Environmental pollution.*

Source: By User:Synth85 – Own work, CC BY-SA 4.0, https://commons.wikimedia.org/w/index.php?curid=78383107.

As mentioned in (Kumar, 2023), initiatives such as the Namami Gange Programme and Sewage Treatment Plants standards demonstrate the Indian government's commitment to mitigating water pollution. Effective enforcement of these regulations is essential to ensure the sustainable management of water resources. By prioritizing clean water initiatives, communities can work towards restoring and protecting vital aquatic ecosystems. Soil pollution, often less visible than air and water contamination, nonetheless poses a substantial risk to environmental integrity and public health. This pollution frequently results from agricultural practices, such as the excessive use of pesticides and fertilizers, as well as industrial waste disposal. Contaminants in the soil can inhibit plant growth, disrupt local ecosystems, and even enter the food chain, posing risks to human health. The ongoing degradation of soil quality is a direct consequence of inadequate waste management and urban sprawl. As highlighted in (Kumar, 2023), the government has recognized the necessity for soil protection through strategies aimed at reducing pollutants and promoting sustainable agricultural techniques. Effective soil management practices, coupled with public awareness campaigns, are crucial for reversing degradation trends. A concerted effort to address soil pollution can not only restore ecosystem balance but also enhance agricultural productivity and ensure food security.

1.6.2. Effects of Pollution on Human Health and Ecosystems

Pollution remains one of the most pressing environmental challenges, significantly affecting human health across the globe. Elevated levels of contaminants, such as nitrogen in drinking water, pose serious risks, with high nitrogen concentrations linked to various health issues, including methemoglobinemia, or blue baby syndrome in infants. According to data, approximately 40% of the European population could be exposed to unsafe nitrate levels without adequate treatment measures in place (Gilles et al., 2010). This scenario highlights the urgent need for stringent policies and efficient water treatment facilities to mitigate these health risks. Ultimately, the intersection of pollution control and public health is critical, as successful interventions can protect vulnerable populations and improve overall community well-being. Ecosystems, too, bear the brunt of pollution, experiencing detrimental changes that threaten biodiversity and the balance of natural habitats. Nitrogen pollution specifically contributes to eutrophication, a process that causes the overgrowth of algae in water bodies, leading to hypoxia—an oxygen-depleted condition detrimental to aquatic life. The toxic algal blooms produced can lead to fish kills and negatively impact habitats, pushing many species to the brink of extinction. The European Union has recognized these challenges and implemented policy tools to combat nitrogen pollution, yet the full effectiveness of these measures has often been hindered by historical accumulations and inadequate planning (Gilles et al., 2010).

Addressing these ecological impacts requires comprehensive strategies that promote ecosystem health and resilience, highlighting the interconnectedness of all environmental components. Achieving ocean sustainability rests heavily on addressing pollution, which is intricately linked to broader global initiatives such as the UN Sustainable Development Goals (SDGs). Specifically, Target 14 aims to protect marine ecosystems from degradation driven by pollutants. The pursuit of marine health contributes to the resilience of coastal communities and supports economic activities reliant on thriving oceans. Implementing measures to attain SDG 14 can create co-benefits that extend beyond marine health, ultimately contributing to enhanced climate resilience and social equity. Thus, addressing pollution, particularly in marine environments, underscores a fundamental aspect of environmental science, as it connects the health of our oceans directly to human prosperity and ecological sustainability (Figure 1.11).

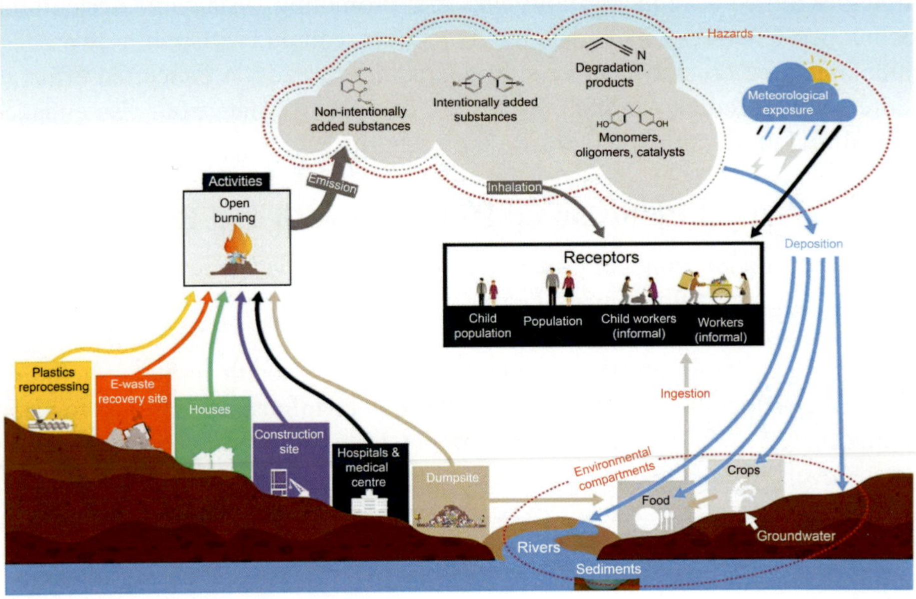

Figure 1.11. *Air quality and health.*

Source: By Cook, E – https://eprints.whiterose.ac.uk/169766/, CC BY-SA 4.0, https://commons.wikimedia.org/w/index.php?curid=140325515.

1.6.3. Waste Management Practices and Their Effectiveness

Implementing effective waste management practices is essential for promoting sustainability and reducing environmental impact. One significant avenue for improvement lies in the adoption of reverse marketing, which focuses on the recycling and management of waste products. This innovative approach not only minimizes landfill usage but also engages consumers in sustainable practices. As noted in the research, an active commitment to recycling communicates a brand's dedication to sustainability, ultimately enhancing its equity within the marketplace (Deivanayagampillai et al., 2019). By fostering a collaborative relationship between businesses and consumers, reverse marketing creates a platform for effective waste management while simultaneously improving the overall brand image. The effectiveness of various waste management strategies can be evaluated through comprehensive metrics, such as the reduction of waste generated and the efficiency of recycling processes. Techniques such as type-2 fuzzy sets and neutrosophic sets are critical in assessing the uncertainties and complexities associated with traffic flow and waste disposal. These mathematical frameworks allow for nuanced decision-making under uncertain conditions, ensuring that the waste management systems can adapt and respond effectively to varying levels of waste generation. By applying methods like the triangular interval type-2 Schweizer and Sklar operators, managers can make informed decisions that enhance the efficacy of waste management (Deivanayagampillai et al., 2019). Such analytical tools provide clarity and precision, ultimately leading to better waste management outcomes. In conclusion, the integration of advanced analytical techniques and innovative marketing strategies can significantly elevate the effectiveness of waste management practices. Businesses can improve their operational efficiencies and positively influence consumer behavior towards recycling and responsible waste disposal by recognizing the inherent value in sustainable practices. Engaging customers through clear communication of sustainability initiatives generates trust and loyalty, contributing to brand equity while working toward a cleaner environment. As industries increasingly adopt comprehensive waste management strategies, the synergy between reverse marketing and analytical methodologies will be pivotal in achieving not only ecological sustainability but also economic viability (Gupta, 2015). Thus, a holistic approach to waste management practices emerges as vital in addressing pressing environmental concerns.

1.7. RENEWABLE ENERGY SOURCES

The transition from fossil fuels to renewable energy sources is driven primarily by the pressing need to address environmental degradation and climate change. Renewable energy sources such as solar, wind, and bioenergy provide sustainable alternatives to traditional energy production and promise significant reductions in greenhouse gas emissions. The concept of cascading utilization of renewable materials amplifies this benefit by highlighting the potential for these resources to be used efficiently and effectively across multiple applications before their eventual conversion into energy (Geldermann et al., 2015). This multi-faceted approach not only enhances resource efficiency but also amplifies economic benefits by creating new markets and industries centered around innovative and eco-friendly technologies. Adopting renewable energy solutions requires substantial policy support and innovative strategies that stimulate both market growth and technological advancements. The European Union's biofuels policy is a prime example, emphasizing the intricate relationship between markets, technoscience, and environmental sustainability. By fostering demand for biofuels, the policy encourages investments in research and development, ultimately leading to innovative solutions that are responsive to sustainability challenges (Birch et al., 2012). This interplay not only drives economic growth but also promotes a shift in consumer behavior towards more sustainable products, aligning market forces with environmental objectives (Figure 1.12).

Figure 1.12. *Renewable energy.*

Source: By Warren Rohner from Cape Town, South Africa – Darling Wind Farm, CC BY-SA 2.0, https://commons.wikimedia.org/w/index.php?curid=15581862.

Ensuring that these policies are effective hinges on interdisciplinary frameworks that integrate environmental science with economic growth. The ongoing evolution of renewable energy is not without its challenges, particularly in terms of resource allocation and technological adaptation. As we explore the potential of various renewable technologies, it becomes critical to understand the lifecycle impacts of these sources. The utilization of by-products and the implementation of advanced technologies can significantly bolster resource efficiency and sustainability. According to recent research, innovative materials derived from renewable resources can be applied across various sectors, enhancing the overall sustainability of products (Geldermann et al., 2015). By embracing a holistic view of resource efficiency and fostering interdisciplinary collaboration, we can pave the way for a resilient energy future that mitigates climate change while harnessing the benefits of renewable energy sources.

1.7.1. Types of Renewable Energy: Solar, Wind, Hydro, and Geothermal

A variety of renewable energy sources are available today, each with distinct advantages and challenges. Solar energy, harnessed through photovoltaic cells, has seen significant advancements in efficiency and affordability, largely due to government incentives such as tax rebates and net metering. These measures have encouraged residential investments in solar installations, which in turn contribute to decreased electricity demand on the grid (Worldwatch Institute, 2005). Wind energy, generated through turbines, complements solar by providing a reliable power source during different times of the day.

Wind farms, which are increasingly prevalent in regions with optimal wind patterns, contribute to lower emissions and create jobs in installation and maintenance, further promoting economic sustainability. Hydropower has long been recognized as one of the most established renewable energy sources. It utilizes flowing or falling water to generate electricity, making it a consistent energy supply that can meet high-demand periods. However, the environmental implications of large hydroelectric dams, such as ecosystem disruption and displacement of communities, necessitate a careful examination of their implementation.

Smaller-scale hydro projects are emerging as viable alternatives, focusing on sustainability and minimal ecological impact. As highlighted in the Global Status Report, many countries are beginning to adopt policies that support diverse renewable energy portfolios, including innovative hydro technologies (Worldwatch Institute, 2005). Geothermal energy, harnessed from the Earth's internal heat, presents another renewable option with significant potential, particularly in volcanic regions. This energy source can provide both electricity and direct heating applications, capitalizing on the Earth's constant temperatures. While the initial costs of geothermal plants can be high, their long-term benefits often include low operating costs and high efficiency. As renewable energy technologies continue to mature, the integration of solar, wind, hydro, and geothermal resources can create a balanced and resilient energy grid. The success of California's Renewable Portfolio Standard, aiming for 33% renewables by 2020, exemplifies the potential for such integration to foster a sustainable future (Atkins et al., 2014) (Figure 1.13).

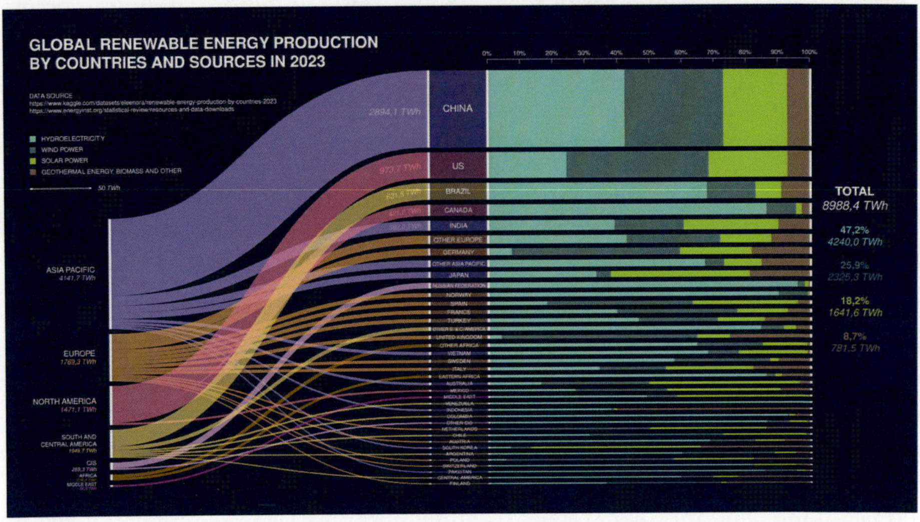

Figure 1.13. *Renewable energy: Sources and types.*

Source: By Eleonora Galliani – https://www.kaggle.com/datasets/eleenora/renewable-energy-production-by-countries-2023Dataset based on data from Energy Institute, Statistical Review of World Energy(https://www.energyinst.org/statistical-review/resources-and-data-downloadsStatistical Review of World Energy Data (2023), Renewables Generation by Source)The Energy Institute is, as of 2023, the home of the Statistical Review of World Energy, which analyses data on world energy markets providing timely, comprehensive, and objective data to the energy community since 1952, CC0, https://commons.wikimedia.org/w/index.php?curid=153480356.

1.7.2. Benefits of Transitioning to Renewable Energy

Transitioning to renewable energy systems has significant economic advantages that are becoming increasingly evident as technological innovations enhance efficiency. One key benefit is the potential for job creation in the renewable energy sector, which encompasses diverse fields ranging from solar panel manufacturing to wind turbine installation. The shift from fossil fuels to cleaner energy fosters the development of new, sustainable industries and revitalizes local economies, particularly in areas previously reliant on traditional energy sources. By embracing this transition, nations can stimulate job growth, as highlighted by good practices from European players who have successfully implemented strategies to create green jobs (Kryk, 2015). These jobs often pay well and contribute to workforce development, thus reinforcing the economic viability and resilience of communities undertaking this essential shift. Another crucial aspect of utilizing renewable energy is its capacity to mitigate environmental degradation. Traditional energy sources, particularly fossil fuels, are significant contributors to air and water pollution, leading to adverse health effects and biodiversity loss. By shifting to renewable sources such as wind, solar, and hydropower, we can drastically reduce carbon emissions and pollution levels, helping combat climate change. This aligns with the vision outlined in the call for a Green Energy Revolution, which emphasizes the importance of decentralized energy supplies that promote sustainability while fostering community empowerment (Kryk, 2015).

Figure 1.14. A solar power station.

Source: By kallerna – Own work, CC BY-SA 4.0, https://commons.wikimedia.org/w/index.php?curid=103494974.

As societies adopt these cleaner technologies, they enhance environmental conditions and pave the way for innovative practices that prioritize ecological health, contributing to a more sustainable future for generations to come. Finally, transitioning to renewable energy sources enhances energy security and resilience. Unlike fossil fuels, which are subject to volatile market fluctuations and geopolitical tensions, renewable resources are inherently abundant and widely distributed. This decentralization allows communities to rely on local energy generation, reducing dependency on imported fuels and mitigating the risks associated with supply chain disruptions. The adoption of smart electricity networks, as suggested in the push for a Green Energy Revolution, could further optimize energy distribution and consumption (Kulatilaka et al., 2010). By making energy systems more robust against disruptions, societies can improve their overall stability, leading to greater confidence in energy availability. Consequently, transitioning to renewable energy not only addresses immediate environmental concerns but also promotes long-term sustainability and security for citizens worldwide (Figure 1.14).

1.7.3. Challenges in Implementing Renewable Energy Solutions

Transitioning to renewable energy solutions presents significant logistical challenges that cannot be overlooked. One of the primary hurdles involves the infrastructure required to support these technologies. For instance, wind, and solar energy installations necessitate extensive physical support systems, including grid upgrades and energy storage solutions, to ensure reliability and efficiency. These infrastructural developments often face bureaucratic delays and high initial costs, which deter investment from both the public and private sectors. Furthermore, many regions lack the existing frameworks needed to efficiently integrate renewable sources into their energy mix, resulting in a lag in implementation. Thus, infrastructural inadequacies emerge as a primary barrier to realizing the full potential of renewable energy. In addition to infrastructure concerns, economic factors continue to be a significant impediment to adopting renewable energy solutions. While the long-term benefits of renewables, such as cost savings and environmental protection, are widely acknowledged, the upfront investment required for technology and installation can be a major deterrent. The inconsistency in government policies and subsidies related to renewable energy further complicates the financial landscape, making it difficult for investors to make informed decisions. According to the World Ocean Council (2016), industries must engage in marine planning to understand the economic implications of new energy solutions. Without clear and stable financial policies, the renewable energy sector struggles to attract the necessary funding, stifling innovation and growth. Lastly, societal attitudes and public perception play a crucial role in the implementation of renewable energy solutions. Misinformation and skepticism surrounding renewable technologies can lead to resistance from communities and policymakers alike. Many individuals may be uninformed about the practical benefits and feasibility of renewable energy, creating friction when initiatives are proposed. This can stifle community support and hinder projects from moving forward, as local buy-in is often essential for success. According to Buckley et al (2014), educational outreach and stakeholder engagement are vital in addressing these concerns. Overcoming these social challenges necessitates targeted communication strategies that clarify the advantages of renewable energy, thereby fostering a more supportive environment for future developments in this essential field.

1.8. ENVIRONMENTAL POLICY AND REGULATION

The relationship between environmental quality and social equity distinctly highlights the role of policy in promoting or hindering sustainable development. As indicated in recent research, socially, and economically deprived populations often inhabit the most polluted areas, exacerbating health disparities and leading to adverse living conditions. To address this concern, regulatory frameworks must evolve to enforce environmental protections and to consider the socio-economic factors at play. For example, the Environment Agency in the UK has recognized the necessity of integrating social justice into its environmental policies, aiming to mitigate inequalities that adversely affect marginalized communities (McNew-Birren et al., 2011). Such integration can lead to more comprehensive and effective environmental strategies that not only improve ecological health but also promote equitable access to a clean environment for all. The concept of environmental policy learning underpins the need for adaptive approaches in regulatory practices. Stakeholders evolving perspectives on environmental issues can greatly influence policy outcomes, as seen in the case of lead contamination management in Herculaneum, Missouri. Local community members, categorized as Community Health Activists and Community Preservation Activists, have offered contrasting views on how to handle lead contamination, showcasing broader community involvement in policymaking (McNew-Birren et al., 2011). This dialogue emphasizes the importance of incorporating diverse voices into environmental regulation, ensuring that policy objectives do not overlook the concerns of those directly affected. By prioritizing inclusive stakeholder engagement, regulatory frameworks can become more responsive, thereby fostering an environment where equitable and sustainable practices can flourish. To achieve effective environmental regulation, continuous evaluation and reflection on policies are essential. Implementing participatory approaches encourages local communities to engage in discussions about their environmental conditions, which can lead to better-tailored solutions. The Environment Agency's commitment to addressing environmental inequalities through pilot projects and community involvement demonstrates a proactive approach to policy development (Chalmers et al., 2006). Furthermore, ongoing research into the effects of environmental hazards, such as those associated with lead exposure, helps refine regulatory strategies. Thus, a robust environmental policy framework prioritizes the ecological health of a community and champions social equity, ensuring that all individuals can thrive regardless of their socio-economic status. Through such comprehensive approaches, the challenges posed by environmental degradation can be effectively addressed.

1.8.1. Overview of Key Environmental Laws and Regulations

Environmental laws and regulations play a crucial role in protecting natural resources and public health. The National Environmental Policy Act (NEPA) of 1969 marked a significant milestone, requiring federal agencies to assess the environmental impacts of their proposed actions through Environmental Assessments (EAs) and Environmental Impact Statements (EISs). This framework ensures that environmental considerations are integrated into the decision-making process, allowing for informed choices that prioritize sustainability. The influence of NEPA extends beyond federal jurisdiction, as it has inspired various state-level initiatives, promoting a more comprehensive approach to environmental governance. Such laws emphasize the necessity of transparency and encourage public participation in governmental processes, fostering a sense of community responsibility and engagement in environmental stewardship. Additionally, the implementation of the Clean Air Act (CAA) and the Clean Water Act (CWA) delineates critical regulatory measures aimed at preserving air and water quality. The CAA, established in 1970, empowers the Environmental Protection Agency (EPA) to set and enforce national air quality standards, addressing pollutants that pose risks to human health and the environment. In parallel, the CWA, enacted in 1972, regulates discharges of pollutants into U.S. waters, requiring states to establish water quality standards. Both laws underscore the federal commitment to combat pollution while allowing states some autonomy in their enforcement strategies. However, as noted, the recent amendment to the Toxic Substances Control Act raises concerns about potential preemption, which could hinder states abilities to address specific contaminants effectively, particularly regarding emerging threats such as PFAS chemicals, which are increasingly linked to serious health risks (McDonald et al., 2020).

A critical review of existing legislation reveals the need for continuous refinement and adaptation of environmental regulations. While comprehensive laws like the Endangered Species Act (ESA) highlight efforts to safeguard biodiversity, the effectiveness of such regulations often depends on public awareness and compliance. Furthermore, as highlighted in the literature, improper implementation can lead to inefficiencies and gaps in achieving intended environmental outcomes (Anthoff and Robert, undated). Policymakers must therefore address these deficiencies while recognizing the importance of collaboration among federal, state, and local governments. Engaging stakeholders and communities can facilitate more robust environmental protections, ensuring that laws evolve in response to emerging challenges such as climate change and pollution. Ultimately, a proactive and integrated regulatory framework is essential for fostering a sustainable future and preserving the planet for generations to come.

1.8.2. Role of Government and Organizations in Environmental Protection

The interplay between government bodies and environmental organizations is crucial in shaping effective policies aimed at environmental protection. Governments create frameworks that define regulatory standards and enforce compliance among industries and communities. Through legislation, such as the Clean Air Act or the Endangered Species Act, officials set parameters that bolster conservation efforts while maintaining economic interests. Simultaneously, these governing entities collaborate with organizations dedicated to environmental advocacy, drawing on their expertise to address complex issues like biodiversity loss and climate change. For instance, the structured interviews conducted within the Central Hungary region highlighted how public institutions interact with private and civil organizations, indicating the importance of cooperative governance in successful resource management efforts (Fleischer et al., 2005). By fostering these partnerships, governments can leverage the specialized knowledge of organizations to inform policymaking, creating a robust response to environmental challenges. Moreover, organizations play a critical role in mobilizing public support and awareness around environmental issues. They act as intermediaries, translating complex scientific information into accessible language that the general public can comprehend. This is particularly important in addressing topics like Colony Collapse Disorder (CCD), where misinformation can hinder effective action. By providing clear, reliable information, organizations empower citizens to advocate for change, influencing both public opinion and governmental priorities. Additionally, they often serve as watchdogs, holding both governments and corporations accountable for their environmental practices. This dynamic is illustrated by ongoing research into CCD and the exploration of effective communication strategies. The ability of organizations to effectively disseminate crucial information is vital in educating stakeholders and supporting informed decision-making processes (Boehm, 2012).

The collaboration between these organizations and government entities thus enhances community engagement and strengthens overall environmental protection efforts. Lastly, funding, and research initiatives often involve significant contributions from both government and organizations, underlining their synergistic relationship. Governments allocate resources to research organizations and academic institutions, ensuring that empirical studies inform policy decisions. Such partnerships lead to the development of innovative solutions for pressing environmental issues, exemplified by collaborative projects targeting pollution control or sustainable resource management. In navigating the complexities of climate-related data and environmental challenges, the alignment of interests between public bodies and organizations becomes essential. The ISPA program mentioned in the Central Hungary case is a prime example of how financial and technical support from governmental sources can enable organizations to implement impactful environmental strategies (Fleischer et al., 2005). Ultimately, robust collaboration fosters your extensive understanding of environmental challenges, presenting a united front against ecological degradation and promoting sustainable practices across multiple sectors.

1.8.3. International Agreements and Their Impact on Global Environmental Issues

The interconnected nature of environmental issues demands coordinated efforts through international agreements which can create frameworks for cooperation among nations. These agreements aim to address common challenges, such as climate change and biodiversity loss, which often transcend national borders. For example, the Paris Agreement represents a significant step in global climate action, where countries commit to reducing greenhouse gas emissions to limit global temperature rise. By establishing enforceable targets and fostering transparency, these international pacts can enhance accountability and facilitate shared technological advancements in renewable energy. Furthermore, such agreements can lead to financial mechanisms that support developing nations in their transition toward sustainable practices, highlighting the importance of global solidarity in addressing environmental degradation. Though the intention behind international agreements is commendable, their effectiveness can be compromised by a lack of compliance among member states. Countries often possess differing economic priorities and capabilities, leading to discrepancies in commitment levels. As noted by the Working Group on Trade, Investment, and Climate Policy, trade agreements like the Trans-Pacific Partnership (TPP) have historically fallen short of adequately addressing environmental concerns (Alam et al., 2016). Many nations prioritize economic growth over environmental regulations, resulting in loopholes and insufficient enforcement of climate objectives. Hence, for international agreements to wield meaningful impact, they must be designed with robust mechanisms for enforcement, regular monitoring, and avenues for accountability that can address non-compliance and foster genuine engagement. Moreover, the interplay between trade policies and environmental agreements is crucial in shaping sustainable practices on a global scale. Trade agreements can either promote or hinder environmental progress, depending on how they are structured. The Working Groups analysis emphasizes the need for trade and investment rules to align with climate goals, suggesting that significant reforms are necessary to curtail negative impacts on the environment stemming from unfettered trade (Alam et al., 2016). As countries engage in negotiations, a holistic approach that intertwines economic growth with environmental stewardship could create pathways for sustainability. By prioritizing climate-friendly policies within trade regimes, nations can collectively mitigate environmental degradation, reaffirming their commitment to shared global values in the fight against climate change.

1.9. SUSTAINABLE DEVELOPMENT

Sustainable development requires a careful balance between economic growth, social equity, and environmental protection. This triad of principles underscores the urgency of addressing a wide range of global challenges, including poverty, climate change, and resource scarcity. A focus on sustainable practices is essential for the preservation of biodiversity and for the well-being of future generations. By integrating sustainable practices into various sectors, from agriculture to energy, society can foster resilience against ecological degradation while promoting economic stability. For instance, the implementation of green technologies and practices can lead to more efficient use of resources, providing a pathway toward a sustainable economy that benefits both people and the planet. Moreover, effective regulatory frameworks play a crucial role in advancing sustainable development, especially in emerging economies where rapid growth can often lead to environmental degradation. The study of swine manure recovery in China illustrates the empirical need for command-and-control regulations, which can initially stimulate farmers' intentions to engage in sustainable practices. However, to bridge the gap between intention and action, incentive-based, and information-driven approaches are vital. Combining these strategies enhances engagement and ensures that sustainable behavior becomes the norm rather than the exception. This multifaceted regulatory approach is essential for advancing agricultural sustainability, thereby contributing to the United Nations Sustainable Development Goals by addressing both environmental impacts and economic viability (Tou et al., 2024). The growing recognition of global engagement and collaboration in sustainability initiatives highlights the interconnectedness of environmental issues and international policy. Research shows that themes like education and cross-cultural learning experiences are central to fostering international cooperation in addressing pressing global challenges (Ifan et al., 2023). Furthermore, as countries like the United States, the United Kingdom, and Australia lead in contributions to sustainable development research, their collective efforts pave the way for more cohesive global strategies. By prioritizing sustainable development in legislative agendas and academic research, nations can better respond to the complexities of environmental degradation and promote practices that are beneficial for all. Such commitments are critical for ensuring that the resources required for sustainable development are managed responsibly, creating a framework that enables present and future generations to thrive alongside a healthy planet (Figure 1.15).

Figure 1.15. *Sustainable development.*

Source: By Melanie Maecker-Tursun – Melanie Maecker-Tursun, www.ponymithorn.com, hello@ponymithorn.com, CC BY-SA 4.0, https://commons.wikimedia.org/w/index.php?curid=88679630.

1.9.1. Principles of Sustainable Development

Sustainable development is fundamentally rooted in the balance between economic, social, and environmental considerations. This tripartite model emphasizes that true progress must not compromise the ecological integrity of the planet or the well-being of its inhabitants. Instead, sustainable development advocates for practices that harmonize resource utilization with the need for conservation. For instance, the principles of sustainable development are reinforced through education systems that integrate these concepts into curricula, enabling future generations to engage with environmental stewardship effectively. Such educational frameworks ensure that students understand the interconnectedness of natural systems and human societies, which is crucial for fostering responsible citizenship in a complex global landscape (Retief et al., 2002).

Another vital principle of sustainable development involves the recognition that environmental management must be guided by philosophical and ethical considerations. A Christian worldview, as noted in certain discussions on the subject, posits that humanity has a moral obligation to care for the environment, suggesting that ecological preservation aligns with spiritual values. This perspective challenges us to reevaluate our approaches to environmental issues, demanding that we balance economic pursuits with sincere respect for biodiversity and ecological systems (Retief et al., 2002). By embracing diverse philosophical understandings, sustainable development can derive richer, more inclusive solutions that address the root causes of ecological degradation while fostering equitable socio-economic growth. To effectively implement the principles of sustainable development, collaboration among various sectors is essential. This collaboration encompasses governmental bodies, non-profit organizations, and the private sector, all working together to create policies and strategies that promote sustainability. For example, integrating sustainable practices into urban planning and infrastructure development can significantly reduce environmental impact while enhancing the quality of life for residents. Research shows that integrating education for sustainable development (ESD) into curricula can equip students with critical skills to address pressing environmental challenges (Avcı et al., 2022). By encouraging an interdisciplinary approach that transcends traditional boundaries, communities can develop holistic solutions that respond to the pressing needs of today without jeopardizing the resources required for future generations. Through such collective efforts, the ideal of sustainable development can move from a theoretical framework to a tangible reality (Figure 1.16).

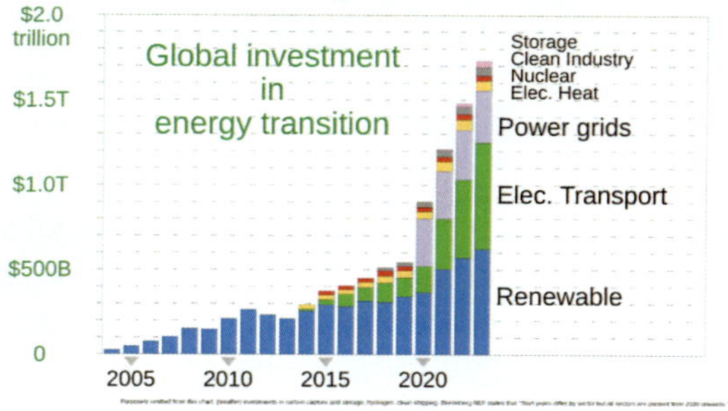

Figure 1.16. *Principles of sustainable development.*

Source: By RCraig09 – Own work, CC BY-SA 4.0, https://commons.wikimedia.org/w/index.php?curid=99302131.

1.9.2. Strategies for Promoting Sustainability in Communities

Creating sustainable communities requires a multifaceted approach that encompasses educational initiatives, community engagement, and policy reform. One effective strategy involves integrating sustainability education into local curricula to empower individuals with knowledge about environmental impacts and resource management. Institutions, such as the Mohamed Bin Zayed University for Humanities, have recognized the need for environmental education, aiming to foster a generation that actively participates in climate action (Aboukhousa, 2024). By providing students with hands-on learning experiences and promoting environmental awareness campaigns, communities can cultivate a strong sense of responsibility and inspire engagement at all levels. Collaboration among various stakeholders, including local governments, non-profit organizations, and businesses, is crucial for promoting sustainability initiatives. Partnerships enable the pooling of resources and expertise, facilitating the development of innovative solutions to address environmental challenges.

Policymakers play a critical role in establishing frameworks that encourage sustainable practices, such as incentivizing green building projects and implementing effective waste management systems. As seen in Lagos State, a comprehensive policy approach can effectively integrate structural and non-structural flood mitigation measures to enhance community resilience against climate-related impacts (Olukunga et al., 2024). These efforts require a commitment to long-term planning and the involvement of community members to ensure that strategies are both practical and culturally relevant. Community-based initiatives represent another vital strategy for promoting sustainability. By fostering grassroots movements, individuals can mobilize resources to address specific environmental challenges within their neighborhoods. This may include establishing community gardens, organizing clean-up drives, or participating in local governance to advocate for sustainable policies. Involving residents not only raises awareness but also empowers them to affect change in their surroundings. The success of such initiatives often hinges on dedicated volunteers and local leadership, which contribute to a shared vision of sustainability and communal responsibility. Overall, these strategies collectively promote a culture of sustainability, illustrating how informed and engaged communities can significantly contribute to environmental stewardship and a more sustainable future.

1.9.3. Role of Technology in Achieving Sustainable Development

In the quest for sustainable development, technology serves as a critical catalyst for innovation and efficiency. For instance, the integration of renewable energy technologies has revolutionized the way we generate power, significantly reducing our reliance on fossil fuels. Solar panels, wind turbines, and other renewable energy systems are becoming increasingly affordable and accessible. This transition lowers greenhouse gas emissions and supports the creation of green jobs, paving the way for economic growth without compromising environmental integrity. Moreover, advanced technologies in transportation, such as electric vehicles and smart traffic systems, enhance energy efficiency and minimize pollution, addressing pressing challenges in urban environments. Embracing these technological advancements is essential for achieving a balance between economic development and ecological preservation. Moreover, the role of technology extends beyond energy production to the realm of resource management and conservation. Environmental monitoring systems powered by satellite imagery and data analytics allow for better assessment and management of natural resources. These technologies enable policymakers and stakeholders to monitor deforestation, water supply, and air quality with unprecedented accuracy, facilitating informed decision-making.

As demonstrated in the environmental life cycle impact assessment of transportation infrastructure, employing technology in assessing and minimizing environmental impacts can significantly reduce emissions associated with infrastructure projects (Zhang et al., 2021). In sectors like agriculture, precision farming technologies optimize resource usage, promoting sustainable practices that minimize waste and enhance productivity. Such innovations not only propel industries towards sustainability but also support food security initiatives globally. Finally, the emergence of the green market underscores the intersection of technology, entrepreneurship, and sustainable development. As consumer awareness around environmental issues grows, there lies an opportunity for technology-driven solutions that align with sustainable practices. Green entrepreneurship fosters the development of eco-friendly products and services, driving economic development while addressing environmental challenges. Research indicates a positive correlation between the emergence of the green market and advancements in green entrepreneurship, highlighting technology's role in facilitating sustainable practices in knowledge-based industries (Lotfi et al., 2018). This symbiotic relationship between technology, entrepreneurship, and sustainable development demonstrates how innovative solutions can meet the needs of consumers while advancing societal goals. Harnessing these technological opportunities is crucial for building a sustainable future that balances economic viability with ecological stewardship.

1.10. CONCLUSION

The exploration of eco-innovation in manufacturing, particularly in the automotive sector, highlights the importance of sustainable practices in today's environmental landscape. Companies are increasingly recognizing that adopting eco-innovation can significantly reduce their ecological footprints while simultaneously improving operational efficiency. As noted in the findings related to the interdependence of product, process, and management innovations, it is clear that a comprehensive approach is crucial for meaningful change. This evidence underlines the necessity for organizations to combine efforts across different aspects of their operations to achieve the desired environmental outcomes. When companies embrace this holistic view, they not only comply with regulations but often enhance their competitive advantage in a market that increasingly values sustainability (Garza-Reyes et al., 2020). Furthermore, the discourse surrounding sustainable consumption provides insight into the broader societal implications of individual and organizational behaviors. The conference proceedings illustrate that understanding consumption patterns is essential for developing strategies that encourage more sustainable practices among consumers. The ambivalence of private consumption, being both a driving force for the economy and a contributor to environmental degradation, necessitates a multifaceted approach to foster change. Engaging in collaborative efforts among researchers and practitioners allows for more effective dissemination of knowledge and innovative strategies to address these challenges. This concerted effort is vital in reshaping norms and expectations concerning consumption, thereby aligning them with sustainability goals (Balmer et al., 2011).

In conclusion, the current research in environmental science underscores a pivotal shift toward sustainability through eco-innovation and mindful consumption. By closely examining the intertwining relationship between eco-innovation practices and sustainable consumption patterns, it becomes apparent that both sectors must work in tandem for effective environmental stewardship. Future advancements in technology and policy must prioritize collaborative frameworks that integrate eco-friendly practices into manufacturing and consumer behavior alike. Ultimately, by embracing these changes, society can make strides toward a more sustainable future, reflecting a collective commitment to preserving our planet for generations to come. The interplay of innovation and informed consumption represents a promising pathway for addressing the pressing environmental challenges of our time.

1.10.1. Summary of Key Points Discussed

In understanding the evolution of library environments, a significant point of discussion has been the development of master plans to enhance these facilities. The initiative undertaken by BDP for UCL highlights the importance of collaborative efforts involving academic representatives, library staff, and design consultants. By engaging diverse stakeholders and conducting thorough consultations with users, the master plan aimed to not only improve the physical environment of the library but also to optimize library services. This approach demonstrates that addressing user needs through direct feedback is crucial for successful library modernization efforts. Such foresight in planning can ensure that library spaces remain relevant and functional in accommodating the changing demands of academic communities. Moreover, the project highlights the essential role of foresight in managing risks, particularly in environmental contexts, as showcased by the Foresight Future Flooding (FFF) project. By researching flood risks up to the year 2100, the FFF initiative underscores the proactive measures that must be adopted in policy formulation and scientific research. The project successfully engaged a panel of experts to provide critical insights on flood management, thereby elevating the discussion within governmental frameworks. Importantly, while the project did not instigate immediate policy changes, it served to legitimize existing proposals and push forward ongoing discussions in flood risk management. This illustrates the value of integrating scientific research with political action to enhance resilience against climate-related challenges. Lastly, ongoing dialogues surrounding environmental adaptation reveal the necessity of embracing innovative solutions and strategies. The collaborative nature of projects like the UCL library master plan and the FFF research demonstrates that engaging multiple perspectives can lead to more comprehensive and effective outcomes. Both initiatives provide compelling evidence that knowledge dissemination is vital for influencing policy and practice. By learning from these examples, future environmental science projects can better tailor their findings to meet policymaker's needs, driving meaningful change. As highlighted in the previous cases, understanding user needs and fostering collaborations are critical components of reshaping our approaches to environmental challenges and community resilience in the face of uncertainty.

1.10.2. The Future of Environmental Science and Its Relevance

The future trajectory of environmental science hinges on its ability to adapt and respond to pressing global challenges, particularly climate change and resource depletion. As nations grapple with diverse energy security needs, the role of innovative low-carbon technologies becomes increasingly paramount. Research has shown that evaluating scenarios in which carbon emissions are contained can illuminate pathways for sustainable development while addressing the socio-economic implications of the energy transition (Staley et al., 2009). Moreover, this scientific discipline must evolve to incorporate interdisciplinary approaches, integrating insights from fields like economics, sociology, and policy-making to develop comprehensive solutions. By fostering collaboration among scientists, policymakers, and communities, environmental science can transform theoretical knowledge into practical strategies that mitigate environmental crises. As environmental issues proliferate, public awareness and engagement with scientific findings will play a crucial role in shaping policy and promoting sustainable practices. The relevance of environmental science extends beyond academia; it permeates everyday decision-making at individual, corporate, and governmental levels. For instance, understanding the impact of consumer behavior on carbon emissions can spur movements toward sustainable living and responsible consumption.

Furthermore, funding from organizations like the Swiss National Science Foundation is vital in sustaining research endeavors and pushing the boundaries of knowledge in this field, as highlighted in M. Kliegel's work (Ballhausen et al., 2016). Consequently, the dissemination of scientific insights into accessible formats can galvanize public support for environmental initiatives and bolster grassroots activism aimed at fostering a more sustainable future. Ultimately, the significance of environmental science in contemporary society cannot be overstated. It serves as the backbone of evidence-based decision-making necessary for addressing climate threats and environmental degradation. With advancements in data collection and analytical technology, such as satellite monitoring and geographic information systems, environmental scientists are better equipped to analyze complex ecosystems and propose effective interventions. The future of this field may also involve greater emphasis on ecological restoration and biodiversity conservation, as these elements are vital for maintaining the resilience of ecosystems. As the urgency for action intensifies, environmental science will remain a crucial discipline, providing the knowledge and tools necessary for society to navigate the challenges ahead and ensure a sustainable planet for future generations.

1.10.3. Call to Action for Individuals and Communities to Engage in Environmental Stewardship

Individuals and communities play a crucial role in the ongoing effort to combat environmental degradation and promote sustainability. Engaging in environmental stewardship not only fosters a sense of responsibility but also cultivates a deeper awareness of the interconnectedness of all living things. Simple actions, such as reducing waste through recycling and composting, can significantly mitigate harmful impacts on the ecosystem. Furthermore, individuals can advocate for policies that support renewable energy, preserve natural habitats, and reduce carbon footprints, effectively transforming their commitment into a collaborative movement. By initiating conversations on environmental issues within their circles, individuals can inspire others to take action, creating a ripple effect that amplifies collective influence on local and national levels. In addition to individual efforts, communities must come together to create vibrant networks focused on environmental stewardship. Local initiatives, such as community gardens and tree-planting events, not only enhance biodiversity but also foster a sense of unity and shared purpose among residents. Collaboratively addressing local environmental challenges, such as pollution or habitat destruction, can lead to innovative solutions tailored to specific community needs. Moreover, establishing partnerships with local schools can integrate environmental education into curricula, inspiring younger generations to become active participants in sustainability efforts. When communities work together, they can effectively leverage resources, knowledge, and skills to address complex environmental issues more efficiently and robustly. Ultimately, the call for environmental stewardship is both a collective and individual responsibility that cannot be ignored. Engaging in sustainable practices today will ensure a healthier and more resilient planet for future generations. Individuals and communities need to recognize their power in this critical movement, as every small action contributes to wider change. Through participation in local environmental initiatives, advocacy for sustainable policies, and personal commitment to eco-friendly habits, we can collectively create a more sustainable future. The urgency of the climate crisis compels us to act now; by embracing our roles as stewards of the Earth, we can forge paths to lasting ecological balance and regeneration.

REFERENCES

1. Aboukhousa, H. (2024). *Promoting Climate Action Through Youth Empowerment and Environmental Education: Insights from Mohamed Bin Zayed University for Humanities*. https://www.semanticscholar.org/paper/fe7958372f6cc4b7a938eba5fdcd1461fb42d263 (accessed on 5 December 2024).
2. Ackerman, F., Bernasconi, N., Cabré, M. M., Corbett, J. J., Gallagher, K. P., Güven, B. S., Hu, T., Janetos, A. C., Johnson, L., Mbengue, M. M., Monasterolo, I., Porterfield, M. C., Raberto, M., Schacherer, S., & Schachter, J. (C.), Selin, H., & Thrasher, R. (2016). Trade in the balance: Reconciling trade and climate policy: Report of the Working Group on Trade, Investment, and Climate Policy. *Boston University Frederick S. Pardee Center for the Study of the Longer-Range Future*. ISBN 978-1-936727-13-1
3. Aggarwal, P. K., Baethegan, W. E., Cooper, P., Gommes, R., Lee, B., Meinke, H., Rathore, L., & Sivakumar, M. (2010). Managing climatic risks to combat land degradation and enhance food security: Key information needs. *Procedia Environmental Sciences, 1*, 305–312. https://doi.org/10.1016/j.proenv.2010.09.019.
4. Alabi, A. O. (2020). Assessment of environmental consciousness among patrons in selected academic and public libraries in Lagos metropolis. *eScholarship, University of California*. https://core.ac.uk/download/287623618.pdf (accessed on 5 December 2024).
5. Alam, N. N. (2016). DIPL 4555 AA/AB economic aspects of international affairs. *eRepository@SetonHall*. https://core.ac.uk/download/216888507.pdf (accessed on 5 December 2024).
6. Anthoff, D., & Hahn, R. (2010). Government failure and market failure: On the inefficiency of environmental and energy policy. *Oxford Review of Economic Policy, 26*(2), 197–224. https://doi.org/10.1093/oxrep/grq004 (accessed on 5 December 2024).
7. Archana, A., Datta, C., & Tiwari, P. (2016). Impact of environmental degradation on human health. *Scientific and Literature Open Access Publishing*. https://media.neliti.com/media/publications/280807-impact-of-environmental-degradation-on-h-b05f4f6d.pdf (accessed on 5 December 2024).
8. Atkins, M. J., Walmsley, M. R. W., & Walmsley, T. G. (2014). California's renewables portfolio standard (RPS) requires 33% renewable electricity generation by 2020—Dream or reality? *SDEWES Centre*. Retrieved from: https://core.ac.uk/download/29202646.pdf (accessed on 5 December 2024).

9. Avcı, G., & Barak, B. (2022). *Comparative Analysis of Turkey and Germany (Bavaria) Secondary Education Curricula in Terms of Education for Sustainable Development*. Walter de Gruyter GmbH. https://core.ac.uk/download/568394020.pdf (accessed on 5 December 2024).

10. Ballesteros-Pérez, P., Ke, Y., Li, H., Lu, Q., Ouyang, Y., Peng, X., & Skitmore, M. (2021). Environmental life cycle impact assessment of transportation infrastructure: A multi-case study in international perspective. *International Journal of Sustainable Transportation, 16*(11), 1020–1031. https://doi.org/10.1080/15568318.2021.1959684.

11. Balmer, I., Defila, R., DiGiulio, A., Kaufmann-Hayoz, R., & Kobel, P. (2011). *Sustainable Consumption: Towards Action and Impact. International Scientific Conference November 6th-8th, 2011, Hamburg–European Green Capital 2011, Germany: Abstract Volume*. https://core.ac.uk/download/17203350.pdf (accessed on 5 December 2024).

12. BDP, Master Planning Team, UCL. (2008). UCL (University College London) libraries masterplan: Masterplanning report. *UCL Library Services*. https://core.ac.uk/download/1673357.pdf (accessed on 5 December 2024).

13. Benkens, A., Buck, B. H., Hoerterer, C., Krause, G., Nickiewicz, D., & Schupp, M. F. (2020). Stakeholder perspectives on opportunities and challenges in achieving sustainable growth of the blue economy in a changing climate. *Frontiers Media SA*. https://core.ac.uk/download/287601633.pdf (accessed on 5 December 2024).

14. Berezina, E., Moiseenko, K., Shtabkin, Y., Skorokhod, A., & Vasileva, D. (2020). Sources and variations of tropospheric ozone in central Siberia: Observations and model simulations. *IOP Publishing*. https://core.ac.uk/download/481106223.pdf (accessed on 5 December 2024).

15. Billen, G., Bouraoui, F., Cardoso, A., Curtis, C., Garnier, J., Grizzetti, B., & Van Grinsven, H. (2010). Nitrogen as a threat to European water quality. *Cambridge University Press (CUP)*. Retrieved from: https://core.ac.uk/download/38622535.pdf (accessed on 5 December 2024).

16. Birch, K., Levidow, L., & Papaioannou, T. (2012). Neoliberalising technoscience and environment: EU policy for competitive, sustainable biofuels. *Ashgate*. https://core.ac.uk/download/2832274.pdf (accessed on 5 December 2024).

17. Boehm, M. K. (2012). An exploratory study of online information regarding colony collapse disorder. *SJSU ScholarWorks*. https://core.ac.uk/download/215412951.pdf (accessed on 5 December 2024).

18. Bostwick, M. (2021). The psychological impacts of climate change education on undergraduate students. *Environmental Studies Electronic*

Thesis Collection, 70. https://scholarworks.uvm.edu/envstheses/70 (accessed on 5 December 2024).

19. Buckley, C., Cloete, E., Godfrey, L., Hildebrandt, D., Makhafola, M., Pouris, A., & Watson, J. (2014). *The State of Green Technologies in South Africa*. Academy of Sciences of South Africa and Ministry of Science and Technology, Republic of South Africa. Retrieved from: https://core.ac.uk/download/30610831.pdf (accessed on 5 December 2024).
20. Bunge, J., Cohen-Rosenthal, E., & Ruiz-Quintanilla, S. A. (1995). Employee participation in pollution reduction: Preliminary analysis of the toxics release inventory.
21. Chalmers, H., & Chalmers, H. (2006). Developing the environment agency's policy position on addressing environmental inequalities. Retrieved from: https://core.ac.uk/download/573861876.pdf (accessed on 5 December 2024).
22. Chew, D., Das, I., Kueh, B. H., & Maryati, M. (2004). Integrated effort in the prioritization of areas of conservation in Borneo. https://core.ac.uk/download/326905712.pdf (accessed on 5 December 2024).
23. Chu, M., Tou, L., & Wang, J. (2024). From intention to action: Enabling sustainable agriculture in emerging economies through decentralized regulations for manure management. *Public Administration and Development*, 1–15. https://doi.org/10.1002/pad.2066.
24. Cisneros-Montemayor, A., Singh, G., Cheung, W., & Ota, Y. (2017). Oceans and the sustainable development goals: Co-benefits, climate change & social equity. *The Nippon Foundation-UBC Nereus Program*. Retrieved from: https://core.ac.uk/download/87085613.pdf (accessed on 5 December 2024).
25. Cowie, A., Schneider, U. A., & Montanarella, L. (2007). Potential synergies between existing multilateral environmental agreements in the implementation of Land Use Change, and Forestry activities. *Working Papers FNU-123, Research Unit Sustainability and Global Change*, Hamburg University.
26. Deivanayagampillai, N., Lathamaheswari, M., Said, B., & Jacob, K. (2019). A new perspective on traffic control management using triangular interval type-2 fuzzy sets and interval neutrosophic sets. *Operations Research Perspectives, 6*, 100099. https://doi.org/10.1016/j.orp.2019.100099.
27. *DigitalCommons@ILR*. https://core.ac.uk/download/5131586.pdf (accessed on 5 December 2024).
28. Fleischer, T., & Futó, P. (2005). EU integration mechanisms affecting Hungarian public policies in the case of waste management. *Institute for World Economics of the Hungarian Academy of Science*. https://core.

ac.uk/download/11856140.pdf (accessed on 5 December 2024).

29. Frediani, K. (2024). Advancing arboriculture through human and tree ecology. Retrieved from: https://core.ac.uk/download/603667568.pdf (accessed on 5 December 2024).
30. Garza-Reyes, J. A., & Maldonado-Guzman, G. (2020). *Eco-Innovation Practices' Adoption in the Automotive Industry*. Emerald. https://core.ac.uk/download/288637710.pdf (accessed on 5 December 2024).
31. Gauthier, S. (2021). Applying ecosystem-based adaptation to protect biodiversity of avian species in coastal communities in the Greater Niagara Region. https://core.ac.uk/download/480267556.pdf (accessed on 5 December 2024).
32. Geldermann, J., Kolbe, L., Krause, A., Mai, C., Militz, H., Osburg, V. S., Schöbel, A., Schumann, M., Toporowski, W., & Westphal, S. (2015). Improved resource efficiency and cascading utilisation of renewable materials. *Journal of Cleaner Production, 110*.
33. Gülhan, F. (2023). *Environmental Sustainability in Science Education in the Anthropocene: A View from STEAM Education*. Bonoi Science Advancement and Education LLC. https://core.ac.uk/download/555483885.pdf (accessed on 5 December 2024).
34. Gupta, S. (2015). Enhancing brand equity through sustainability: Waste recycling. *Wiley*. https://core.ac.uk/download/30712270.pdf (accessed on 5 December 2024).
35. Ifan, M., Purwanto, E., Ate, J., & Simanjuntak, M. (2023). Bibliometric analysis of global engagement research: Trends, prolific authors, and impact in environmental science and business & economics fields. *Widyakala Journal: Journal of Pembangunan Jaya University, 10*(2), 51. https://doi.org/10.36262/widyakala.v10i2.804.
36. Kaushal, S. S., McDowell, W. H., Wollheim, W. M., Newcomer Johnson, T. A., Mayer, P. M., Belt, K. T., & Pennino, M. J. (2015). Urban evolution: The role of water. *Water, 7*(8), 4063–4087. https://doi.org/10.3390/w7084063.
37. Kinnebrew, E., Shoffner, E., Farah-Pérez, A., Mills-Novoa, M., & Siegel, K. (2021). Approaches to interdisciplinary mixed methods research in land-change science and environmental management. *Conservation Biology, 35*(1), 130–141. https://doi.org/10.1111/cobi.13642.
38. Kliegel, M., Ballhausen, N., Hering, A., Ihle, A., Schnitzspahn, K. M., & Zuber, S. (2016). Prospective memory in older adults: Where we are now and what is next. *Gerontology, 62*(4), 459–466. https://doi.org/10.1159/000443698.

39. Kryk, B. (2015). Green jobs–good practices. In M. Burchard-Dziubińska (Ed.), *Towards a Green Economy. From Ideas to Practice* (pp. 107–124). Wydawnictwo Uniwersytetu Łódzkiego.

40. Kulatilaka, N. (2010). *Green Revolution 2.0: A Sustainable Energy Path*. Boston University Frederick S. Pardee Center for the Study of the Longer-Range Future. https://open.bu.edu/bitstream/2144/22688/1/UNsdkp006-Green-Revolution-2.0.pdf (accessed on 5 December 2024).

41. Kumar, A. (2023). The Indian government's environmental stewardship: A review of pollution control efforts. https://core.ac.uk/download/587416644.pdf (accessed on 5 December 2024).

42. Lomas, P., & Giampietro, M. (2017). Environmental accounting for ecosystem conservation: Linking societal and ecosystem metabolisms. *Ecological Modelling, 346*, 10–19. https://doi.org/10.1016/j.ecolmodel.2016.12.009.

43. Lotfi, M., Yousefi, A., & Jafari, S. (2018). The effect of emerging green market on green entrepreneurship and sustainable development in knowledge-based companies. *Sustainability, 10*(7), 2308. https://doi.org/10.3390/su10072308.

44. Maes, J., Zulian, G., Thijssen, M., Castell, C., Baró, F., Ferreira, A., & Teller, A. (2016). *Mapping and Assessment of Ecosystems and their Services: Urban Ecosystems (4th report, final)*. Publication Office of the European Union. Retrieved from: https://publications.jrc.ec.europa.eu/repository/handle/JRC101639 (accessed on 5 December 2024).

45. McDonald, F. A. (2020). Omnipresent chemicals: TSCA preemption in the wake of PFAS contamination. *DigitalCommons@Pace*. https://core.ac.uk/download/286538257.pdf (accessed on 5 December 2024).

46. McNew-Birren, J. (2011). *The Impacts of Lead Contamination on the Community of Herculaneum, Missouri*. Washington University Open Scholarship. Retrieved from: https://core.ac.uk/download/233208186.pdf (accessed on 5 December 2024).

47. Meilinda, M., Rustaman, N. Y., & Tjasyono, B. (2017). The perceptions of pre-service science teachers and science teachers about climate change. *Jurnal Pendidikan IPA Indonesia, 6*(2), 292–397.

48. Musakwa, W., Rouget, M., Musakwa, W., Smit, A. J., Slotow, R., Roberts, D., Boon, R., Douwes, E., O'Donoghue, S., Downs, C. T., Murkherjee, S., Mwabvu, T., Odindi, J., Odindo, A., Prochess, S., Ramdhani, S., Murkherjee, J. R., Naidoo, S., Schoeman, M. C., & Willows-Munro, S. (2016). Implementation of a science-action partnership to manage a threatened ecosystem in an urban context. https://ujcontent.uj.ac.za/esploro/outputs/journalArticle/Implementation-of-a-science-action-

partnership-to/9913513907691#file-0 (accessed on 5 December 2024).

49. Ogbuigwe, A. (2009). Sigtuna think piece 1: Climate change education in Africa. *African Journals Online (AJOL).* https://core.ac.uk/download/478460641.pdf (accessed on 5 December 2024).

50. Olukunga, O. A., Adewumi, B. J., & Hassan, T. (2024). Comparative analysis of flood mitigation strategies for residential housing in Lagos State. *African Journal of Environmental Sciences and Renewable Energy, 15*(1), 209–225. https://doi.org/10.62154/efah7e88.

51. Pavao-Zuckerman, M. A., & Byrne, L. B. (2009). Scratching the surface and digging deeper: Exploring ecological theories in urban soils. *Urban Ecosystems, 12*(1), 9–20. https://doi.org/10.1007/s11252-008-0078-3.

52. Penning Rowsell, E. C., Evans, E. P., Hall, J. W., & Borthwick, A. G. L. (2013). From flood science to flood policy: The Foresight Future Flooding project seven years on. *Foresight, 15*(3), 190–210. https://doi.org/10.1108/fs-06-2012-0046.

53. Pilgrim, J., & Langhammer, P. (2015). Lower Mekong portfolio: Interim evaluation. *Terra Consilium.* https://core.ac.uk/download/80511031.pdf (accessed on 5 December 2024).

54. Retief, F. P. (2002). *Omgewingsbestuur Vanuit 'n Christelike Lewensbeskouing–Riglyne vir Volhoubare ontwikkeling*. AOSIS. https://core.ac.uk/download/335036289.pdf (accessed on 5 December 2024).

55. Rosiek, K. (2015). *Modifications to the Scope of the Term 'Ecological Security' in Poland Using the Example of Water Management*. University of Lodz. https://core.ac.uk/download/71991248.pdf (accessed on 5 December 2024).

56. Staley, B. C., Goodward, J., Zyla, K., & Ladislaw, S. (2009). *Evaluating the Energy Security Implications of a Carbon-Constrained U.S. Economy.* World Resources Institute. https://core.ac.uk/download/71351146.pdf (accessed on 5 December 2024).

57. Szejwald, H. B., & Goble, R. L. (1990). *The Role of Scientists in Risk Assessment.* University of New Hampshire Scholars Repository. https://core.ac.uk/download/72056967.pdf (accessed on 5 December 2024).

58. Thomson, A. I., Archer, F. I., Coleman, M. A., et al. (2021). Charting a course for genetic diversity in the UN Decade of Ocean Science. *Evolutionary Applications, 14*(8), 1497–1518. https://doi.org/10.1111/eva.13224.

59. World Ocean Council. (2016). *Marine Spatial Planning: Case Studies.* Retrieved from: https://core.ac.uk/download/75761082.pdf (accessed on 5 December 2024).

60. Worldwatch Institute. (2005). *Renewables 2005: Global Status Report*. Retrieved from: https://core.ac.uk/download/71338835.pdf (accessed on 5 December 2024).

61. Yang, J., Zhang, X., Jin, X., Seymour, M., Richter, C., Logares, R., Khim, J. S., & Klymus, K. (2021). Recent advances in environmental DNA-based biodiversity assessment and conservation. *Diversity and Distributions, 27*, 1876–1879. https://doi.org/10.1111/ddi.13415.

CHAPTER 2

MATTER, ENERGY, AND LIFE

Contents

INTRODUCTION

The interplay between matter and energy forms the foundation of all natural processes, including those that sustain life on Earth. At its core, matter refers to anything that occupies space and has mass, while energy is defined as the ability to do work or produce change. Together, they create a dynamic system that influences everything from microscopic cellular reactions to the vast movements of celestial bodies. By examining the fundamental principles governing matter and energy, we can gain invaluable insights into the mechanisms that underpin biological systems, as well as ecological interactions. This relationship is not merely academic; it is essential for understanding how living organisms thrive in their environments, adapt to challenges, and evolve over time. To further comprehend how matter and energy intersect with life, one must explore the role of biological processes at the molecular level. Cellular respiration, photosynthesis, and metabolic pathways are key examples of how organisms harness energy and transform matter for growth and survival. In photosynthesis, plants convert sunlight into chemical energy, utilizing carbon dioxide and water to produce glucose and oxygen—an elegant process that illustrates how energy inputs can generate complex organic molecules from simple inorganic compounds. Conversely, cellular respiration allows organisms to break down those molecules to release energy necessary for cellular functions. The circular flow of energy and matter through ecosystems not only highlights the intricacy of life but also underscores the interdependence of various species within their habitats. Understanding these processes is crucial for addressing contemporary environmental challenges and leveraging sustainability. As human activities disrupt natural systems and deplete resources, the delicate balance of energy and matter in ecosystems faces unprecedented threats. Climate change, pollution, and habitat loss are symptomatic of a broader disconnect between our understanding of these fundamental principles and their application in real-world contexts. By studying matter and energy flows, researchers can develop innovative strategies for conservation and restoration, informing policy decisions that protect biodiversity and ecological integrity. Through a meticulous examination of these relationships, we can appreciate the complexity of life and motivate collective efforts toward a more sustainable future that honors the intricate dance of matter and energy.

2.1. DEFINITION OF MATTER

Understanding the nature of matter is pivotal within the realms of both science and philosophy. Matter is fundamentally defined as anything that possesses mass and occupies space. This definition encapsulates a vast array of substances, ranging from familiar solids and liquids to gaseous forms that may challenge our everyday perceptions. The state of matter can fluctuate based on temperature and pressure, showing that matter is not static but rather dynamic in its interactions with energy and external influences. For instance, heating a solid can lead to its transformation into a liquid and ultimately a gas, demonstrating how energy can manipulate the state of matter and its properties. Such transitions highlight the intrinsic relationship between matter and energy, forming a foundation for understanding broader concepts in physical sciences. Moving beyond the classical definitions, it is crucial to acknowledge the subatomic realm, where matter reveals its complexity at the fundamental particle level. Within this context, matter is composed of atoms, which are themselves made up of smaller particles like protons, neutrons, and electrons. This atomic theory fundamentally reshapes our perception of matter, illustrating that it is not merely a collection of physical substances but an intricate arrangement of particles governed by electromagnetic forces and quantum mechanics. As discussed in the realm of cognitive sciences, understanding the structure of matter extends into a deeper comprehension of interactions that underpin all forms of life and energy processes. Thus, matter serves as a vital component in the synthesis of biological systems, reinforcing the interdependence of matter, energy, and life itself. Significantly, the interaction between matter and energy is a physical phenomenon and a philosophical exploration of existence itself. The fluidity with which matter interacts with energy forms the basis of many processes essential for life, such as metabolism and homeostasis. For example, biological organisms embody various forms of matter, utilizing energy derived from their environment to sustain life processes. The exploration of matter extends to philosophical implications regarding existence and reality, as seen in Haken and Portugali's work on self-organization and the interplay of structural and symbolic forms of information within complex systems (Haken and Juval, 2016). Such an understanding fosters a more profound appreciation of the interconnectedness of all entities, displaying how matter, energy, and life coalesce into a unified framework that underpins the existence and functioning of the universe.

2.2. DEFINITION OF ENERGY

Energy is fundamentally understood as the capacity to perform work or produce change in a system. This definition encapsulates a wide range of physical phenomena, from the kinetic energy of a moving object to the potential energy stored in an elevated mass. In contemporary scientific discourse, energy is not simply a measurable quantity but a crucial concept that governs interactions across the universe. As reflected in the law of conservation of energy, energy can neither be created nor destroyed; it only transforms from one form to another. This principle highlights the interconnectedness of different energy types and emphasizes the importance of energy in driving physical processes, thereby underpinning both the inanimate and living aspects of matter. Moreover, various forms of energy—such as thermal, electrical, chemical, and nuclear—illustrate the diversity of energy and its manifestations in different contexts. For instance, chemical energy, stored in molecular bonds, becomes kinetic energy when a reaction occurs, resulting in motion or heat. This transformative ability makes energy a central focus in multiple scientific fields, including physics, chemistry, and biology. The relationship between matter and energy is further explicated within the framework of thermodynamics, where the flow and transformation of energy within systems dictate both the state of matter and the direction of natural processes. Thus, understanding energy requires an appreciation of its forms and its dynamic role in facilitating changes and interactions in the physical world. In the context of living systems, the definition of energy expands to encompass biophysical processes that sustain life. Energetic principles govern growth, reproduction, and homeostasis, as organisms constantly convert energy from one form to another. For example, photosynthesis in plants transforms solar energy into chemical energy, which is then utilized to fuel cellular processes. This concept of energy is vital in understanding ecological dynamics and nutrient cycles within ecosystems. It is evident that life's complexity cannot be decoupled from energy transformations, as these processes enable organisms to maintain order and complexity amidst the natural tendency toward entropy. Thus, within the broader narrative of matter, energy plays an essential role in defining interactions between physical entities and in creating and sustaining the intricate systems that are characteristic of life itself.

2.3. IMPORTANCE OF THE RELATIONSHIP BETWEEN MATTER AND ENERGY IN LIFE

The relationship between matter and energy is foundational to all biological processes that sustain life. In essence, matter provides the structural components of living organisms, while energy facilitates the transformations necessary for biochemical reactions. For instance, the process of photosynthesis exemplifies this relationship; plants convert light energy into chemical energy stored in glucose molecules. This conversion is essential not only for the growth and development of plants but also for animals and humans that rely on plants as food sources. Without the intricate interplay of matter and energy, the cycles of life, including metabolism and reproduction, would be impossible. Therefore, understanding how energy is derived and utilized from the matter enables deeper insights into biological functions and the overall sustainability of ecosystems.

Moreover, the microbiota-gut-brain axis is a compelling illustration of this relationship, where the matter of microbial communities interacts with the energy derived from nutrients to influence brain function and behavior. This connection underscores how microbial metabolites, generated from the interaction between dietary matter and gut bacteria, can impact mental health and neurological conditions. Significant research has been undertaken to investigate how stress and other factors that affect the composition of gut microbiota can lead to changes in energy metabolism and, subsequently, alter brain chemistry. These findings highlight the crucial role that both matter—through microbial composition—and energy—through metabolic processes—play in enhancing our understanding of mental and physical health. As the exploration of this axis continues, it points to potential interventions that could harness this relationship for therapeutic purposes. In conclusion, the relationship between matter and energy is integral to the functioning of all living systems, spanning from basic metabolic processes to complex interactions like the gut-brain axis. Current research indicates that disruptions in this relationship can lead to significant health implications, as demonstrated by the mental health outcomes associated with changes in microbiota composition linked to stress and lifestyle factors (Holmes et al., 2020). Furthermore, the understanding of how energy, originating from the matter within our diet, affects our biological systems can lead to innovative approaches in preventive and therapeutic healthcare strategies (Cryan et al., 2019). By prioritizing this relationship in research and practice, we deepen our comprehension of life processes and may enhance the well-being of individuals within their biological and environmental contexts.

2.4. THE NATURE OF MATTER

Examining the intricate composition of matter reveals its essential role within our universe. Matter is defined as anything that possesses mass and occupies space, existing in various states such as solid, liquid, gas, and plasma. Each of these states is characterized by the arrangement and motion of its constituent particles—atoms, molecules, or ions. For instance, solids maintain a fixed shape due to closely packed particles, whereas gases expand to fill their containers, showcasing a significant variation in energy levels and interactions among particulate matter. Understanding these distinctions is critical, as they highlight the relationship between matter and energy, which governs physical transformations and chemical reactions. As noted in discussions of complex systems, these processes underline the significance of information in structuring interactions, emphasizing the nuanced nature of matter as it interacts with energy and environment (Nightingale et al., 2018). The examination of matter goes beyond its physical states; it delves into the fundamental building blocks of life itself. At the most basic level, matter is composed of atoms, which combine to form molecules, the essential units of biological structures and functions. The abundance of elements like carbon, hydrogen, and oxygen gives rise to the organic compounds fundamental to life, including proteins, nucleic acids, and carbohydrates. These molecules not only serve as structural components but also facilitate the myriad chemical reactions necessary for metabolism and growth. Moreover, the principles of self-organization shed light on how complex biological systems arise from simpler matter. This relational viewpoint sees organisms as dynamic entities shaped by interactions involving both human and non-human elements, implicating matter in a continuous dialogue with energy flows and environmental factors that enhance biological complexity (Nightingale et al., 2018). To grasp the nature of matter fully, it is essential to appreciate its dual role as both a participant in and a product of complex interactions. Matters behavior can be described by various scientific principles, underscoring its interplay with energy and information. For instance, the transition from solid to liquid and then gas reflects the energy changes associated with heat absorption or release, revealing how matter responds to energetic influences. Additionally, the concept of complex systems emphasizes that matter does not exist in isolation but is part of larger networks where the qualities of one component can influence the whole. This interconnectedness aligns with insights from synergetic theories that illustrate how self-organizing properties of matter facilitate emergent behaviors and structures in life forms. In this way, the nature of matter emerges as a tapestry woven from the interactions of energy, information, and the myriad forces that govern life (Haken and Juval, 2016).

2.4.1. States of Matter: Solid, Liquid, Gas

Understanding the diverse states of matter—solid, liquid, and gas—plays a foundational role in both scientific inquiry and daily life. Solids are characterized by tightly packed particles that maintain fixed shapes and volumes, offering stability and structure. The integrity of solids is essential in numerous applications, from building materials to essential tools. Their fixed arrangement allows solids to resist deformation, a feature that supports a variety of engineering and manufacturing processes. The knowledge of solid-state physics can also guide innovations in material science, helping researchers develop lighter and stronger composites. This exploration into the solid state exemplifies the broader relationship between states of matter and technological advancement, demonstrating how our grasp of physical properties can lead to enhanced life quality and resource efficiency. Conversely, liquids present a fascinating contrast to solids in their ability to flow and take the shape of their containers while maintaining a constant volume. The intermolecular forces in liquids are strong enough to keep particles close, yet sufficiently weak to allow them to slide past one another. This unique characteristic makes liquids indispensable in biological systems and industrial processes alike. For instance, the transport of nutrients through blood is a critical application of liquid dynamics in living organisms. Furthermore, insights drawn from the study of liquid states can facilitate advancements in chemical engineering, where understanding the behavior of liquid reactants is vital. Many complex phenomena, including phase transitions and thermodynamic processes, hinge on the properties of liquids, underscoring their importance in scientific research and their applications in technology and health. Gases bring an entirely different set of characteristics, marked by a lack of definite shape or volume. In this state, particles are spaced far apart and move freely, which allows gases to expand and compress. This fluidity makes gases vital for various forms of energy generation and plays a crucial role in ecological systems. For example, the action of gas exchange in respiration and photosynthesis demonstrates the interconnectedness of gases in sustaining life. Moreover, the advancements in understanding gas behaviors lead to pivotal technologies, such as those employed in powering engines or optimizing chemical reactions. As we continue to explore the properties and interrelations of different states of matter, our comprehension of the fundamental principles governing matter and energy transforms our understanding of life itself, revealing intricate connections that enhance our capabilities in various scientific and practical domains (Figure 2.1).

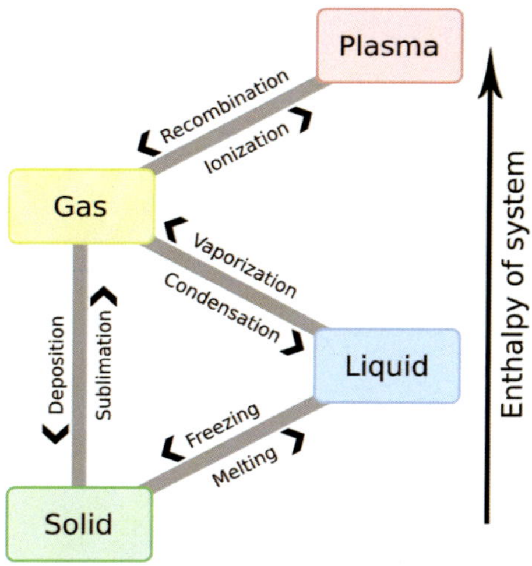

Figure 2.1. *The particulate nature of matter.*

Source: By F l a n k e r, penubag – Own work, Public Domain, https://commons.wikimedia.org/w/index.php?curid=5098161.

2.4.2. Atomic Structure and Elements

Understanding atomic structure is fundamental to grasping the nature of matter and the elements that constitute our universe. Atoms, the building blocks of all matter, consist of three primary subatomic particles: protons, neutrons, and electrons. Protons and neutrons, which reside in the nucleus, determine the atomic number and overall mass of an element. Electrons, on the other hand, occupy energy levels surrounding the nucleus and are crucial in forming chemical bonds between atoms. This arrangement not only dictates the properties of elements but also their reactivity, leading to the vast diversity of compounds we observe in nature. Thus, the study of atomic structure lays the groundwork for understanding chemical interactions that are vital for life itself. The periodic table organizes all known elements based on their atomic structure and characteristics. Each element is classified according to its atomic number, which reflects the number of protons in its nucleus, providing inherent ordering to its physical and chemical properties. Elements are broadly categorized into metals, nonmetals, and metalloids, each exhibiting distinct behaviors. Metals, known for their conductivity and malleability, play critical roles in technology and industry. Nonmetals are essential for various biological processes; for instance, they are integral to molecular formation and energy transfer. Furthermore, as environmental pollutants such as heavy metals accumulate in ecosystems, they pose significant risks to health and biodiversity, as described in (Hazrat Ali et al., 2019). Thus, understanding the periodic classification of elements allows for deeper insights into both chemical behavior and ecological interactions (Figure 2.2).

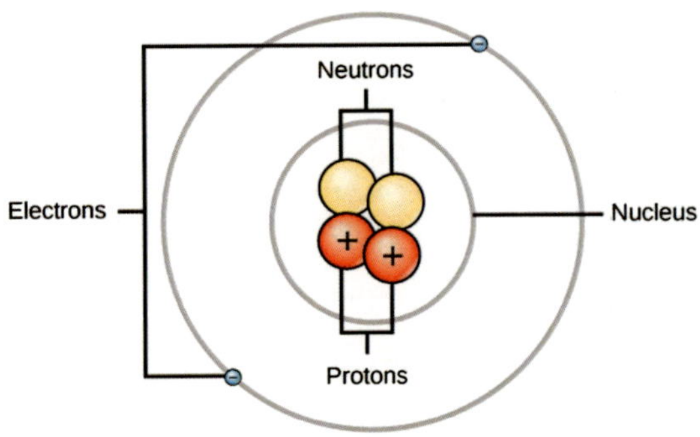

Figure 2.2. *Elements, such as helium, depicted here, are made up of atoms. Atoms are made up of protons and neutrons located within the nucleus, with electrons in orbitals surrounding the nucleus.*

Source: https://uen.pressbooks.pub/introductorychemistry/chapter/the-structure-of-the-atom/.

Elements can heavily influence life on Earth, serving as the foundational substances in biological systems. For example, carbon, hydrogen, oxygen, and nitrogen are the four essential elements in organic molecules that form the basis of proteins, nucleic acids, and carbohydrates, all crucial for maintaining life. The manipulation of these elements at the atomic level leads to the formation of complex molecules with varied functions in cellular processes. Moreover, as technology evolves, applications of our knowledge about atomic structure are becoming increasingly significant. The advent of paradigms such as Software-Defined Networking (SDN), as noted in Kreutz et al (2014), exemplifies the intersection of atomic science and technological advancement, allowing for improved network management that can better support global communications and data transfer systems. Consequently, an understanding of atomic structure not only fosters comprehension of the microcosm of life but also informs broader technological innovations that shape our world.

2.4.3. The Role of Molecules in Biological Systems

Molecules serve as the fundamental building blocks of life, playing critical roles in the structure and function of biological systems. At the cellular level, biomolecules, including proteins, nucleic acids, lipids, and carbohydrates, contribute to various essential processes. For instance, proteins are not only structural components of cells but also serve as enzymes that catalyze biochemical reactions, regulating crucial pathways within living organisms. The intricate interactions between these molecules facilitate life at an incredibly complex level, allowing systems to maintain homeostasis and adapt to their environments. This interplay forms the basis of cellular communication and metabolic pathways, underscoring the essential nature of molecular interactions in sustaining life. The importance of molecules extends beyond mere structural roles; they also participate in the regulation of cellular functions and phenomena. Recent advances in synthetic biology demonstrate how artificial constructs can be created by manipulating molecular components, leading to outcomes that mimic biological functions. As highlighted in recent research, bottom-up synthetic biology is investigating the reconstitution of cellular processes using minimal sets of purified proteins and nucleic acids, which reveals insight into the mechanisms governing life (Göpfrich et al., 2018). By understanding these molecular roles, scientists can unlock new potential for applications in medicine, biotechnology, and environmental sustainability. This demonstrates that the study of molecules is not just a theoretical exercise but has profound implications for the future of health and technology. Moreover, the interactions among molecular components can give rise to emergent properties in biological systems. For example, liquid-liquid phase separation, a process not commonly associated with living organisms, is now recognized as a significant mechanism for the organization of biomolecules within cells. This phenomenon allows proteins and RNA to form dynamic droplets that facilitate various cellular functions, such as signal transduction and gene expression (Shin et al., 2017). Understanding these complex behaviors is vital, as they can both support cellular efficiency and contribute to pathological conditions when dysregulated. By critically analyzing the roles of molecules in biological systems, researchers can deepen their understanding of disease mechanisms, leading to innovative therapeutic strategies based on restoring proper molecular interactions (Figure 2.3).

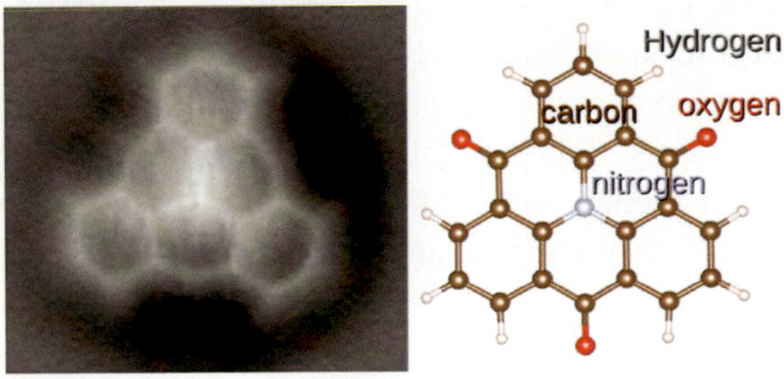

Figure 2.3. *Molecules.*

Source: By Prokop Hapala et al. – [1], CC BY 4.0, https://commons.wikimedia.org/w/index.php?curid=55297341.

2.5. THE CONCEPT OF ENERGY

The interplay between energy and matter fundamentally shapes the dynamics of our universe. Energy, defined as the capacity to perform work, exists in various forms, including kinetic, potential, thermal, and chemical. These diverse manifestations enable essential processes, from the movement of physical objects to the biochemical reactions that sustain life. Understanding energy transfer—how it moves from one form to another—is crucial in grasping natural phenomena. For example, photosynthesis exemplifies this concept, where plants convert solar energy into chemical energy stored in glucose. This transformation not only serves as a source of energy for the plants themselves but also forms the foundation of the food chain, underscoring the complex interdependencies of life and energy. Energy functions extend beyond mere mechanical processes to encompass social and technological dimensions that influence societal development. Recent initiatives, such as Germany's Industry 4.0 and China's Made-in-China 2025, reveal the critical integration of advanced energy systems into industrial practices (Xu et al., 2018). These projects underscore an ongoing transition toward more efficient energy use, reflecting growing awareness of sustainability issues. The research into sustainability transitions, as indicated by the push for cleaner energy sources and more responsible consumption patterns, exemplifies how modern society's challenges are deeply intertwined with energy. This perspective not only highlights the significance of energy in fostering innovation but also calls for a coherent approach to governing energy transitions that considers ethical implications and the involvement of diverse stakeholders. Ultimately, energy is both a driving force and a lens through which we can understand the intricacies of existence. It aids in unraveling the grand challenges of sustainability while urging us to reconsider our consumption habits, governance structures, and technological advancements. As the ongoing research indicates, bridging the gap between energy and societal needs is imperative for achieving lasting solutions to environmental dilemmas (Köhler et al., 2019). Through a continuous exploration of energy's various dimensions—from its fundamental role in life processes to its strategic importance in industrial transitions—we can better navigate the complexities of our ever-evolving world. In this way, energy not only fuels our physical reality but also shapes the frameworks within which we address the pressing issues of our time (Figure 2.4).

Figure 2.4. *Understanding the concept of energy.*

Source: By Dbc334 (first version); Jynto (second version) – Water-3D-vdW.png, Public Domain, https://commons.wikimedia.org/w/index.php?curid=1332739.

2.5.1. Types of Energy: Kinetic, Potential, Thermal

When examining the different forms of energy, kinetic energy stands out as a crucial element in understanding physical interactions. Defined as the energy possessed by an object due to its motion, kinetic energy is proportional to the mass and the square of the velocity of that object. For example, a rolling ball, a speeding car, or even the wind all exemplify kinetic energy in action. This form of energy plays a vital role in various systems and processes, particularly in the realms of mechanics and dynamics. As kinetic energy transfers from one object to another, it can change forms, illustrating the principle of energy conservation. This adaptability makes kinetic energy essential not just in theoretical physics but in practical applications, such as engineering and ecological sustainability, where the management of moving systems is crucial for efficiency and innovation. On the other hand, potential energy serves as a counterbalance to kinetic energy, representing the stored energy in an object due to its position or configuration. This form of energy can manifest in various ways, such as gravitational potential energy, which relies on the height of an object relative to the ground, or elastic potential energy, stored in materials like rubber bands when stretched. Understanding potential energy is critical in fields ranging from architecture to environmental science, as it affects the stability and behavior of structures and ecosystems. For instance, the concept of potential energy is integral in designing buildings to withstand forces acting upon them, ensuring safety and functionality. Therefore, potential energy is not merely a passive form of energy; it plays an active role in the dynamics of matter and life, impacting how organisms adapt to their environments and utilize energy for various functions.

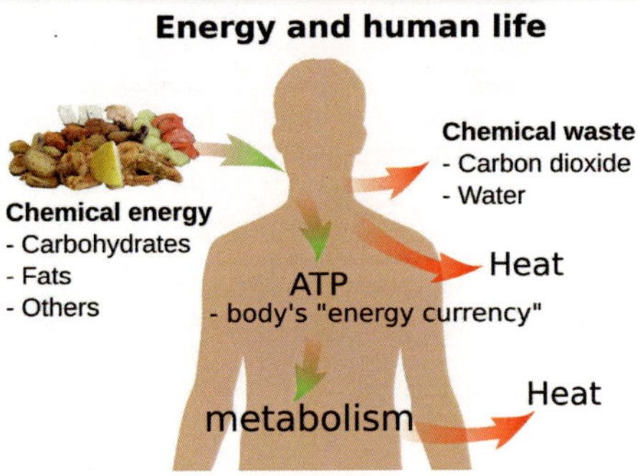

Figure 2.5. *Types of energy.*

Source: By Mikael Häggström – All images are of Public Domain license. File:Pikilia.JPG, Public Domain, https://commons.wikimedia.org/w/index.php?curid=6343781.

Thermal energy, the third type of energy in our exploration, refers to the energy that arises from the temperature of an object, which is a measure of the kinetic energy of its particles. When heat is added to a system, the particles move faster, increasing the thermal energy and resulting in temperature changes. This form of energy is fundamental to various biological and physical processes.

For example, the human body relies on thermal energy regulation to maintain homeostasis, while industries harness thermal energy in processes like electricity generation and manufacturing. Furthermore, the principles surrounding heat transfer and thermal dynamics have led to advancements in technologies such as thermal insulation and energy-efficient systems. By connecting thermal energy with both natural processes and technological innovation, we gain insight into how energy influences the fabric of life and matter in a constantly changing world, where the dynamics of energy remain central to understanding sustainability and environmental impact (Figure 2.5).

2.5.2. The Law of Conservation of Energy

One of the most fundamental principles governing the interactions within our universe is the law of conservation of energy. This law posits that energy cannot be created or destroyed; it can only be transformed from one form to another. This idea permeates various aspects of science, underpinning both classical mechanics and thermodynamics. For instance, the transformation of potential energy into kinetic energy during a falling object's descent illustrates this concept. This equilibrium is not only crucial for physical systems but also applies to biological systems, where energy transitions occur through processes such as cellular respiration. Hence, understanding this law provides a foundational framework for analyzing how energy flows and transforms within both living and non-living entities. In the context of life sciences, the law of conservation of energy is particularly significant. Biological systems typically rely on energy transformations to sustain metabolic processes essential for survival. Plants, for example, convert solar energy into chemical energy through photosynthesis, which sets the stage for more complex energy transformations within ecological systems. The subsequent transfer of energy through food chains illustrates how energy is conserved, as energy flows from producers to consumers while maintaining the overall energy balance within ecosystems.

By recognizing these patterns, we gain insights into the delicate interdependence between organisms and their environments, allowing a deeper understanding of ecological dynamics and the sustainability of life on Earth. While the law of conservation of energy is a well-established principle in physics, its implications extend beyond mere theory into practical applications in the field of synthetic biology. Researchers are increasingly aiming to construct synthetic life forms through understanding and manipulating energy pathways at the molecular level. These efforts highlight the intricate relationships between energy consumption, biosynthesis, and cellular function. As detailed in recent studies, simple reaction routes can operate essential functions by efficiently generating ATP and maintaining electrochemical gradients (Sikkema et al., 2019). Moreover, as ethnobotany bridges human interactions and plant life, it showcases how energy dynamics are essential for sustaining traditional practices and medicinal uses of various species (Chavda et al., 2022). Therefore, appreciating the law of conservation of energy in both natural and synthetic contexts enhances our understanding of life processes and the interconnectedness of matter and energy in our universe (Figure 2.6).

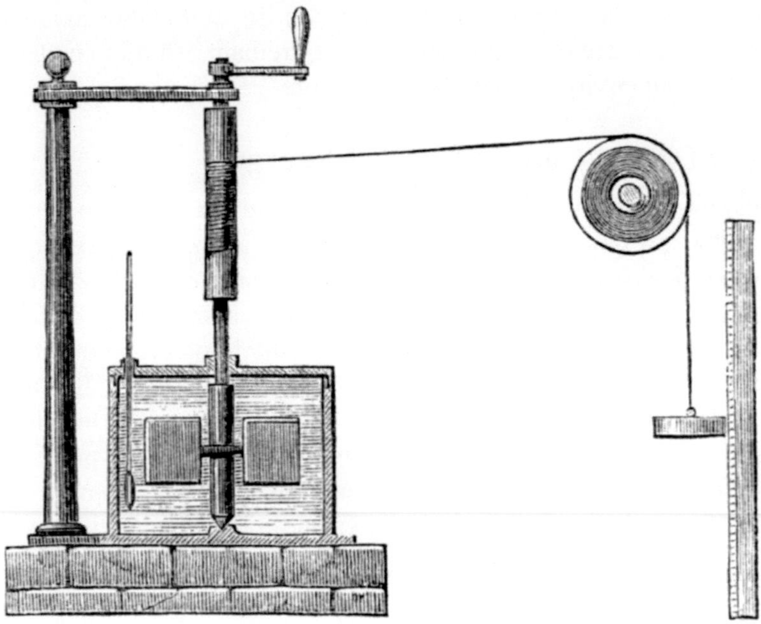

Figure 2.6. *The Law of conservation of energy: Joule's apparatus.*

Source: By Jacob Abbott – Harper's New Monthly Magazine, Vol. 39, No. 231, p. 327, August 1869, Public Domain, https://commons.wikimedia.org/w/index.php?curid=1527228.

2.5.3. Energy Transformations in Living Organisms

Living organisms rely on energy transformations to sustain life processes, which fundamentally hinges on the conversion of one form of energy into another. At the cellular level, this metabolic machinery facilitates the transition of nutrients into usable energy. For instance, during cellular respiration, glucose is metabolized to produce adenosine triphosphate (ATP), the energy currency of cells. This process underscores the connection between food intake and energy availability, showcasing how energy from the sun is captured through photosynthesis in plants, which, in turn, becomes a vital source of energy for animals through consumption. Hence, a complex web of energy transfer underpins the survival of all living organisms, emphasizing the significance of understanding these transformations in the context of ecological interactions and food webs. Energy transformations in living organisms are not only restricted to physical processes; they also underpin critical biochemical interactions, influencing behavior and health. Recent studies highlight the intricate relationship between the gut microbiota and energy metabolism. For example, microbial metabolites such as short-chain fatty acids, which are produced during the fermentation of dietary fibers, have been shown to impact host energy homeostasis and immune responses, thus affecting the overall well-being of the organism. This interdependence is a clear indication of how energy transformations can extend beyond individual organisms, demonstrating their collective impact on ecosystems and the environment. Addressing these connections contributes to a deeper understanding of energy efficiency in biological systems, paving the way for potential applications in nutrition and health interventions. As the Anthropocene era presents unprecedented challenges, the role of microorganisms in energy transformations remains crucial for sustaining ecosystems under climate change. Microorganisms are foundational to biogeochemical cycles—transforming energy and matter in ways that support higher trophic levels. For instance, the breakdown of organic matter through microbial processes plays a pivotal role in nutrient cycling and energy flow within ecosystem dynamics. As outlined in (Ricardo Cavicchioli et al., 2019), these unseen majority organisms not only contribute to the resilience of environmental processes but also influence human applications in biotechnology and climate mitigation strategies. Recognizing and addressing the interactions between microorganisms and energy transformations will be critical for achieving sustainability in our rapidly changing world. Thus, understanding these dynamics is essential for both ecological conservation and addressing societal challenges.

2.6. THE INTERPLAY BETWEEN MATTER AND ENERGY

The relationship between matter and energy is foundational to understanding the processes that govern life on Earth. Matter, in its various forms, serves as the building blocks for all biological entities, while energy acts as the catalyst that sustains the mechanisms of life. This interplay is evident in the way cells utilize energy to convert matter into usable forms. For instance, during the process of photosynthesis, plants absorb sunlight, a form of energy, to transform carbon dioxide and water into glucose and oxygen. This transformation not only illustrates the conversion of energy into a chemical form but also highlights how matter becomes integral to the life-sustaining processes of living organisms. The intricate balance between the two allows for biological systems to thrive, adapt, and evolve in their environments. When examining the molecular dynamics underlying cellular function, the phase separation of biomolecules reveals significant insights into this interplay.

Research indicates that the condensation of proteins and nucleic acids into liquid-like droplets within cells exemplifies how energy influences matter at a microscopic level (Yongdae Shin et al., 2017). These droplets, driven by liquid-liquid phase separation, organize cellular components without the need for membranes. This organization is vital for regulating various biochemical reactions necessary for life. Such phenomena underscore how energy can dictate the spatial and temporal organization of matter, allowing for efficient cellular processes. The behavior of biomolecules in these droplets also suggests that energy-driven phase transitions are crucial for maintaining homeostasis in living systems, further illustrating the dynamic nature of matter and energy in biological contexts. Ultimately, the interplay between matter and energy encapsulates the essence of life's complexity. Matter provides the structural basis for biological organisms, while energy powers the myriad processes that sustain them. The continuous flow of energy through ecosystems, from the sun to plants and subsequently to consumers, showcases this relationship's cyclic nature. This interconnectedness is profound; as energy transfers between organisms, it transforms matter into various forms—from organic compounds to cellular structures—enabling life's continuity. Understanding this interplay not only enhances our grasp of biological processes but also raises critical questions about the origins of life and the nature of existence itself. By interlinking matter and energy, we can begin to appreciate the intricate dance that sustains life on Earth (Göpfrich et al., 2018) (Figure 2.7).

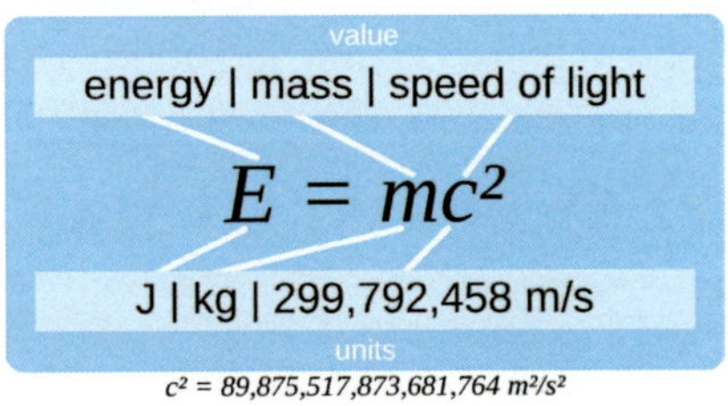

Figure 2.7. *Relations between matter, energy, and information.*

Source: By JTBarnabas – Own work, CC BY-SA 3.0, https://commons.wikimedia.org/w/index.php?curid=25129322.

2.6.1. Chemical Reactions and Energy Changes

Chemical reactions are fundamental processes that involve the transformation of substances through the breaking and forming of chemical bonds. These reactions can be classified into several types, including combination, decomposition, and displacement reactions, each characterized by unique energy changes. For instance, when a reaction occurs, it may either absorb energy from the surroundings, indicating an endothermic process, or release energy, denoting an exothermic reaction. Understanding these energy changes is essential, as they govern the spontaneity and feasibility of reactions under different conditions. The intricate balance between energy absorbed and released during a reaction reflects the driving forces of chemical processes and underpins the principles of thermodynamics vital in both natural phenomena and industrial applications. In addition to the classification of reactions, the role of chemical energy in biological systems is pivotal. Organisms rely on the transformation of energy stored in chemical bonds to sustain life processes. For example, during cellular respiration, glucose undergoes oxidation, releasing energy that is captured in the form of ATP, a molecule crucial for cellular work. The successful transfer and transformation of energy through various metabolic pathways, as seen in both aerobic and anaerobic pathways, demonstrate the interconnectedness of chemical reactions with life. This is paralleled in the sulfur cycle, where microorganisms partition energy from sulfide through diverse metabolic processes, supporting their survival and function in ecosystems (Jørgensen, 2021). The delicate balance of these energy changes highlights the vital role of chemical reactions in sustaining life on Earth. Moreover, the implications of chemical reactions extend beyond biological systems to encompass environmental and health considerations. Certain chemical pollutants, such as particulate matter and nitrogen oxides, result from anthropogenic activities and pose significant risks to human health and global ecosystems.

Figure 2.8. *A thermite reaction using iron(III) oxide. The sparks flying outwards are globules of molten iron trailing smoke in their wake.*

Source: https://en.wikipedia.org/wiki/Chemical_reaction.

The energy changes associated with the formation and breakdown of these pollutants are crucial for understanding their long-term effects. For example, when volatile organic compounds (VOCs) undergo chemical reactions in the atmosphere, they can lead to increased ozone formation, impacting air quality and public health (Manisalidis et al., 2020). Thus, the interplay of chemical reactions and energy changes not only shapes the processes of life but also determines the state of our environment, underscoring the need for a multifaceted understanding of chemistry as it relates to energy dynamics and its broader implications on health and sustainability.

2.6.2. The Role of Energy in Metabolic Processes

Metabolic processes are fundamentally driven by energy, which is essential for maintaining the structural and functional integrity of living organisms. Energy is harnessed primarily through chemical reactions during metabolism, allowing cells to convert nutrients into usable forms. At the heart of this process lies ATP, the primary energy currency of the cell. Through cellular respiration, organisms break down glucose and other organic molecules, facilitating the transfer of energy necessary for various biochemical reactions. This relationship between energy and metabolism highlights the interconnectedness of biochemical processes and life itself. The efficient conversion and management of energy resources are critical; when metabolic pathways are disrupted, as seen in various diseases, overall cellular function declines, demonstrating the vital role energy plays in sustaining life. Understanding the implications of energy in metabolic processes involves recognizing how energy deficiencies can lead to severe health issues. For instance, insufficient energy can hinder nutrient absorption and utilization, creating a cascade effect that results in systemic problems. More than 820 million people currently experience energy malnutrition, which starkly illustrates a global health crisis linked to energy imbalance within metabolic systems (Willett et al., 2019). Moreover, conditions such as obesity and related non-communicable diseases originate from disrupted energy balance and inefficient metabolic pathways, emphasizing the importance of energy in maintaining metabolic homeostasis. Thus, the study of energy within metabolic processes is essential not only for understanding individual health but also for addressing broader public health challenges.

Ultimately, the role of energy in metabolic processes underscores an intricate relationship between energy intake, utilization, and health. Through mechanisms such as neurovascular coupling, energy supply and demand are finely tuned to ensure that metabolic processes align with the physiological needs of the organism (Iadecola, 2017). An optimal balance is crucial, as both energy excess and deficiency can lead to significant metabolic disturbances. This dynamic equilibrium is particularly evident in the brain, where the precise regulation of energy delivery affects cognitive function and overall brain health. As we deepen our understanding of these metabolic processes and their reliance on energy, we can better develop strategies to promote health, prevent disease, and enhance our comprehension of the intricate web of life, underscoring the foundational theme of Matter, Energy, and Life (Figure 2.9).

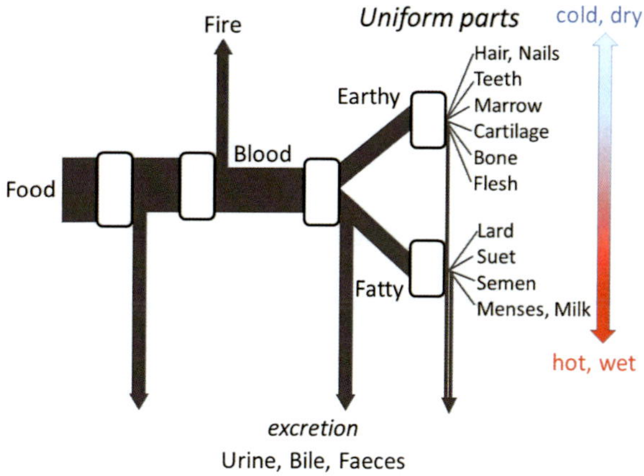

Figure 2.9. *Metabolism.*

Source: By Chiswick Chap – Own work, CC BY-SA 4.0, https://commons.wikimedia.org/w/index.php?curid=55539096.

2.6.3. Matter-Energy Equivalence in Biological Systems

The concept of matter-energy equivalence, famously encapsulated by Einsteins equation ($E=mc^2$), has profound implications for biological systems. At the most fundamental level, this principle underscores that all biological processes, from cellular metabolism to physiological functions, rely on strict conversions and transfers of energy and matter. In living organisms, energy is obtained from food sources through chemical reactions that break down macromolecules, releasing energy stored in chemical bonds. For example, the process of cellular respiration converts glucose and oxygen into carbon dioxide, water, and energy in the form of ATP, highlighting the transformation of matter into usable energy. This interplay illustrates that energy does not simply flow into biological systems; it is intricately bound up with mass, as every organism is a complex assembly of atoms and molecules undergoing constant transformation and energy exchange. Furthermore, the relationship between matter and energy extends to the growth and reproduction of an organism, characterized by biological processes that embody the principles of thermodynamics. An organism requires a continuous input of energy to maintain homeostasis and support life processes, with matter serving as both a substrate and a product in these energetic exchanges. For instance, plants utilize photosynthesis to convert light energy into chemical energy while simultaneously embodying the transformation of light into the organic matter of glucose, which forms the basis of the food chain. As highlighted in the realm of methanol as an alternative fuel, efficient energy transfer mechanisms can serve as models for understanding how biological systems optimize energy usage. By generating energy in a manner analogous to fuel cell technology, living organisms efficiently harness and convert energy, demonstrating a remarkable sophistication inherent in biological systems (Verhelst et al., 2018). This intricate dance between matter and energy culminates in the understanding that all living systems are subject to the same physical laws governing energy transformations. The guidelines drawn upon in pharmacology also reflect this interconnectedness, as understanding biological mechanisms relies on a clear framework of how energy is utilized and converted within living organisms. Studying bipolar disorder or any biological condition necessitates recognizing the fundamental energetic exchanges in cellular systems, where imbalances may lead to dysfunctions (Goodwin et al., 2016). The significance of matter-energy equivalence extends beyond mere biochemical reactions, revealing a broader spectrum of life's processes, including signaling pathways, metabolic reactions, and even neurological functions. As such, exploring matter-energy equivalence unveils profound insights into the sustainability of life and the underlying principles that dictate the behavior of living systems, emphasizing the harmonious relationship between matter and energy throughout biological existence.

2.7. MATTER AND ENERGY IN ECOSYSTEMS

Ecological systems are fundamentally driven by the interplay of matter and energy, with each impacting the other in a continuous cycle. The transformation of energy through various trophic levels illustrates this dynamic. Producers, such as plants, convert solar energy into chemical energy via photosynthesis, forming the foundation of the food web. This process not only captures energy but also facilitates the cycling of essential matter, like carbon and nitrogen, through biological communities. As energy flows through the ecosystem—from producers to consumers and decomposers—matter is also recycled, underscoring the interdependence of biotic and abiotic components. Understanding these relationships is vital, especially given the challenges posed by human activities that threaten ecosystem stability and function. Moreover, the challenges facing global food systems exemplify the complex relationship between matter, energy, and human health. The pressing need to provide healthy diets for a growing population underscores the necessity of sustainable food production methods, which require both energy efficiency and material conservation. Food production processes must be optimized to ensure that energy used for cultivation, processing, and distribution is balanced with the nutritional quality of the food produced. As highlighted by Willett et al (2019), while calories are theoretically abundant, the quality of these calories directly impacts human health, emphasizing the critical role of sustainable practices in harnessing energy and matter effectively. Transitioning to agricultural systems that prioritize sustainability is essential to preserve both the environment and the health of populations. To comprehensively address the interactions of matter and energy within ecosystems, incorporating advanced scientific methodologies such as metagenomics becomes increasingly important. This innovative approach allows researchers to explore the complex microbial communities that play crucial roles in nutrient cycling and energy flow (Garza et al., 2015). By understanding the various microorganisms' contributions, scientists can uncover critical pathways through which energy is transferred and the matter is decomposed within ecological systems. Such insights into microbial networks enhance our understanding of ecosystem dynamics and resilience, particularly as we face ecological challenges such as climate change and biodiversity loss. Thus, a deeper exploration of these microbial processes not only enriches our knowledge of matter and energy but also presents pathways for more sustainable management of ecosystems.

2.7.1. Energy Flow Through Food Chains

In ecosystems, energy flow is fundamentally governed by the principles of food chains, which delineate the interactions among producers, consumers, and decomposers. At the base of these chains are primary producers, typically plants or phytoplankton, that harness solar energy through photosynthesis. This process not only provides energy for their growth but also forms the foundation of energy transfer within the ecosystem. As illustrated by the intricate relationships between various trophic levels, the energy captured by producers is passed on to herbivores and subsequently to carnivores, highlighting the linear progression of energy flow. However, energy is not conserved; each transfer results in significant loss, primarily as heat, due to the laws of thermodynamics, which limit the number of trophic levels in a given food chain. The impact of energy flow can be observed through the dynamics of the food web, where interdependencies among species influence overall ecosystem stability. As energy diminishes with each successive level, apex predators are typically fewer in number compared to primary producers. These predator-prey dynamics are crucial for regulating populations, ensuring that no single species dominates the ecosystem. Moreover, the introduction of a new species or the extinction of a current one can dramatically alter the balance, affecting energy distribution. Thus, understanding these relationships provides insight into ecological responses to environmental changes, emphasizing the importance of biodiversity for maintaining energy flow and ecosystem resilience. Recent studies have begun to reveal the underlying mechanisms of the microbiota-gut-brain axis and its role in regulating biological systems, which could have broader implications for understanding energy flow in ecological contexts. The gut microbiota, which varies among individuals and species, is influenced by factors such as diet, stress, and genetics, as outlined in research highlighting the complex interactions that affect health and behavior (Cryan et al., 2019). These interactions suggest that nutrient availability and energy allocation within organisms may also be influenced by microbial communities, further complicating traditional views of energy flow through food chains. As we advance our understanding of these interactions, it becomes evident that the study of energy flow must consider not just direct transfers between trophic levels, but also the intricate biological and ecological interdependencies that govern life on Earth (François Mach et al., 2019) (Figure 2.10).

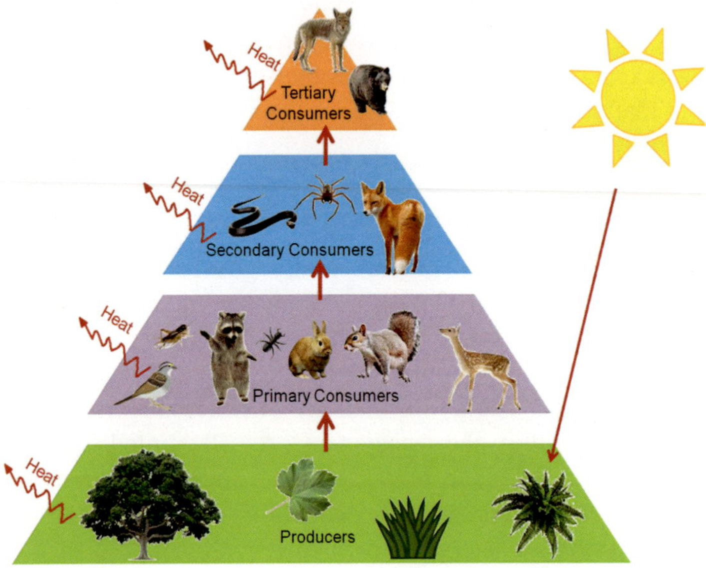

Figure 2.10. *Energy flow.*

Source: https://www.mrgscience.com/42-energy-flow.html.

2.7.2. The Cycling of Matter in Ecosystems

The cycling of matter in ecosystems is a fundamental process that enables life to thrive on Earth. Matter, including essential nutrients like carbon, nitrogen, and phosphorus, continually moves through various biotic and abiotic components of the ecosystem. This cycling occurs in several interconnected processes such as photosynthesis, decomposition, and nutrient uptake. Plants absorb inorganic matter from the soil and atmosphere, transforming it into organic forms through photosynthesis. These organic materials then move through food webs as they are consumed by herbivores and subsequently transformed by higher trophic levels. The efficiency of these cycles is vital; disruption in any phase can lead to ecosystem imbalances, demonstrating the intricate interdependence of life forms within their habitats. Moreover, human activity poses significant threats to the natural cycling of matter. The overuse of fertilizers in agriculture, for instance, disrupts nitrogen and phosphorus cycles, leading to eutrophication in aquatic ecosystems. This process increases nutrient loads in water bodies, causing excessive growth of algae which can deplete oxygen levels, harming aquatic life. The consequences extend beyond the immediate environment, affecting food systems that rely on healthy ecosystems to function properly. As discussed, "more than 820 million people have insufficient food and many more consume low-quality diets" partly due to disruptions in these natural cycles (Willett et al., 2019). Sustainable practices in agriculture and industry are increasingly necessary to restore balance and ensure the continued cycling of matter, which supports both environmental health and human well-being. Furthermore, the physical environment itself plays a critical role in facilitating the cycling of matter. Oceanic systems, in particular, are experiencing notable changes due to human-induced factors like climate change and pollution, which contribute to deoxygenation in marine environments. As observed, "rising nutrient loads coupled with climate change.. are changing ocean biogeochemistry and increasing oxygen consumption," leading to a decline in viable habitats for marine life (Breitburg et al., 2018). The loss of oxygen not only affects fish stocks but threatens broader ecological systems that rely on marine biodiversity. With diminishing resources, both human communities and ecosystems face potential collapses, underscoring the urgency of addressing our impact on the cycling of matter. Taking action to mitigate human influences will be essential for maintaining the resilience of ecosystems essential for life on Earth.

2.7.3. The Impact of Human Activity on Matter and Energy Cycles

Human activity has significantly altered the natural cycles of matter and energy, leading to profound environmental consequences. Industrialization, urbanization, and agricultural expansion have disrupted biogeochemical cycles, particularly the carbon, nitrogen, and phosphorus cycles. For instance, the excessive burning of fossil fuels has increased the concentration of carbon dioxide in the atmosphere, which exacerbates the greenhouse effect and contributes to climate change. This alteration not only influences weather patterns but also disrupts ecosystems, as species must adapt to rapidly changing conditions. Furthermore, nitrogen, and phosphorus runoff from agricultural fertilizers has led to eutrophication in aquatic systems, resulting in dead zones where marine life cannot survive. These changes highlight the urgent need for sustainable practices that minimize our impact on these vital cycles. In addition to their direct effects on ecological processes, human activities also impact the energy flow within ecosystems. The conversion of land for agriculture and urban development alters habitats, affecting local flora and fauna and consequently disrupting energy transfer within food webs. The depletion of resources through unsustainable practices, such as overfishing and deforestation, not only affects biodiversity but also diminishes the amount of energy available to support various trophic levels in an ecosystem. Essentially, as energy flow is impaired, the stability of ecosystems becomes jeopardized, resulting in decreased resilience to environmental changes.

Figure 2.11. *Humans: A force of nature.*

Source: https://en.wikipedia.org/wiki/File:Akha_cropped_hires.JPG.

Recognizing the interconnectedness of human activities with energy dynamics emphasizes the importance of adopting practices that promote ecological balance and resource sustainability. Moreover, our consumption habits reflect a deeper interplay with the cycles of matter and energy, ultimately shaping our environmental footprint. Industrial food systems, which dominate global food production, exemplify how human-induced changes can lead to unsustainable practices that threaten both human health and environmental integrity. The current global food landscape produces excessive waste and generates more greenhouse gas emissions than necessary, thereby amplifying the crisis of climate change (Willett et al., 2019). Simultaneously, the reliance on resource-intensive agricultural practices has been linked to biodiversity loss, soil erosion, and the depletion of water resources (Azzalini et al., 2019). To mitigate these negative impacts, societies need to reassess consumption patterns and implement more sustainable food systems that nurture human health while fostering environmental sustainability. This shift not only addresses the intricate dynamics of matter and energy cycles but also promotes a healthier planet for future generations (Figure 2.11).

2.8. THE ROLE OF MATTER AND ENERGY IN CELLULAR PROCESSES

Understanding the intricate relationship between matter and energy is fundamental to the biological processes that sustain life at the cellular level. Matter, composed of atoms and molecules, forms the building blocks of all living organisms. Cells utilize various macromolecules—like proteins, lipids, carbohydrates, and nucleic acids—for structural integrity and functional roles. For instance, proteins, formed from amino acids, play critical roles in catalyzing biochemical reactions as enzymes and facilitating cellular communication. Simultaneously, matter undergoes transformation through metabolic pathways, which are energized processes that convert substrates into usable forms. This constant exchange underlines the connection between matter and energy as essential components driving cellular function. Energy is primarily derived from the chemical bonds in nutrients and is harnessed through metabolic processes, notably cellular respiration and photosynthesis. During cellular respiration, glucose molecules are broken down, releasing energy stored in their bonds to produce ATP, the energy currency of the cell. This energetic transformation is vital for sustaining cellular activities, such as biosynthesis, transport, and cellular division. Moreover, energy from ATP powers various cellular mechanisms, ensuring proper function and adaptability in response to environmental stimuli.

In plant cells, photosynthesis captures sunlight, converting solar energy into chemical energy stored within glucose molecules. This process illustrates the dual role of energy as both a resource for cellular activity and a driver of matter conversion, underscoring the need for energy input in biological systems. The interplay between matter and energy is not only limited to sustaining individual cellular functions but also extends to broader physiological processes and homeostasis within organisms. For example, the energy and matter derived from food intake must be balanced with energy expenditure for an organism to maintain a stable internal environment. Importantly, disruptions in this balance can lead to metabolic disorders or diseases that impact both cellular and systemic functions. Furthermore, as Uhlén et al (2015) emphasize, protein expression across various tissues reflects how energy and matter contribute to organismal complexity and health. Understanding the intricacies of how matter and energy interact within cells enhances our grasp of life itself, illustrating their critical role in both individual cells and entire biological systems.

2.8.1. Cellular Respiration and Energy Production

One of the fundamental processes that sustain life is cellular respiration, which is the method through which cells convert nutrients into energy. This intricate biochemical process primarily occurs in the mitochondria, often referred to as the powerhouse of the cell. Cellular respiration involves several stages—glycolysis, the citric acid cycle, and oxidative phosphorylation. During glycolysis, glucose is broken down into pyruvate, generating a net gain of two ATP molecules and two NADH electron carriers. The subsequent citric acid cycle further decomposes pyruvate, releasing carbon dioxide while producing more electron carriers in the form of NADH and FADH2. Finally, oxidative phosphorylation utilizes these carriers to produce the bulk of ATP through a series of redox reactions and a proton gradient, culminating in the production of water. This multi-step process showcases the efficient transformation of energy found in organic molecules into a usable form for cellular activities. Energy production through cellular respiration is not merely a linear pathway; it encompasses critical regulatory mechanisms that integrate with cellular demands. ATP, the energy currency of the cell, is tightly regulated to ensure a balanced supply in response to varying metabolic needs. For instance, high levels of ADP and low energy state within the cell stimulate pathways associated with ATP production, indicating a need for increased respiration. In contrast, excessive ATP can inhibit certain enzymatic steps, such as those in the citric acid cycle, showcasing negative feedback mechanisms that ensure homeostasis. Additionally, cellular respiration is connected to other metabolic processes, such as fermentation, providing alternative routes for energy production under anaerobic conditions. The interplay between these pathways, alongside the dynamic regulation of ATP, reflects the intricate adaptability of cellular respiration in response to environmental changes and energy requirements. Moreover, the implications of cellular respiration extend beyond mere energy production; they play a pivotal role in ecological and evolutionary contexts. The ability of organisms to efficiently extract energy from carbohydrates has shaped the evolution of metabolic pathways across different species. Aerobic respiration yields approximately 36 to 38 ATP molecules per glucose molecule, significantly higher than anaerobic processes like fermentation, which generates only 2 ATP. This efficiency underscores why aerobic organisms have flourished in oxygen-rich environments. Furthermore, the byproducts of cellular respiration—namely carbon dioxide and water—are integral to various biogeochemical cycles, notably the carbon cycle. The respiration of living organisms contributes to atmospheric CO_2 levels, which plants subsequently utilize in photosynthesis, forming a symbiotic relationship that sustains life on Earth. Thus, the significance of cellular respiration extends well beyond individual cellular activity, highlighting its essential role in both biological integrity and ecosystem functionality (Figure 2.12).

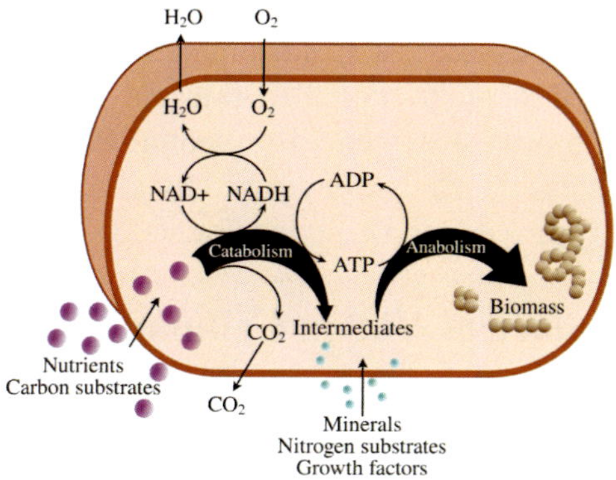

Figure 2.12. *Cellular respiration.*

Source: By D. Ilyin: vectorization – File:Metabolism.png by Linares-Pastén, J. A, CC0, https://commons.wikimedia.org/w/index.php?curid=150135192.

2.8.2. Photosynthesis and the Conversion of Energy

The process of photosynthesis is fundamentally vital for life on Earth as it converts light energy into chemical energy, specifically glucose, which serves as a primary energy source for many organisms. Plants, algae, and certain bacteria harness sunlight through pigments like chlorophyll to initiate this complex sequence of reactions. During photosynthesis, carbon dioxide from the atmosphere and water absorbed by the roots are transformed into glucose and oxygen, a process that underpins the food web and sustains various life forms. This conversion not only fuels plant growth but also indirectly supports herbivores and carnivores, emphasizing its critical role in biospheric energy dynamics. Moreover, photosynthesis contributes significantly to the Earth's carbon cycle, acting as a natural regulator for atmospheric composition. Through the uptake of carbon dioxide, it helps mitigate the greenhouse effect, which is increasingly threatened by anthropogenic activities and climate change. The role of photosynthesis in sequestering carbon has become a focal point for environmental studies aimed at understanding how biological processes can offset human-induced climate impacts. As stated in recent literature, the interplay between photosynthetic activity and climate variations highlights the importance of this process in maintaining ecological balance (Cavicchioli et al., 2019). In concert with the nitrogen cycle, which is essential for synthesizing amino acids and proteins, the role of microorganisms in photosynthesis and energy conversion complements the broader biogeochemical cycles on Earth. Microorganisms, such as nitrogen-fixing bacteria, are crucial for making nitrogen bioavailable to plants, thereby enhancing their capacity for photosynthesis and energy conversion (Kuypers et al., 2018). This synergy between microbial activity and plant processes underscores a sophisticated network of life, where energy conversion is intricately linked to nutrient cycling. Recognizing the interdependencies within these systems is essential for developing sustainable practices that can address the challenges posed by climate change and resource depletion, as humanity strives for an environmentally sustainable future.

2.8.3. The Significance of ATP in Cellular Functions

ATP serves as the primary energy currency within biological systems, playing a crucial role in driving numerous cellular functions. At its core, ATP is composed of adenosine and three phosphate groups, which together store potential energy that can be readily harnessed for cellular activities. When the high-energy bonds between the phosphate groups are broken—most notably during hydrolysis—this energy is released and can be employed to power essential processes such as muscle contraction, active transport, and biosynthesis. Without ATP, cells would be unable to perform the myriad functions necessary for life, exhibiting the fundamental significance of this molecule in sustaining cellular activities and overall organismal health. Furthermore, its role extends beyond mere energy provision; it is intricately involved in signaling and regulation within cells. The molecule participates in various enzymatic reactions, often functioning as a cofactor to facilitate the action of enzymes. For instance, in metabolic pathways, ATP acts as a substrate for enzymes that catalyze glycolysis and the citric acid cycle, linking metabolic energy production directly to cellular function. Additionally, ATP mediates signaling pathways within cells, influencing processes such as cell growth, differentiation, and apoptosis. This dual role as both an energy carrier and a regulatory molecule underscores the complexity of the functions in maintaining cellular homeostasis amid constantly changing environmental conditions. The importance of ATP can also be observed in relation to health and disease. For example, cellular impairment resulting from insufficient ATP production can lead to conditions such as insulin resistance, which is linked to cardiovascular disease and other metabolic disorders (Ormazábal et al., 2018). Similarly, the disruption of ATP-dependent processes can adversely affect nitrogen cycling in ecosystems, highlighting the biochemical interconnections between ATP and broader environmental systems (Kuypers et al., 2018). Addressing these implications necessitates an understanding of how ATP is integral not only to individual cellular functions but also to the functioning of entire biological systems. Hence, comprehending the significance of ATP is essential for advancing both cellular biology and medical research, as it helps to unveil the intricate relationship between energy, matter, and life itself.

2.9. THE IMPLICATIONS OF MATTER AND ENERGY IN EVOLUTION

In the intricate dance of evolution, the roles of matter and energy cannot be understated. Matter serves as the substrate upon which life evolves, providing the necessary building blocks for biological structures. Variations in molecular composition, such as changes in the arrangement of carbon atoms in organic compounds, can lead to distinct life forms possessing unique traits and abilities. Meanwhile, energy is fundamental for biological processes, driving the biochemical reactions necessary for survival and reproduction. As organisms harness energy—from sunlight in photosynthesis to chemical energy in food—these processes enable the development and variability that form the basis of natural selection. The balance and flow of matter and energy within ecosystems directly influence evolutionary pathways, leading to adaptations that enhance survival in fluctuating environments. The evolution of species also hinges on the dynamic interplay between matter and energy at a cellular level, influencing genetic variation and the mechanisms by which traits are passed on. As outlined in the research on **self-organization** and **information dynamics** (Haken and Juval, 2016), complex systems evolve from simpler forms through a series of energy-rich interactions that promote diversification. For instance, genetic mutations arising from environmental stresses can be viewed as adaptations driven by energy fluctuations within ecosystems. These mutations, in turn, shape the genetic landscape of populations over generations. Thus, the availability of energy sources not only dictates the survival of existing species but also fosters the emergence of new ones, reflecting a critical relationship where energy access shapes evolutionary trajectories in response to both abiotic and biotic pressures. Moreover, the implications of understanding matter and energy extend beyond biological evolution to societal and technological advancements. As our comprehension of these foundational concepts deepens, parallels can be drawn to resource management and environmental sustainability. The work on the shifting contributions of risk factors to health outcomes illustrates how changes in energy consumption and material use impact societal health (Lim et al., 2012). Consequently, recognizing the evolutionary significance of matter and energy allows for an integrated approach to solving both environmental issues and human health challenges. By applying these principles across different domains, societies can not only address present challenges but also enact policies that foster resilience against future ecological and evolutionary shifts, suggesting that the fundamental aspects of matter and energy truly underpin the fabric of life itself.

2.9.1. Natural Selection and Energy Efficiency

In the ongoing struggle for survival, organisms find themselves constantly adapting to their environments through the process of natural selection. This mechanism favors traits that enhance an organism's energy efficiency, directly impacting its ability to thrive and reproduce. For example, species that successfully maximize energy extraction from their surroundings, whether through specialized metabolic pathways or behavioral adaptations, are more likely to pass on their genes. Such energy-efficient traits can also lead to competitive advantages in resource-limited settings, where the ability to obtain and use energy smoothly can dictate survival and reproductive success. The evolutionary implications of energy efficiency are further underscored by the 2023 call for sustainable practices in renewable energy production, highlighting that species utilizing resources effectively often mirror these principles in natural ecosystems (Mata et al., 2009). Efficient energy use is not limited to individual organisms but also plays a crucial role in shaping entire ecosystems. Natural selection acts on various traits that influence how energy flows within these complex systems. Some species evolve to occupy specific ecological niches, allowing them to exploit available resources optimally. In this context, the relationships between herbivores and vegetation, for instance, can reveal how energy transfer through food webs is influenced by natural selection. Additionally, as observed in research on biofuels derived from microalgae, the adaptation of these organisms to different environmental factors aids in maximizing their energy output. As such, the adaptations that promote energy efficiency contribute not only to the survival of individual species but also to the overall productivity of ecosystems, creating a dynamic balance that is essential for biodiversity (Mata et al., 2009). The effects of energy efficiency extend beyond individual organisms and ecosystems to broader implications for societal development and technology. Our understanding of natural selection sheds light on how energy-efficient designs can inspire innovations, particularly in producing renewable resources. Graphene, as a material exhibiting unique electronic properties, exemplifies how efficiency can be harnessed in technology and energy storage solutions. Emphasizing the interconnectedness between natural systems and human innovations, we must recognize that applying the principles of natural selection to energy use can promote sustainability and resilience in our societies. By valuing energy efficiency in both biological and technological contexts, we are equipped to tackle pressing challenges such as climate change and resource conservation, ensuring our future aligns with the natural efficiencies that have evolved over eons.

2.9.2. The Evolution of Metabolic Pathways

Metabolic pathways are intricate networks of biochemical reactions that have evolved over billions of years, facilitating life's dependence on energy and matter. Initially, the most primitive life forms on Earth utilized simple pathways such as glycolysis, which allowed them to convert glucose into ATP without the need for oxygen. These pathways are believed to be conserved due to their efficiency, showing that evolutionary processes favor metabolic functions that optimize energy extraction from available substrates. As organisms evolved and diversified, so too did their metabolic capabilities, adapting to various environmental niches and transforming available resources into usable energy forms. This diversification highlights the relationship between life forms and their surroundings, showcasing an evolutionary trend where complexity is interwoven with ecological adaptability. The emergence of more complex metabolic pathways illustrates another key aspect of evolution: the integration of various biochemical processes. As organisms transitioned from anaerobic to aerobic environments, pathways such as the citric acid cycle developed, reflecting a significant adaptation to the utilization of oxygen in energy metabolism. The ability to harness more energy from substrates not only gave rise to multicellular organisms but also enabled ecosystems to flourish and diversify. Additionally, the development of specialized pathways, such as those for nitrogen fixation and photosynthesis, underscores the dynamic interplay between life forms and their biogeochemical environments.

Contemporary research continues to illuminate these complex interactions, revealing how biochemical networks have been fine-tuned over millennia, ultimately shaping the biosphere we inhabit today. Finally, understanding metabolic evolution is crucial for addressing contemporary challenges in biology and medicine. As highlighted by recent studies, the resilience of microbial communities and their metabolic pathways, particularly in soil ecosystems, demonstrates their importance in nutrient cycling and ecological health (Nannipieri et al., 2017). Similarly, the balance between oxidative stress and redox signaling mechanisms illustrates how metabolic pathways are integral to regulating biological processes at the cellular level (Sies et al., 2017). By examining the intricate evolution of these pathways, scientists can gain insights into how organisms adapt and respond to environmental pressures, which can inform fields ranging from environmental science to biotechnology. Therefore, the study of metabolic evolution reveals not only the historical context of life's complexity but also its implications for future research and technological advancement in understanding biological systems.

2.9.3. Adaptations to Energy Availability in Different Environments

In various ecological contexts, organisms have evolved distinct adaptations that enable them to optimize energy utilization based on available resources. For instance, some plants have developed specific root architectures to enhance nutrient absorption from the soil, directly addressing the challenge of limited mineral availability, particularly phosphorus and nitrogen. Such adaptations are crucial in nutrient-poor environments where conventional growth strategies may fail. In contrast, soil-dwelling microorganisms can mobilize and process these nutrients effectively, thereby enriching the surrounding environment and improving their own survival chances. This mutualistic interaction underscores how energy acquisition mechanisms in plants and microbes can co-evolve, demonstrating the intricate balance of energy dynamics within ecosystems. Animals, too, exhibit remarkable adaptations in response to energy scarcity or abundance. For example, species residing in nutrient-rich environments may exhibit rapid growth rates and high reproductive outputs that increase their population size, while their counterparts in energy-deficient settings often adopt slower growth rates and reproductive strategies focused on longevity. These differences arise from the necessity of conserving energy and resources amidst competing demands.

In this regard, the mammalian gut microbiota exemplifies how symbiotic relationships can enhance digestion and nutrient extraction from food, optimizing energy availability and contributing to overall metabolic efficiency. The role of gut microbes in breaking down complex carbohydrates into simpler forms illustrates a critical adaptation that underpins dietary versatility in diverse habitats. Conversely, the relationship between energy availability and organismal adaptation is clearly pronounced in extreme environments, such as deserts and polar regions. In these habitats, both flora and fauna exhibit specialized metabolic pathways that enable them to thrive despite energy constraints. For instance, desert plants may employ methods such as crassulacean acid metabolism (CAM) to minimize water loss while maximizing photosynthetic efficiency during cooler nighttime hours. Meanwhile, animals in colder climates have developed insulating layers of fat and fur, alongside metabolic adaptations that allow for energy conservation during scarce food availability. These evolutionary strategies reflect a fundamental aspect of survival: the demand for organisms to adapt their physiological and behavioral traits to meet the challenges posed by varying energy landscapes effectively. Such adaptations are critical for maintaining the delicate balance outlined in the interactions of matter and energy that govern life on Earth (Figure 2.13).

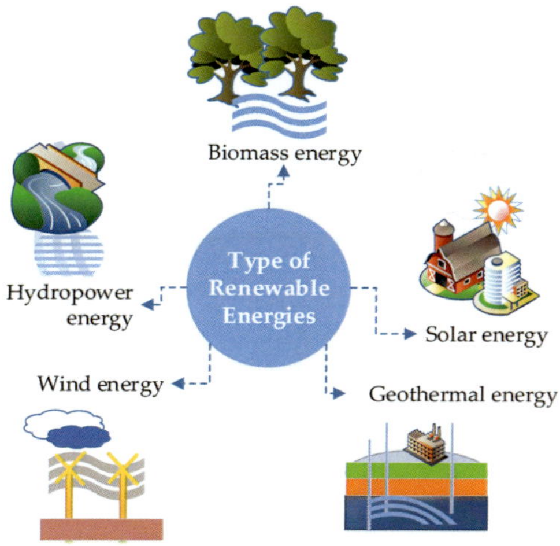

Figure 2.13. *Role of renewable energy sources in environmental protection.*

Source: Avtar R, Sahu N, Aggarwal AK, Chakraborty S, Kharrazi A, Yunus AP, Dou J, Kurniawan TA. Exploring Renewable Energy Resources Using Remote Sensing and GIS—A Review. Resources *2019; 8(3):149. https://doi.org/10.3390/resources8030149.*

2.10. CONCLUSION

In summation, the intricate interplay among matter, energy, and life underscores the complexity of biological systems and their dependence on environmental factors. Matter forms the structural foundation of all living organisms, while energy drives the processes that sustain life. The relationship between these elements is crucial in understanding how organisms adapt and evolve. For instance, dietary choices directly influence gut microbiota, illustrating how the composition and quality of our food—not only a source of matter and energy—can profoundly affect our health. As highlighted in (Makki et al., 2018), dietary fibers interact with gut microbes, affecting their composition and producing essential metabolites that influence host physiology. This exemplifies how the fundamental principles of matter and energy converge in biological contexts, reinforcing the idea that life is a dynamic balance maintained within a complex matrix of interactions. Moreover, the implications of energy and matter conservation extend beyond mere biological processes; they play a significant role in social and organizational realms as well. The Conservation of Resources (COR) theory, as discussed by Hobfoll et al (2017), elucidates that managing resources—both material and energetic—can influence workplace dynamics and employee well-being. By applying the principles of COR theory, organizations can enhance productivity and reduce stress, highlighting the universal relevance of energy conservation. This not only demonstrates the interconnectedness of individual and organizational health but also reflects a broader principle: that careful stewardship of both matter and energy leads to sustainable outcomes. Consequently, recognizing the critical nature of these relationships across different contexts encourages a holistic understanding of life. In reflecting on the dynamics of matter, energy, and life, it becomes evident that these concepts are not isolated but are richly intertwined. The insights gleaned from the interplay of biological systems and organizational behavior reveal common threads that underscore the importance of resource conservation in both realms. As we continue to seek balance in our lives—fostering healthy relationships with the environment and our communities—it is vital to embrace the lessons drawn from various fields of study. By integrating knowledge from biology, psychology, and sociology, we can forge a deeper understanding of the mechanisms that govern life itself. Ultimately, the ongoing exploration of how matter and energy influence life encourages a comprehensive approach to both personal and societal well-being, paving the way for a more resilient future.

2.10.1. Summary of the Relationship between Matter and Energy

The intricate relationship between matter and energy is grounded in the foundational principles of physics, particularly those articulated by Einstein's theory of relativity. This theory revolutionized our understanding by establishing that matter can be transformed into energy and vice versa, encapsulated in the famous equation $E=mc^2$. In this context, energy is not an independent entity but rather a manifestation of the potential of matter to perform work or generate change. This principle has far-reaching implications, from the fission reactions in nuclear power plants, where mass is converted to energy, to the biochemical processes within living organisms, where food is transformed into usable energy to sustain life. Understanding this dynamic interchange is crucial for advancing scientific endeavors across various fields, including medicine, environmental science, and technology. Within biological systems, the interplay between matter and energy is evident at every level, from cellular functions to ecological interactions. Living organisms utilize energy derived from matter, particularly nutrients, to maintain homeostasis, grow, and reproduce. The process of photosynthesis exemplifies this relationship, wherein plants convert light energy into chemical energy stored in glucose, thereby transforming inorganic matter into organic compounds. This energy transfer is pivotal for life as it supports food webs and underpins the flow of energy through ecosystems. Additionally, various scientific studies highlight the necessity of this relationship for understanding mental and physical health in light of recent challenges like the COVID-19 pandemic, which has emphasized the importance of energy dynamics in maintaining overall well-being and psychological resilience (Holmes et al., 2020). The concept of this transformative role extends beyond biological systems and into technological and societal advancements, demonstrating the centrality of energy in modern life. Artificial intelligence (AI), for example, is predicated on the efficient conversion of vast amounts of data (matter) into meaningful insights and predictions (energy), underscoring the importance of explainability in AI systems (Arrieta et al., 2019). This relationship illustrates the broader implications of matter-energy dynamics not only in computational fields but also in addressing global challenges such as climate change and resource management. As we explore and exploit these connections, it becomes evident that a comprehensive understanding of the relationship between matter and energy is paramount for progress in scientific research, technological innovation, and enhancing life itself.

2.10.2. The Significance of Understanding this Relationship for Life Sciences

A nuanced grasp of the interplay between matter, energy, and life is fundamental for advancing our understanding of biological systems and their intricate functioning. This relationship underscores how living organisms interact with their environment, employing various forms of energy to sustain metabolic processes. For example, the concept of the neurovascular unit (NVU) (Iadecola, 2017) illustrates the collaboration between neurons, astrocytes, and blood vessels, emphasizing that neural activity necessitates a precise coupling with cerebral blood flow. Understanding these dynamic interactions allows researchers in life sciences to develop targeted treatments for neurodegenerative diseases, improving patient outcomes by addressing the root physiological issues rather than merely managing symptoms. The challenges posed by environmental degradation and climatic change also highlight the importance of understanding the relationship between matter, energy, and life. As emphasized in discussions about planetary health, changes such as resource depletion and loss of biodiversity have profound implications for human health and ecosystems alike (Whitmee et al., 2015). Recognizing that human health is intimately linked to the health of ecological systems helps scientists formulate innovative solutions to reduce the environmental footprint of our activities. Addressing these challenges through a holistic understanding can lead to sustainable practices that not only enhance public health but also ensure the longevity of our ecological heritage. Moreover, advancements in life sciences reveal that the manipulation of matter and energy at the molecular and cellular levels can yield substantial benefits, such as developing new biotechnologies or improving agricultural practices. As scientists deepen their knowledge of how cells utilize energy and materials, potential applications in medicine, environmental restoration, and bioengineering become increasingly viable. For instance, understanding cellular energy paths can lead to breakthroughs in regenerative medicine or the enhancement of crop resilience against climate change effects. By fostering an educated approach to these relationships, researchers empower society to make informed decisions that will shape sustainable futures—not just for current populations, but for generations to come.

2.10.3. Future Directions for Research in Matter, Energy, and Life

Advancements in material science remain crucial as we look to the future for innovations that can enhance energy efficiency and sustainability. Researchers are increasingly focused on developing new materials with exceptional properties, such as superconductors that operate at higher temperatures, which could revolutionize energy transmission. Additionally, the pursuit of nanomaterials has opened doors to applications in various sectors, including medicine, electronics, and renewable energy. These microscopic structures can be tailored for specific functions, such as targeted drug delivery or solar energy conversion. By bridging fundamental physics with engineering, future studies will deepen our understanding of matter and pave the way for groundbreaking applications that could alleviate pressing global challenges related to energy consumption and environmental impact. The interplay of energy systems and biological processes is another fertile ground for future research, particularly in understanding how energy transfer mechanisms can inform sustainable practices. Exploring bioenergetics can yield insights into optimizing energy use in agricultural systems, potentially enhancing crop yields while minimizing ecological footprints. This avenue of inquiry may lead to advancements in biofuels and waste-to-energy technologies, tapping into the inherent efficiencies of natural systems. Moreover, integrating energy systems into the framework of ecosystem services can guide policy development aimed at sustainable resource management. By fostering collaboration between biologists, energy scientists, and environmental policymakers, we can develop multifaceted solutions that address energy challenges while promoting the health of ecosystems and communities. As we delve deeper into the intricate relationship between matter, energy, and life, interdisciplinary approaches will be paramount in shaping future research directions. The convergence of fields such as quantum biology, biomimetics, and synthetic biology offers a unique lens to examine how life exploits energy and materials at the molecular level. This approach not only illuminates the fundamental principles governing biological systems but also inspires innovative technologies that mimic these processes. For instance, advancements in synthetic biology could lead to engineered organisms capable of producing renewable energy or bioplastics, while biomimetic designs might yield more efficient solar cells. Ultimately, fostering an environment where diverse scientific domains intersect will be essential in unlocking new avenues of inquiry, guiding us toward a more sustainable and resilient future.

REFERENCES

1. Abe, J. O., Popoola, A. P. I., Ajenifuja, E., & Popoola, O. (2019). Hydrogen energy, economy and storage: Review and recommendation. *International Journal of Hydrogen Energy, 44*(30), 15072–15086. https://doi.org/10.1016/j.ijhydene.2019.04.068.
2. Ali, H., Khan, E., & Ilahi, I. (2019). Environmental chemistry and ecotoxicology of hazardous heavy metals: Environmental persistence, toxicity, and bioaccumulation. *Journal of Chemistry, 2019*, 1–14. https://doi.org/10.1155/2019/6730305.
3. Arrieta, A. B., Díaz-Rodríguez, N., Del Ser, J., Bennetot, A., Tabik, S., Barbado, A., & Herrera, F. (2019). Explainable Artificial Intelligence (XAI): Concepts, taxonomies, opportunities and challenges toward responsible AI. *Information Fusion, 58*, 82–115. https://doi.org/10.1016/j.inffus.2019.12.012.
4. Azzalini, D., Rebollo, I., & Tallon-Baudry, C. (2019). Visceral signals shape brain dynamics and cognition. *Trends in Cognitive Sciences, 23*(6), 488–509. https://doi.org/10.1016/j.tics.2019.03.007.
5. Bechinger, C., Di Leonardo, R., Löwen, H., Reichhardt, C., Volpe, G., & Volpe, G. (2016). Active particles in complex and crowded environments. *Reviews of Modern Physics, 88.* https://doi.org/10.1103/revmodphys.88.045006.
6. Breitburg, D. L., Levin, L. A., Oschlies, A., Grégoire, M., Chávez, F. P., Conley, D. J., Garçon, V., Gilbert, D., Gutiérrez, D., Isensee, K., Jacinto, G. S., Limburg, K. E., Montès, I., Naqvi, S. W. A., Pitcher, G. C., Rabalais, N. N., Roman, M. R., Rose, K. A., Seibel, B. A., & Zhang, J. (2018). Declining oxygen in the global ocean and coastal waters. *Science, 359.* https://doi.org/10.1126/science.aam7240.
7. Castro Neto, A. H., Guinea, F., Peres, N. M. R., Novoselov, K. S., & Geim, A. K. (2009). The electronic properties of graphene. *Reviews of Modern Physics, 81*, 109–162. https://doi.org/10.1103/revmodphys.81.109.
8. Cavicchioli, R., Ripple, W. J., Timmis, K. N., Azam, F., Bakken, L. R., Baylis, M., & Webster, N. S. (2019). Scientists' warning to humanity: Microorganisms and climate change. *Nature Reviews Microbiology, 17*(9), 569–586. https://doi.org/10.1038/s41579-019-0222-5.
9. Cavicchioli, R., Ripple, W. J., Timmis, K. N., Azam, F., Bakken, L. R., Baylis, M., Behrenfeld, M. J., Boetius, A., Boyd, P. W., Classen, A. T., Crowther, T. W., Danovaro, R., Foreman, C. M., Huisman, J., Hutchins, D. A., Jansson, J. K., Karl, D. M., Koskella, B., Mark Welch, D. B., & Webster, N. S. (2019). Scientists' warning to humanity: Microorganisms

and climate change. *Nature Reviews Microbiology, 17*(9), 569–586. https://doi.org/10.1038/s41579-019-0222-5.

10. Chavda, N. J., Rathore, R., & Solanki, H. (2022). An overview of the historical background of ethnobotany and indigenous culture of India. *IJCRT, 10*(4). ISSN: 2320–2882.
11. Cryan, J. F., O'Riordan, K. J., Cowan, C. S. M., Sandhu, K. V., Bastiaanssen, T. F. S., Boehme, M., Codagnone, M. G., Cussotto, S., Fulling, C., Golubeva, A. V., Guzzetta, K. E., Jaggar, M., Long-Smith, C. M., Lyte, J. M., Martin, J. A., Molinero-Perez, A., Moloney, G., Morelli, E., Morillas, E., & Dinan, T. G. (2019). The microbiota-gut-brain axis. *Physiological Reviews, 99*(4), 1877–2013. https://doi.org/10.1152/physrev.00018.2018.
12. Faraco, G., & Iadecola, C. (2013). *Hypertension, 62*, 810–817. https://doi.org/10.1161/hypertensionaha.113.01063.
13. Garza, D. R., & Dutilh, B. E. (2015). B?*Cellular and Molecular Life Sciences, 72*(22), 4287–4308. https://doi.org/10.1007/s00018-015-2004-1.
14. Goodwin, G. M., Haddad, P. M., Ferrier, I. N., Aronson, J. K., Barnes, T., Cipriani, A., & Young, A. H. (2016). Evidence-based guidelines for treating bipolar disorder: Revised third edition recommendations from the British Association for Psychopharmacology. *Journal of Psychopharmacology, 30*(6), 495–553. https://doi.org/10.1177/0269881116636545.
15. Göpfrich, K., Platzman, I., & Spatz, J. P. (2018). Mastering complexity: Towards bottom-up construction of multifunctional eukaryotic synthetic cells. *Trends in Biotechnology, 36*(9), 938–951. https://doi.org/10.1016/j.tibtech.2018.03.008.
16. Hacquard, S., Garrido Oter, R., González, A., Spaepen, S., Ackermann, G., Lebeis, S. L., McHardy, A. C., Dangl, J. L., Knight, R., Ley, R. E., & Schulze Lefert, P. (2015). Microbiota and host nutrition across plant and animal kingdoms. *Cell Host & Microbe, 17*, 603–616. https://doi.org/10.1016/j.chom.2015.04.009.
17. Haken, H., & Portugali, J. (2016). Information and self-organization. *Entropy, 19*(1), 18. https://doi.org/10.3390/e19010018.
18. Hobfoll, S. E., Halbesleben, J. R. B., Neveu, J. P., & Westman, M. (2017). Conservation of resources in the organizational context: The reality of resources and their consequences. *Annual Review of Organizational Psychology and Organizational Behavior, 5*(1), 103–128. https://doi.org/10.1146/annurev-orgpsych-032117-104640.
19. Holmes, E. A., O'Connor, R. C., Perry, V. H., Tracey, I., Wessely, S., Arseneault, L., Ballard, C., Christensen, H., Cohen Silver, R., Everall, I.,

Ford, T., John, A., Kabir, T., King, K., Madan, I., Michie, S., Przybylski, A. K., Shafran, R., Sweeney, A., & Bullmore, E. (2020). Multidisciplinary research priorities for the COVID-19 pandemic: A call for action for mental health science. *The Lancet Psychiatry, 7*(6), 547–560. https://doi.org/10.1016/S2215-0366(20)30168-1.

20. Iadecola, C. (2017). /lb*Neuron, 96*(1), 17–42. https://doi.org/10.1016/j.neuron.2017.07.030.
21. Jørgensen, B. B., Findlay, A. J., & Pellerin, A. (2019). The biogeochemical sulfur cycle of marine sediments. *Frontiers in Microbiology, 10*, 849. https://doi.org/10.3389/fmicb.2019.00849.
22. Köhler, J., Geels, F. W., Kern, F., Markard, J., Onsongo, E., Wieczorek, A., & Wells, P. E. (2019). An agenda for sustainability transitions research: State of the art and future directions. *Environmental Innovation and Societal Transitions, 31*, 1–32. https://doi.org/10.1016/j.eist.2019.01.004.
23. Kreutz, D., Ramos, F. M. V., Veríssimo, P. E., Rothenberg, C. E., Azodolmolky, S., & Uhlig, S. (2014). Software-defined networking: A comprehensive survey. *Proceedings of the IEEE, 103*, 14–76. https://doi.org/10.1109/jproc.2014.2371999.
24. Kuypers, M. M. M., Marchant, H. K., & Kartal, B. (2018). The microbial nitrogen-cycling network. *Nature Reviews Microbiology, 16*, 263–276. https://doi.org/10.1038/nrmicro.2018.9.
25. Kuypers, M., Marchant, H., & Kartal, B. (2018). The microbial nitrogen-cycling network. *Nature Reviews Microbiology, 16*(5), 263–276. https://doi.org/10.1038/nrmicro.2018.9.
26. Lim, S. S., Vos, T., Flaxman, A. D., Danaei, G., Shibuya, K., Adair-Rohani, H., & Ezzati, M. (2012). A comparative risk assessment of burden of disease and injury attributable to 67 risk factors and risk factor clusters in 21 regions, 1990–2010: A systematic analysis for the Global Burden of Disease Study 2010. *The Lancet, 380*(9859), 2224–2260. https://doi.org/10.1016/S0140-6736(12)61766-8.
27. Mach, F., Baigent, C., Catapano, A. L., Koskinas, K. C., Casula, M., Badimon, L., & Wiklund, O. (2020). 2019 ESC/EAS guidelines for the management of dyslipidaemias: Lipid modification to reduce cardiovascular risk. *European Heart Journal, 41*(1), 111–188. https://doi.org/10.1093/eurheartj/ehz455.
28. Makki, K., Deehan, E. C., Walter, J., & Bäckhed, F. (2018). The impact of dietary fiber on gut microbiota in host health and disease. *Cell Host & Microbe, 23*(6), 705–715. https://doi.org/10.1016/j.chom.2018.05.012.
29. Manisalidis, I., Stavropoulou, E., Stavropoulos, A., & Bezirtzoglou,

E. (2020). Environmental and health impacts of air pollution: A review. *Frontiers in Public Health, 8*, 14. https://doi.org/10.3389/fpubh.2020.00014.

30. Mata, T. M., Martins, A. A., & Caetano, N. S. (2010). Microalgae for biodiesel production and other applications: A review. *Renewable and Sustainable Energy Reviews, 14*(1), 217–232. https://doi.org/10.1016/j.rser.2009.07.020.
31. McDougal Littell. (2005). *Matter and Energy: Molecular Biology of the Cell* (ISBN 0618334459). McDougal Littell.
32. Nannipieri, P., Ascher Jenull, J., Ceccherini, M. T., Landi, L., Pietramellara, G., & Renella, G. (2017). Microbial diversity and soil functions. *European Journal of Soil Science, 68*(1), 12–26. https://doi.org/10.1111/ejss.4_12398.
33. Nightingale, A. J., & Ahlborg, H. (2018). Theorizing power in political ecology: The 'where' of power in resource governance projects. *Journal of Political Ecology, 25*(1). https://doi.org/10.2458/v25i1.22804.
34. Obleser, J., & Kayser, C. (2019). Neural entrainment and attentional selection in the listening brain. *Trends in Cognitive Sciences, 23*(10), 913–926. https://doi.org/10.1016/j.tics.2019.08.004.
35. Ormazábal, V., Nair, S., Elfeky, O., Aguayo, C., Salomón, C., & Zúñiga, F. (2018). Association between insulin resistance and the development of cardiovascular disease. *Cardiovascular Diabetology, 17*. https://doi.org/10.1186/s12933-018-0762-4.
36. Polkovnikov, A., Sengupta, K., Silva, A., & Vengalattore, M. (2011). *Colloquium*: Nonequilibrium dynamics of closed interacting quantum systems. *Reviews of Modern Physics, 83*(3), 863–883. https://doi.org/10.1103/revmodphys.83.863.
37. Ragaert, K., Delva, L., & Van Geem, K. M. (2017). Mechanical and chemical recycling of solid plastic waste. *Waste Management, 69*, 24–58. https://doi.org/10.1016/j.wasman.2017.07.044.
38. Shin, Y., & Brangwynne, C. P. (2017). Liquid phase condensation in cell physiology and disease. *Science, 357.* https://doi.org/10.1126/science.aaf4382.
39. Sies, H., Berndt, C., & Jones, D. P. (2017). Oxidative stress. *Annual Review of Biochemistry, 86*, 715–748. https://doi.org/10.1146/annurev-biochem-061516-045037.
40. Sikkema, H. R., Gaastra, B. F., Pols, T., & Poolman, B. (2019). Cell fuelling and metabolic energy conservation in synthetic cells. *Chembiochem, 20*(20), 2581–2592. https://doi.org/10.1002/cbic.201900398.

41. Uhlén, M., Fagerberg, L., Hallström, B. M., Lindskog, C., Oksvold, P., Mardinoglu, A., & Pontén, F. (2015). Tissue-based map of the human proteome. *Science, 347*(6220), 1260419. https://doi.org/10.1126/science.1260419.

42. Verhelst, S., Turner, J., Sileghem, L., & Vancoillie, J. (2018). Methanol as a fuel for internal combustion engines. *Progress in Energy and Combustion Science, 70*, 43–88. https://doi.org/10.1016/j.pecs.2018.10.001.

43. Whitmee, S., Haines, A., Beyrer, C., Boltz, F., Capon, A. G., de Souza Dias, B. F., & Yach, D. (2015). Safeguarding human health in the Anthropocene epoch: Report of The Rockefeller Foundation–Lancet Commission on planetary health. *The Lancet, 386*(10007), 1973–2028. https://doi.org/10.1016/S0140-6736(15)60901-1.

44. Willett, W. C., Rockström, J., Loken, B., Springmann, M., Lang, T., Vermeulen, S. J., Garnett, T., Tilman, D., DeClerck, F., Wood, A., Jonell, M., Clark, M., Gordon, L., Fanzo, J., Hawkes, C., Zurayk, R., Rivera, J. A., de Vries, W., Sibanda, L. M., & Murray, C. J. L. (2019). Food in the Anthropocene: The EAT–Lancet Commission on healthy diets from sustainable food systems. *The Lancet, 393*, 447–492. https://doi.org/10.1016/s0140-6736(18)31788-4.

45. Xu, L. D., Xu, E., & Li, L. (2018). Industry 4.0: State of the art and future trends. *International Journal of Production Research, 56*, 2941–2962. https://doi.org/10.1080/00207543.2018.1444806.

CHAPTER 3

ECOSYSTEMS AND THE BIOSPHERE

Contents

INTRODUCTION TO ECOSYSTEMS

Within the intricate web of life on Earth, ecosystems play a fundamental role in the survival of various organisms, demonstrating the delicate balance that sustains biological communities. An ecosystem encompasses all living organisms, referred to as biotic factors, and their physical environment, known as abiotic factors, which include elements such as climate, soil, and water. Together, these components interact to form a functional unit where energy flows and matter cycles, fostering a diversity of life. The relationships among species, encompassing predators, prey, and symbiotic affiliations, contribute to the stability and resilience of an ecosystem, underscoring the interdependence of life forms and their habitats. As such, the study of ecosystems reveals valuable insights into environmental dynamics and the essential processes that support life on the planet. Examining the various types of ecosystems reveals the vastness of biological interactions that shape our planet.

Ecosystems can be categorized into terrestrial types, such as forests, grasslands, and deserts, and aquatic types, including freshwater and marine environments. Each of these ecosystems harbors unique species adapted to specific conditions, forming a complex tapestry of life characterized by distinct adaptations and interactions. For instance, in a rainforest ecosystem, high biodiversity supports a multitude of ecological niches, contributing to resilience against disturbances. Conversely, desert ecosystems document remarkable adaptations to extreme conditions, highlighting the diversity of survival strategies. Understanding these unique ecosystems is crucial, as they provide essential services, such as carbon sequestration, water filtration, and habitat provision, that directly impact global health and sustainability.

The study of ecosystems also emphasizes the critical importance of ecological balance and the threats posed by human activity. Changes resulting from pollution, deforestation, and climate change can disrupt these natural systems, leading to biodiversity loss and decreased ecosystem function. For example, the introduction of invasive species can outcompete native organisms, altering food webs and diminishing local biodiversity. Conservation efforts aimed at preserving and restoring ecosystems are imperative for maintaining the health of the planet. By promoting sustainable practices and protecting vulnerable habitats, we can help mitigate the adverse impacts of anthropogenic influences and ensure that ecosystems continue to thrive. In this way, understanding ecosystems is not only an academic pursuit but a vital necessity for the future of our planet, underscoring the connection between human well-being and environmental stewardship.

3.1. DEFINITION OF ECOSYSTEMS

At the core of ecological study lies the concept of ecosystems, which consist of dynamic interactions between living organisms and their physical environment. These systems encompass a range of communities, including plants, animals, and microorganisms, all functioning within specific geographic areas and conditions. An ecosystem is not merely a collection of species; it is a network of relationships that shapes energy flow and nutrient cycling vital for sustaining life. This interconnectedness forms the basis of ecosystem services, which are the benefits provided to humans, such as clean water, pollination, and climate regulation. Understanding these relationships is essential for effective biodiversity and resource management, most notably in the face of climate-induced changes that threaten these intricate systems, as outlined in (Hummel et al., 2011).

The intricacies of ecosystems reveal the importance of their defined structures and processes. Each ecosystem is characterized by its abiotic components, such as soil, water, and temperature, alongside biotic factors, including the diversity of species and their interactions. Within this framework, individual species play specific roles, known as ecological niches, which contribute to the overall functioning of the ecosystem. Changes in one component can significantly impact others, demonstrating the delicate balance that ecosystems maintain. Recognizing this balance is crucial, especially in the context of rapid environmental changes, as highlighted in (Frank et al., 2015). Consequently, comprehensive research focusing on the sustainability of these relationships is necessary for both conservation efforts and understanding how ecosystems can adapt to shifts in climate and biodiversity. A critical aspect of defining ecosystems involves acknowledging the socio-economic values they provide. These values are not just ecological but extend to cultural and economic dimensions, underscoring the intricate links between human societies and their natural environments.

Ecosystem services, driven by biodiversity, enhance societal well-being and economic stability, reinforcing the need for policies that promote sustainable use and conservation. By evaluating the societal values associated with ecosystems, researchers can better understand how to implement sustainable adaptations to biodiversity losses caused by climate change. The integration of societal factors into ecological research, as emphasized by Hummel et al (2011) provides a rounded perspective essential for informed decision-making in biodiversity and resource management. Thus, a comprehensive definition of ecosystems must consider both ecological and socio-economic dimensions to foster resilience in an ever-changing biosphere.

3.2. IMPORTANCE OF ECOSYSTEMS IN THE BIOSPHERE

In addition to their intrinsic value, ecosystems are vital for their role in supporting biodiversity, which in turn undergirds the stability and resilience of the biosphere. Biodiversity allows ecosystems to adapt to changes, whether those are natural shifts or anthropogenic pressures. Diverse biological communities are better equipped to handle disturbances, such as climate fluctuations or disease outbreaks, thereby ensuring the continuity of essential ecological processes. This interconnectedness underscores the importance of maintaining diverse ecosystems, as declines in one area can have cascading effects. Effective management strategies centered around the preservation of ecosystem functions and their associated biodiversity are crucial for securing both ecological and economic benefits globally, illustrating the need to elevate the discourse around sustainable practices in environmental science. Moreover, the socio-economic dimensions of ecosystems cannot be overlooked. As highlighted by multi-actor transdisciplinary research, particularly in the context of Almería, Spain, agricultural systems that rely heavily on intensive practices must evolve to mitigate negative environmental impacts while still ensuring food security. Current models of agriculture often create tensions between short-term economic gains and long-term sustainability, leading to challenges in biodiversity conservation and resource management. Almerías experience emphasizes the necessity for a governance structure grounded in shared responsibility that reconciles agricultural productivity with environmental integrity. This approach fosters community engagement and aligns diverse stakeholder interests, paving the way for the integration of ecosystem services into economic frameworks. Finally, the interconnectedness of ecosystems and human well-being is increasingly evident in discussions surrounding sustainable development. Ecosystems provide critical services—such as clean air, water filtration, and pollination—that are foundational to human health and prosperity. The degradation of these natural services can directly impact societal structures, leading to food insecurity and loss of livelihoods. As research indicates, understanding the socio-ecological relationships within ecosystems is essential for devising effective responses to climate change and resource depletion. By prioritizing ecosystem conservation within policy agendas, societies can foster not only the resilience of natural environments but also their own socio-economic stability. Continuous dialogue and collaborative efforts across sectors will be imperative in meeting the complex challenges of balancing ecosystem health with human needs (Figure 3.1).

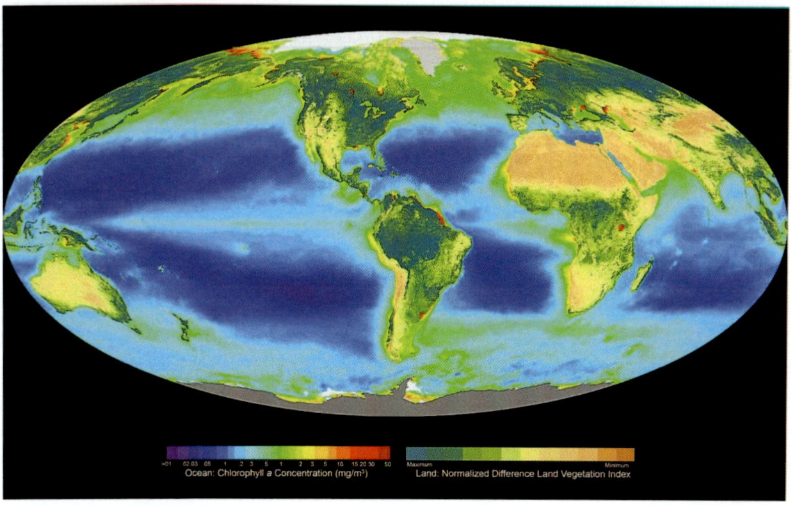

Figure 3.1. *Biosphere.*

Source: By Provided by the SeaWiFS Project, Goddard Space Flight Center and ORBIMAGE – http://oceancolor.gsfc.nasa.gov/SeaWiFS/BACKGROUND/Gallery/index.html and from en:Image:Seawifs global biosphere.jpg, Public Domain, https://commons.wikimedia.org/w/index.php?curid=387228.

3.3. OVERVIEW OF ECOSYSTEM COMPONENTS

Ecosystems are complex networks formed by interactions among various components, each playing a pivotal role in maintaining ecological balance. At the foundation are abiotic factors, which include sunlight, water, temperature, and soil. These non-living elements create the conditions necessary for life. For instance, sunlight is essential for photosynthesis, enabling plants to produce energy, which in turn supports herbivores and, subsequently, carnivores. The physical environment also influences species distribution and abundance, affecting everything from plant growth to animal behaviors. Thus, these abiotic components serve as the underlying structure upon which biotic components—living organisms—interact and thrive. Biotic components of ecosystems are typically categorized into producers, consumers, and decomposers. Producers, or autotrophs, like plants, harness energy from the sun to create organic material through photosynthesis. Consumers, including herbivores and carnivores, depend on these producers for energy, forming intricate food webs that illustrate the flow of energy. Additionally, decomposers, such as fungi and bacteria, break down dead organic matter, recycling nutrients back into the ecosystem. This cycle is crucial for maintaining the health of an ecosystem, as it ensures that vital nutrients are available for producers once again. Understanding these dynamics unveils how disruptions in one component can cascade through an ecosystem, highlighting the significance of biodiversity and interdependence among species. The relationships among these ecosystem components underscore the importance of resilience and adaptability in the face of environmental changes. As ecosystems undergo shifts, driven by factors like climate change or human activity, the interactions that facilitate biodiversity and stability can become threatened. The concept of social-ecological systems, as discussed by Hummel et al (2011), emphasizes the necessity of integrating societal behaviors and structures with ecological dynamics. Addressing these intertwined systems is crucial for developing effective management strategies that bolster both ecosystem health and human well-being. Moreover, recognizing the unpredictability inherent in ecosystems (Kauffman et al., 2012), leads to a deeper understanding of how life evolves amid changing contexts. Therefore, a comprehensive overview of ecosystem components not only reveals their intricate interconnections but also highlights the need for sustainable practices that cherish both ecological and human systems.

3.3.1. Types of Ecosystems

Among the various types of ecosystems, terrestrial ecosystems are particularly significant due to their role in carbon storage and the impact of human activities on their functionality. Forests, grasslands, and deserts each have unique characteristics that contribute to their ecological dynamics. For instance, forests act as substantial carbon sinks, absorbing vast amounts of CO_2 from the atmosphere. However, changes in land use, such as deforestation for agriculture, disrupt this balance, leading to net carbon emissions. The interplay between increasing atmospheric CO_2 and climate variability underscores the importance of these ecosystems in mitigating climate change. As noted, "the long-term effect of climate on carbon storage has been small relative to the effects of increasing atmospheric CO_2 and land use" (A. D. McGuire et al., 2001). This highlights the critical need for sustainable management practices to preserve terrestrial ecosystems, thereby ensuring their functionality as carbon sinks in an era of climate uncertainty. Aquatic ecosystems, encompassing freshwater and marine environments, exhibit complex interactions that contribute to global biodiversity and ecological balance. Freshwater ecosystems, such as rivers, lakes, and wetlands, provide essential services including water filtration, habitat for countless species, and flood regulation. Conversely, marine ecosystems, including coral reefs and open oceans, are vital for global fish populations and play a pivotal role in the Earth's carbon cycle. Marine organisms, particularly phytoplankton, contribute significantly to carbon uptake, further emphasizing the interconnectedness of ecosystems. Through various feedback mechanisms, changes in these environments also influence atmospheric carbon levels, exemplifying how "higher global temperature will promote a rise of greenhouse gas levels" (Drégelyi-Kiss et al., 2008). As climate change raises concerns over habitat degradation and species extinction, understanding the types and functions of aquatic ecosystems is paramount for effective conservation strategies. Lastly, the delineation between terrestrial and aquatic ecosystems gives way to more specialized systems such as wetlands, which embody features from both environments. Wetlands, including swamps, marshes, and mangroves, are crucial for maintaining biodiversity, providing habitats, and regulating water cycles. They act not only as flood buffers but also as effective carbon sinks, with their unique vegetation capable of sequestering substantial amounts of carbon. This ecological diversity underscores the importance of preserving wetlands amidst increased pressures from urban development and climate change. A comprehensive understanding of wetlands' functions supports the notion that all ecosystems contribute to a larger ecological network, reinforcing the need for integrated conservation efforts. Engaging in the protection and restoration of these habitats is essential for sustaining the health of the biosphere, addressing both biodiversity loss and climate change effects concurrently.

3.3.1.1. Terrestrial Ecosystems

The interplay between terrestrial ecosystems and atmospheric conditions is increasingly significant as global climate change progresses. Rising concentrations of atmospheric CO_2 have been shown to enhance carbon uptake by vegetation, particularly since the mid-20th century. This phenomenon occurs largely due to the physiological responses of plants to elevated CO_2 levels, which can drive photosynthesis and growth. Consequently, while early analyses indicated a net release of carbon from terrestrial ecosystems due to extensive cropland establishment, later findings confirm that ecosystems have transformed into a net carbon sink post-1958, highlighting the dual impact of anthropogenic activity and natural processes on carbon dynamics (McGuire et al., 2001). Understanding these intricate interactions is crucial for predicting future ecosystem responses and managing carbon storage effectively. However, the role of climate variability in influencing terrestrial carbon storage remains ambiguous. While terrestrial models suggest that climate has had limited long-term impact relative to the effects of rising CO_2, there is substantial disagreement about whether short-term climate fluctuations, such as those driven by the El Niño/Southern Oscillation (ENSO), foster carbon release or storage (McGuire et al., 2001). The varying performance of models in simulating interannual CO_2 concentration changes illustrates the complexity of these dynamics. As such, the interaction between climate systems and terrestrial ecosystems requires further investigation to unravel the mechanisms behind carbon flux, particularly in the face of increasing anthropogenic pressures and climatic extremes that may alter local and global carbon budgets. The feedback mechanisms between terrestrial ecosystems and climate are critical for understanding potential future scenarios. Higher global temperatures are likely to exacerbate greenhouse gas emissions from these ecosystems, leading to positive feedback loops that intensify climatic changes. As temperatures rise, certain ecosystems may reach tipping points, resulting in shifts in species composition and function, further influencing carbon dynamics. The integration of more precise data on land use, disturbance events, and ecosystem management into modeling efforts is necessary to portray a comprehensive picture of this feedback (McGuire et al., 2001). By addressing these complexities, researchers can better assess how terrestrial ecosystems will respond to ongoing environmental changes, ultimately informing conservation and sustainability strategies aimed at mitigating climate impacts and preserving biodiversity in the biosphere (Figure 3.2).

Figure 3.2. *Terrestrial ecosystem.*

Source: By Wsiegmund – Self-published work by Wsiegmund, CC BY 2.5, https://commons.wikimedia.org/w/index.php?curid=336095.

3.3.1.2. Aquatic Ecosystems

Water bodies serve as essential components of aquatic ecosystems, providing habitats for a diverse array of flora and fauna. These ecosystems, which include freshwater lakes, rivers, wetlands, and marine environments, function as dynamic systems that interact with terrestrial habitats and influence global biodiversity. The health of these ecosystems is crucial, as they are integral to numerous ecological processes, such as nutrient cycling and carbon sequestration. However, human activities have led to significant habitat degradation, threatening the delicate balance that sustains aquatic life. Overexploitation of resources and pollution have resulted in the decline of many species, highlighting the urgent need for sustainable management practices to protect these invaluable ecosystems. As stated, natural resources are not inexhaustible, and the responsible stewardship of aquatic environments is necessary to ensure their longevity for future generations (Petru et al.(n.d)). Moreover, the interconnectivity of aquatic ecosystems with terrestrial and atmospheric systems underscores their importance within the broader biosphere. These ecosystems are not isolated; rather, they are influenced by and contribute to larger ecological dynamics. For instance, nutrient runoff from agricultural lands can lead to harmful algal blooms in freshwater and marine environments, which threaten aquatic life and disrupt food webs. Effective ecosystem-based management requires an understanding of these complex interactions. The justification of EAF is evident in the characteristics of an exploited ecosystem, emphasizing the need for an integrated approach that considers both ecological health and resource utilization in fisheries (Zerbi et al., 2003). By adopting strategies that recognize these relationships, we can better protect aquatic ecosystems, promoting resilience in the face of environmental challenges. Finally, preserving aquatic ecosystems requires concerted efforts at local, national, and global levels. This entails implementing policies that prioritize the sustainability of water resources, habitat restoration, and the protection of endangered species. Community engagement and education are crucial for raising awareness about the importance of aquatic ecosystems and fostering responsible conservation practices. The challenges posed by climate change, pollution, and habitat destruction call for innovative solutions grounded in scientific research and ecological principles. Enhancing our understanding and management of these ecosystems is not merely an environmental imperative but also a necessary endeavor for sustaining human livelihoods reliant on healthy aquatic systems. Ultimately, our capacity to maintain the integrity of these ecosystems will play a significant role in ensuring biodiversity and ecological stability within the biosphere (Figure 3.3).

Figure 3.3. *Aquatic ecosystems.*

Source: By Copyright (c) 2004 Richard Ling, CC BY-SA 3.0, https://commons.wikimedia.org/w/index.php?curid=78275.

3.3.1.3. Urban Ecosystems

Urban areas are often perceived as concrete jungles, yet they also function as unique ecosystems that harbor complex interactions among various species and their environment. The integration of green spaces, such as parks and community gardens, plays a crucial role in maintaining the ecological balance within these densely populated regions. These spaces not only provide habitat for numerous species but also contribute to the overall health of the urban ecosystem by improving air quality, mitigating urban heat, and managing stormwater runoff. This dual functionality of urban green spaces emphasizes the need for thoughtful urban planning that prioritizes biodiversity alongside human development, ensuring that cities are not only livable but also ecologically sound. Understanding the intricate relationships in urban ecosystems is essential to developing sustainable practices that support both human and environmental health. Cities are facing the challenges of rapid urbanization, which can lead to habitat destruction and increased resource consumption. It is here that models of integrated assessment can play a pivotal role, as proposed in recent studies. As noted in the Northern Eurasia Future Initiative, these models can help analyze the feedback loops between human actions and environmental changes, allowing policymakers to craft more effective strategies for resource management and urban resilience (Albrechtova et al., 2017). By recognizing the interconnectedness of urban and natural systems, stakeholders can implement solutions that enhance ecological integrity while also addressing social equity and economic viability. The framework for sustainability in urban ecosystems necessitates a balance between ecological limits and social needs. A comprehensive approach considers the safe and just operating space for humanity, as delineated in ecological studies. This concept posits that sustainable urban development must operate within an environmental ceiling while ensuring that basic social foundations are met, such as access to clean water and sanitation (Brown et al., 2014). In practice, this means that urban planners and community leaders must work collaboratively to monitor environmental health and social equity in tandem, ensuring that urban growth does not come at the cost of ecological degradation. Through this integrated lens, it becomes evident that thriving urban ecosystems are vital not only for biodiversity but also for cultivating communities that can thrive within sustainable boundaries (Figure 3.4).

Figure 3.4. *Urban ecosystems.*

Source: By User:Diliff – Own work, CC BY-SA 3.0, https://commons.wikimedia.org/w/index.php?curid=267594.

3.4. BIOTIC AND ABIOTIC FACTORS

Understanding ecosystems requires a holistic examination of the interplay between biotic and abiotic factors, as both elements jointly shape the environment and influence the organisms that inhabit it. Biotic factors encompass all living components within an ecosystem, including plants, animals, fungi, and microorganisms. These factors establish interactions such as predation, competition, and symbiosis, which significantly affect population dynamics and community structure. For instance, the presence of keystone species can stabilize ecosystems, fostering biodiversity and resilience against external perturbations. These relationships highlight the intricate web of life that characterizes natural systems, revealing how living organisms continuously adapt and respond to one another. Conversely, abiotic factors refer to the non-living elements that support life, comprising aspects such as climate, soil composition, water availability, and sunlight. These components lay the groundwork for ecosystems by determining which biotic factors can thrive in a given environment. For example, variations in temperature and precipitation dictate plant types, ultimately influencing the herbivores and predators that depend on them. As outlined in recent studies, the stability within Earth's ecosystems often hinges on these abiotic conditions, which interplay with biotic responses. This stability is critical, as it allows for the vital processes of energy flow and nutrient cycling, reinforcing the necessity for a balanced relationship between living and non-living factors that underpin the biosphere.

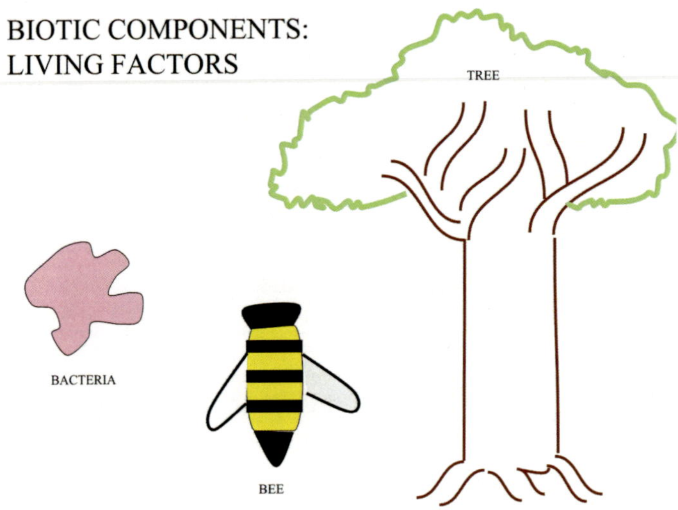

Figure 3.5. *Biotic factors.*

Source: https://en.wikipedia.org/wiki/File:Biotic_Components_Living_Factors.svg.

The complex dynamics between biotic and abiotic factors are not merely theoretical; they illustrate ongoing evolutionary processes that define the Earth's systems. As indicated in recent research, the presence of a biosphere—

where life interacts with its inorganic environment—can lead to self-regulatory mechanisms that contribute to ecosystem stability. For instance, diverse biological communities enhance the resilience of an ecosystem by providing various energy pathways and contributing to nutrient cycling, influencing weather patterns and climate over time. The emergence of these interactions can drastically alter the ecological landscape, marking transitions such as the Anthropocene era, where anthropogenic influences affect both biotic and abiotic factors. Thus, understanding this intricate relationship is essential for effectively managing ecosystems in a rapidly changing world, ensuring that both biodiversity and the abiotic environment can sustain each other (Figure 3.5).

3.4.1. Definition of Biotic Factors

Interactions among living organisms shape the delicate balance of ecosystems, and these interactions are classified as biotic factors. Biotic factors encompass all the living components within ecosystems, including plants, animals, fungi, and microorganisms. These elements have a profound influence on one another, creating a complex web of relationships that support life. For example, predator-prey dynamics dictate population levels, while symbiotic relationships enhance resource use efficiency among species. Additionally, biotic factors can be affected by environmental changes, which may disrupt established interactions and lead to cascading effects throughout the ecosystem. Understanding biotic factors is crucial for recognizing their role in ecosystem stability and productivity. The competitive interactions among species for resources, such as food and habitat, play a fundamental role in shaping community structures. Moreover, biotic factors contribute to nutrient cycling, a process vital for sustaining life. For instance, decomposers break down organic materials, returning essential nutrients to the soil and making them available for plants. McElroy (1983) highlights the importance of biotic factors in nutrient distribution, emphasizing the interconnectedness of the atmosphere, biosphere, and upper ocean. Such interactions illustrate how biotic factors impact ecosystem health and resilience, underscoring their significance in environmental studies. A thorough examination of biotic factors contributes to a deeper appreciation of ecosystem dynamics and the importance of biodiversity. The presence and diversity of species within an ecosystem can influence its functionality and resilience to external pressures, such as climate change and habitat destruction. For example, ecosystems with high biodiversity often exhibit greater stability, as a wider array of organisms may provide functional redundancy, ensuring that essential processes continue even if specific species decline. The FAO (2010) discusses how stability in ecosystems parallels economic structures, highlighting how both rely on diverse interactions. Hence, addressing biotic factors not only enriches our understanding of ecological systems but also informs conservation strategies essential for sustaining life on Earth in the face of anthropogenic threats.

3.4.2. Definition of Abiotic Factors

In ecological studies, understanding the various components that make up an ecosystem is essential. Among these components, abiotic factors play a crucial role in shaping the environment and influencing the living organisms within it. Abiotic factors refer to the non-living physical and chemical elements of an ecosystem, including temperature, light, water, soil composition, and atmospheric gases. These factors set the stage for biological processes, determining which organisms can thrive in specific habitats. For example, temperature affects metabolic rates, while the availability of sunlight impacts photosynthesis in plants.

Consequently, abiotic factors help establish the boundaries of ecosystems and dictate the diversity of life within them, illustrating their fundamental importance in ecosystem dynamics. The persistence of abiotic factors and their interactive tendencies further complicate ecosystems and influence biogeochemical cycles. Changes in one abiotic factor can lead to cascading effects across the ecosystem, affecting all biological entities. For instance, increased levels of heavy metals in soil, resulting from human activities like mining and industrial processes, can render the environment hazardous, thereby affecting plant and animal life (Hazrat Ali et al., 2019). Such changes can disrupt the balance within ecosystems, making resilience and stability contingent upon the maintenance of these abiotic parameters.

The interactivity among abiotic factors also plays a role in determining climate conditions, which in turn modifies habitats and the distribution of species, exemplifying how crucial these factors are to the understanding of ecological systems. Analyses of abiotic factors contribute significantly to the comprehensive study of biomes and understanding of broader environmental issues, including planetary health. For instance, excessive heat or moisture can impact microbial communities in soils, altering the microbiome, which is critical to nutrient cycling (Berg et al., 2020). This highlights the intricate relationships between abiotic factors and biotic components, emphasizing the need for a holistic approach to ecosystem management. An integrated assessment of abiotic factors is essential for addressing anthropogenic challenges, such as promoting biodiversity and ensuring sustainable practices in agriculture and conservation. By recognizing the importance of abiotic factors, researchers, and policymakers can develop strategies that protect ecosystems and promote their resilience against environmental changes.

3.4.3. Interaction Between Biotic and Abiotic Factors

In ecosystems, the interplay between living organisms (biotic factors) and their physical surroundings (abiotic factors) is crucial for shaping the overall dynamics and functioning of the environment. Biotic factors, including plants, animals, and microorganisms, rely on abiotic components such as sunlight, water, minerals, and climate for survival and reproduction. For instance, plants utilize sunlight for photosynthesis, a process that directly influences the availability of energy within an ecosystem. Without adequate sunlight and water, plant growth diminishes, subsequently affecting herbivores and, by extension, the carnivores that depend on them. Understanding this complex interaction helps clarify how shifts in abiotic factors—such as climate change or pollution—can disrupt food webs and biodiversity, ultimately leading to ecological imbalances that may threaten the survival of various species. The reciprocal influence between biotic and abiotic elements also manifests through feedback mechanisms vital for maintaining ecological balance.

One striking example can be seen in the concept of nutrient cycling, where organisms play a significant role in enriching their environments. For instance, organic matter from decomposing plants and animal waste introduces vital nutrients back into the soil, enhancing its fertility and thereby supporting diverse plant life. As highlighted in the conceptual model of homeostasis, increased biodiversity can strengthen stability within ecosystems. The interdependence between biotic and abiotic factors fosters a resilient system, capable of adapting to environmental changes. Without these interactions, ecosystems are at risk of destabilization, demonstrating that the health of biotic communities is inextricably linked to their abiotic surroundings.

Furthermore, the interaction between biotic and abiotic factors can lead to evolutionary adaptations that guide species development in response to environmental pressures. As seen in the classification of planets and their evolutionary states, this relationship reflects broader ecological dynamics, wherein life forms adapt to the energy landscapes defined by their physical environments (Alberti et al., 2017). For example, certain species have evolved physiological mechanisms to survive extreme temperatures or drought conditions, showcasing the capacity of life to endure and thrive in diverse habitats. Additionally, human-induced changes often result in significant shifts in these interactions, impacting the sustainability of ecosystems. The complexities of these relationships underscore the importance of both abiotic and biotic factors in maintaining ecological integrity, as evidenced by the feedback loops that contribute to the stability and functionality of vital life-sustaining systems (Dyke et al., 2013) (Figure 3.6).

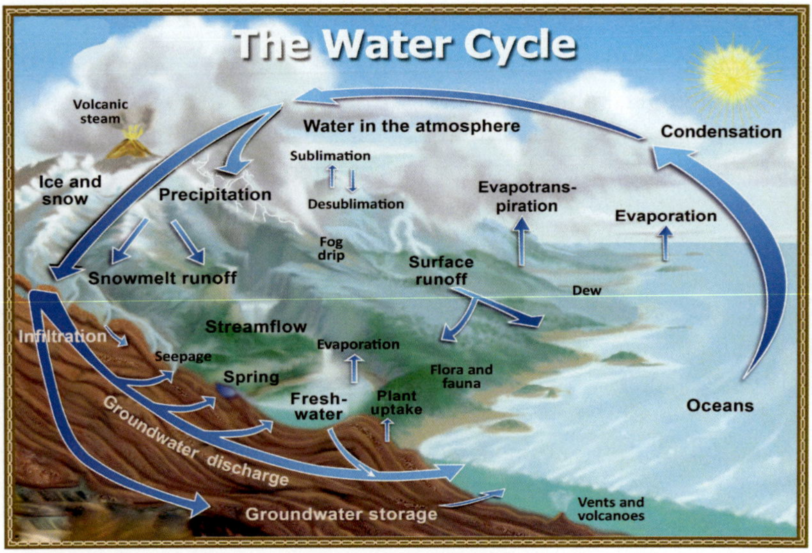

Figure 3.6. *Abiotic and biotic interactions make up a cycle like the water cycle.*

Source: By John Evans and Howard Periman, USGS – http://ga.water.usgs.gov/edu/watercycle.html, Public Domain, https://commons.wikimedia.org/w/index.php?curid=26818355.

3.5. ENERGY FLOW IN ECOSYSTEMS

The intricate network of energy flow in ecosystems is fundamentally driven by the roles of producers, consumers, and decomposers. At the base of the ecological pyramid, producers, primarily plants and photosynthetic organisms, harness solar energy through photosynthesis to create biomass. This process not only fuels their growth but also forms the foundation for energy transfer to higher trophic levels, including herbivores and carnivores. As energy is passed from one trophic level to another, it is diminished due to the second law of thermodynamics, which states that energy transformations are not 100% efficient. Only a fraction of energy, typically around 10%, is transferred to the next level, resulting in less biomass and energy availability for higher trophic levels. This cascading effect illustrates the delicate balance within ecosystems and highlights the importance of each component in maintaining ecological health and function. The complexities of energy flow are further underscored by the feedback mechanisms that exist between ecosystems and climate. Changes in biodiversity, often driven by climate variations, can lead to significant shifts in energy dynamics. For instance, a decline in biodiversity can compromise ecosystem resilience, weakening the production of services that underpin energy pathways. This is critical as biodiversity not only contributes to the stability of food webs but also supports the regeneration of biomass, facilitating the continuation of energy transfer. Research into such interactions underscores the multiplicity of factors affecting energy flow; understanding these complexities is essential for effective resource management and environmental conservation. Since societal actions significantly impact these dynamics, acknowledging the interplay between biodiversity, ecosystem services, and climate change becomes vital (Hummel et al., 2011). Moreover, the consequences of altered energy flow can be amplified through feedback loops within ecological systems. Rising global temperatures, primarily due to anthropogenic greenhouse gas emissions, potentially enhance the frequency and intensity of biodiversity loss, leading to further energy imbalance. As biotic components decline, their ability to sequester carbon diminishes, creating a vicious cycle of increasing atmospheric CO_2 levels, which in turn exacerbates climate change. The interaction between ecosystems and the climate system highlights the urgent need for transdisciplinary approaches to understand these feedback mechanisms. Collectively, these insights underscore the critical importance of preserving biodiversity and maintaining ecosystem services as a strategy for mitigating climate impacts and ensuring sustainable energy flow within the biosphere (Drégelyi-Kiss et al., 2008) (Figure 3.7).

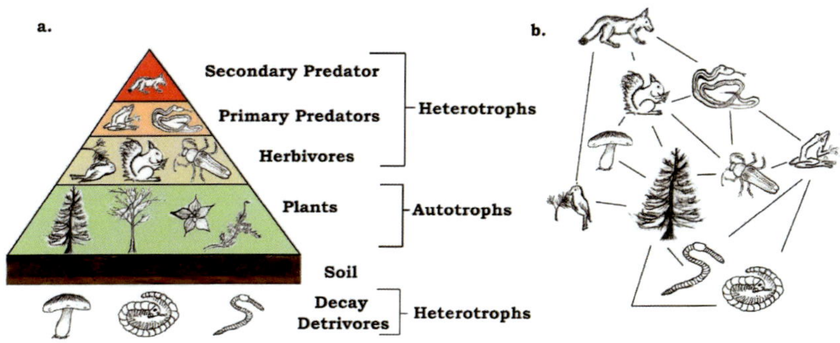

Figure 3.7. *A food pyramid and a corresponding food web, demonstrating some of the simpler patterns in a food web.*

Source: https://en.wikipedia.org/wiki/Energy_flow_%28ecology%29.

3.5.1. The Role of Producers

Producers play a foundational role in ecosystems, serving as the primary source of energy and nutrients for all other organisms within the food web. Through the process of photosynthesis, plants transform sunlight into chemical energy, making them essential for supporting life on Earth. This role extends beyond mere energy provision; producers are integral to the stability of ecosystems. They contribute to soil formation, water retention, and air quality, thereby enabling a diverse range of species to thrive. The complex interactions between producers and their environment illustrate the intricate balance necessary for maintaining biodiversity. When producers are compromised due to climate change or habitat destruction, the entire ecosystem faces potential collapse, as seen in the exploration of social-ecological systems using biodiversity-driven ecosystem services (Hummel et al., 2011). Additionally, the impact of producers on societal structures cannot be overstated, as they form the basis of agricultural and economic systems worldwide. The reliance of human societies on natural resources highlights the importance of sustainable practices in ensuring the long-term viability of these producers. By utilizing a circular business model, communities can innovate and adapt their economic frameworks to promote sustainability and conservation, which further benefits both producers and consumers (Aguayo-González et al., 2020). As societies continue to evolve, fostering an understanding of the relationships among producers, consumers, and the larger ecosystem becomes vital. Sustainable practices in resource management can lead to enhanced ecosystem services, which directly relate to human well-being and economic stability. Ultimately, the role of producers encompasses not just ecological functions but also significant socio-economic dimensions. Their contributions to biodiversity, ecosystem health, and human livelihoods underline the necessity for integrated approaches to environmental management that consider the intertwined fates of producers and human societies. By aligning societal actions with the natural rhythms and requirements of ecosystems, we can cultivate a resilient biosphere that supports both biodiversity and human needs. Understanding this interdependence is essential for crafting policies and practices that mitigate the impacts of climate change and promote sustainable resource utilization, thereby ensuring the survival and prosperity of both ecosystems and the human populations that rely on them.

3.5.2. The Role of Consumers

In the intricate web of ecosystems, consumers play a vital role in sustaining the balance of biological systems. These organisms, which include herbivores, carnivores, and omnivores, are essential for transferring energy between different trophic levels. By consuming producers, such as plants and phytoplankton, consumers facilitate the flow of energy that supports complex ecosystems. Furthermore, the relationship between consumers and producers highlights the interdependence and interactions within an environmental system, which, as outlined in educational standards, is critical for understanding ecological dynamics (TEKS 12A). This intricate relationship underscores the necessity of consumers in maintaining the health and stability of ecosystems. Additionally, consumers influence population dynamics and species diversity within their habitats. Their feeding behaviors can lead to population control of certain species, which in turn can affect competitive dynamics and resource availability. By regulating populations of primary producers and other consumers, they contribute to ecological succession and resilience against external factors. The role of consumers is also evident in how they interact with microorganisms, which are key players in maintaining ecological balance (TEKS 11C) (Mendes et al., 2020). These interactions, whether through predation or mutualism, emphasize the integral part consumers play not only in feeding but also in contributing to the overall health of ecosystems as they navigate shifting environmental conditions. Moreover, today consumers are increasingly recognized for their impact on environmental sustainability through their consumption choices. The understanding that individual choices can influence broader ecological outcomes has led to an emphasis on responsible consumption. This shift echoes the principles of circular economy discussed in the literature, where consumers are seen as active participants in reducing waste and promoting sustainability (Mendes et al., 2020). As awareness grows regarding the implications of consumer behavior on ecosystems, it becomes imperative for individuals to consider their roles in fostering environmental health. Ultimately, consumers are not just passive recipients of ecological processes; they are influential agents of change within the biosphere, underscoring the need for informed decision-making to ensure ecological balance.

3.5.3. The Role of Decomposers

The intricate web of life in ecosystems heavily relies on a group of organisms often overlooked—decomposers. These microorganisms, fungi, and other detritivores play a critical role in nutrient cycling by breaking down organic matter from deceased plants and animals. This decomposition process not only liberates nutrients such as nitrogen and phosphorus back into the soil but also transforms dead matter into valuable organic materials, promoting soil health. Without the action of decomposers, ecosystems would become overwhelmed with waste, and essential nutrients would remain locked within dead organisms, preventing their reuse by living plants. By facilitating these processes, decomposers ensure that ecosystems function efficiently, contributing to the overall balance and sustainability of the biosphere. The relationship between decomposers and other components of ecosystems demonstrates a complex interplay essential to maintaining ecological health. For instance, as pine and spruce tree species absorb nutrients, the eventual decay of their organic matter enriches the soil, thus supporting new plant growth [(cite30)]. Decomposers bridge the gap between producers and consumers, making them indispensable in food webs. They recycle nutrients that plants require for photosynthesis, thereby sustaining primary production. Moreover, their activities significantly influence carbon cycling, as decomposition releases carbon dioxide back into the atmosphere, which can then be utilized by plants during photosynthesis. This intricate cycle showcases the indispensable nature of decomposers as they connect the biosphere's various components and promote ecological resilience. In addition to nutrient recycling, decomposers significantly affect ecosystem dynamics and composition. Their role is particularly enhanced in environments characterized by high plant diversity, where variations in resource availability foster a complex network of relationships among decomposers and other organisms. Consequently, the presence or absence of certain decomposer species can lead to shifts in biomass and nutrient cycling patterns, affecting plant community structures. For example, in cases where nitrogen is abundantly recycled by decomposers, plant growth can flourish, leading to increased biodiversity and productivity in forest ecosystems. Thus, understanding the role of decomposers not only sheds light on their ecological functions but also highlights their importance in addressing broader environmental concerns, such as soil degradation and climate change.

3.6. NUTRIENT CYCLING

Within ecosystems, the process of nutrient cycling serves as a fundamental mechanism that sustains life by facilitating the transfer of essential elements such as carbon, nitrogen, and phosphorus through various biological and geological processes. This cycle comprises several key stages, including the assimilation of nutrients by organisms, their incorporation into food webs, and subsequent decomposition after the death of these organisms. The cycling of nutrients not only governs plant growth and productivity but also influences the overall health and stability of ecosystems. For example, excess nutrients, often resulting from human activities such as fertilizer runoff, can lead to harmful algal blooms and subsequent ecosystem degradation. Thus, understanding, and managing nutrient cycling is critical for maintaining ecological balance and preventing environmental crises. The importance of feedback mechanisms in nutrient cycling can be explored through physical models that simulate the various biotic interactions within ecosystems. Such simulations allow researchers to ascertain how changes in nutrient availability affect atmospheric CO_2 concentrations, ultimately influencing global temperatures. By examining these interactions, scientists can identify how biotic feedbacks contribute to, or mitigate climate change impacts, a notion supported by recent research that acknowledges the role of a hybridized planet in the Anthropocene era (Alberti et al., 2017).

The realization that human-induced changes can significantly alter nutrient fluxes compels us to reconsider traditional ecological models, offering a nuanced understanding of how ecosystems adapt and respond to anthropogenic pressures. Thus, nutrient cycling is not merely a set of processes; it is a dynamic interplay shaped by both natural systems and human interventions. Efforts to preserve and enhance nutrient cycling are essential for restoring and sustaining ecosystems. Initiatives aimed at improving agricultural practices, such as crop rotation and the use of organic fertilizers, can enhance soil fertility and reduce nutrient loss through erosion. Furthermore, maintaining biodiversity is fundamental in ensuring resilient nutrient cycling, as diverse ecosystems are better equipped to adapt to changes and disturbances. Integrating knowledge from studies that utilize physically based models can further inform conservation strategies and provide benchmarks for sustainable practices (Ineson et al., 2011). As we advance our understanding of nutrient dynamics, it becomes increasingly clear that fostering a symbiotic relationship between humanity and nature is vital. This interconnectedness reminds us that nutrient cycling is not only crucial for ecological health but also for ensuring a sustainable future for generations to come (Figure 3.8).

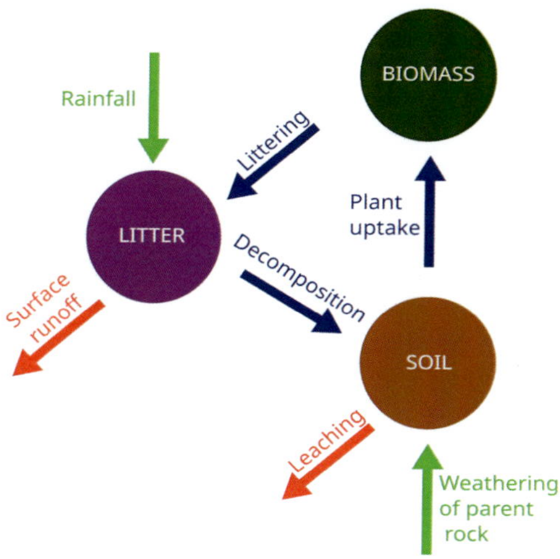

Figure 3.8. *Nutrient cycling in the environment.*

Source: By Kayau – Own work, CC BY-SA 3.0, https://commons.wikimedia.org/w/index.php?curid=23790222.

3.6.1. The Carbon Cycle

The intricacies of the carbon cycle illustrate how integral this process is to sustaining life on Earth. Carbon exists in various forms, circulating through the atmosphere, lithosphere, hydrosphere, and biosphere. Plants play a pivotal role in this cycle by absorbing carbon dioxide during photosynthesis, thereby converting it into organic compounds that support their growth and contribute to the food chain. As organisms consume these plants, carbon is transferred through trophic levels, ultimately reaching decomposers, which break down organic matter and return carbon to the soil and atmosphere. This continuous movement of carbon maintains the delicate balance required for ecosystems to function effectively, highlighting the interconnectedness of life forms and their environment. Understanding the impact of natural and anthropogenic factors on the carbon cycle is essential for assessing its role in maintaining the Earth's habitability. As described in a recent research program, the balance of carbon depends on a complex interplay of chemical, biological, and physical processes within the coupled system of the atmosphere, biosphere, and upper ocean (McElroy, 1983.). Human activities, particularly fossil fuel combustion and deforestation, have disrupted this cycle, leading to elevated carbon dioxide levels and contributing to climate change.

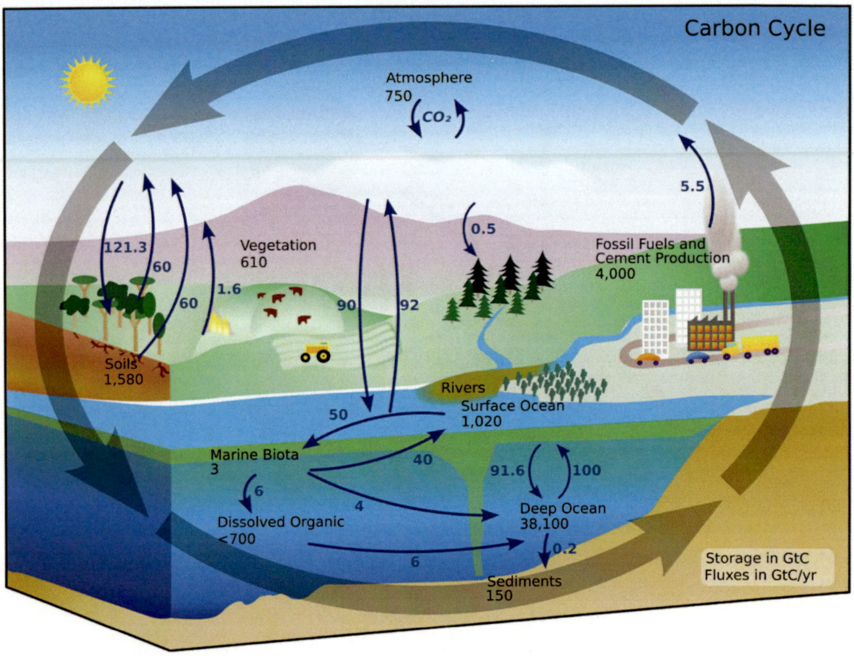

Figure 3.9. *The carbon cycle.*

Source: By Original: Kevin Saff Vector: FischX – Own work based on: Carbon cycle-cute diagram.jpeg (i.e., https://earthobservatory.nasa.gov/features/CarbonCycle), Public Domain, https://commons.wikimedia.org/w/index.php?curid=4134181.

This dynamic underscores the need for sustainable practices that preserve and enhance the natural carbon cycle, enabling us to mitigate the adverse effects of anthropogenic disturbances while promoting ecological health. Efforts to restore the natural carbon cycle can significantly benefit ecosystems and mitigate climate change effects. Strategies such as reforestation not only sequester carbon dioxide but also improve biodiversity and enhance soil health. By adopting a systems approach, industries can align their practices with the carbon cycle principles, fostering ecological, economic, and social sustainability (Conti et al., 2012). Understanding and implementing these strategies is crucial not only for maintaining the planet's life-support system but also for achieving a sustainable future. The necessity for integrating carbon management into environmental policies cannot be overstated, as it directly influences the stability and resilience of ecosystems within the biosphere (Figure 3.9).

3.6.2. The Nitrogen Cycle

Nitrogen is a critical element for all living organisms, as it is a fundamental component of amino acids and nucleic acids. The nitrogen cycle describes the series of processes by which nitrogen is converted between its various chemical forms, facilitating its movement through the atmosphere, and terrestrial and marine ecosystems. This cycle includes several key steps: nitrogen fixation, in which atmospheric nitrogen is converted into ammonia by specific bacteria; nitrification, where ammonia is transformed into nitrates; and denitrification, which returns nitrogen to the atmosphere as gaseous nitrogen. This dynamic cycle ensures the continual availability of nitrogen, promoting plant growth and supporting higher trophic levels in food webs. As terrestrial ecosystems are influenced by these nitrogen dynamics, the productivity and health of vegetative communities can be significantly affected, highlighting the interconnectedness of biological and chemical processes within ecosystems. Human activity has increasingly altered the natural nitrogen cycle, primarily through the intensive use of nitrogen-based fertilizers and fossil fuel combustion. These practices lead to elevated levels of reactive nitrogen in the environment, resulting in various ecological consequences. For instance, the excess nitrogen can stimulate algal blooms in aquatic systems, which subsequently deplete oxygen levels and harm aquatic life. Additionally, the alteration of natural nitrogen dynamics may contribute to soil degradation and reduced biodiversity in terrestrial ecosystems. Research indicates that nitrogen limitation varies across different ecosystems; for instance, while it remains weak in tropical forests, it intensifies in temperate and boreal forests and is particularly strong in tundra ecosystems (McGuire et al., 1992). Understanding these variations is crucial for developing effective management strategies that address the ecological challenges posed by excessive nitrogen. As scientists strive to assess the ramifications of altered nitrogen dynamics on global carbon storage, a clearer picture of the cycle's interaction with other elemental cycles emerges. The interconnectedness of carbon and nitrogen cycles is vital in understanding the overall health of ecosystems and their contribution to climate regulation. For example, studies on terrestrial carbon storage highlight the influence of nitrogen availability on plant productivity, particularly in the context of climate change (McGuire et al., 2001). Elevated levels of atmospheric CO_2 can enhance nitrogen availability through increased moisture and nutrient cycling, ultimately fostering plant growth. However, without proper management, the synergistic effects of excessive nitrogen could undermine these benefits, leading to negative consequences for carbon uptake and overall ecosystem stability. Thus, addressing the implications of the nitrogen cycle is essential for sustainable environmental practices and for mitigating the impacts of climate change on ecosystems (Figure 3.10).

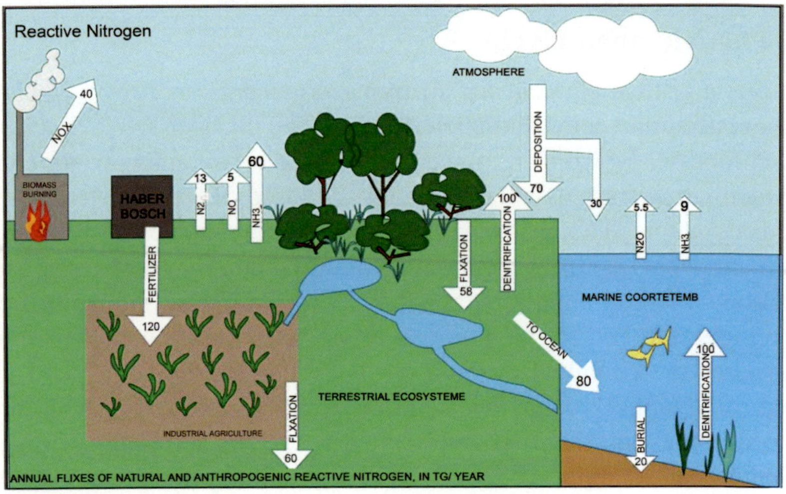

Figure 3.10. *Nitrogen cycle.*

Source: By M maraviglia – Own work, CC BY-SA 4.0, https://commons.wikimedia.org/w/index.php?curid=77954029.

3.6.3. The Phosphorus Cycle

Among the essential biochemical cycles that sustain life on Earth, the phosphorus cycle plays a critical role in maintaining ecosystem health and productivity. Unlike carbon or nitrogen, phosphorus does not exist as a gas in the atmosphere and primarily cycles through soil, water, and living organisms. Phosphate ions derived from weathering rocks or released through decomposition are taken up by plants, forming the foundation of this cycle. Herbivores consume these plants, and through the food web, phosphorus is transferred to higher trophic levels. When organisms die, their bodies break down, returning phosphates to the soil and aquatic systems, which can then be reused by plants. This recycling process is vital for DNA and ATP formation, underpinning energy transfer within ecosystems and demonstrating phosphorus's integral role in sustaining life. Human activity has significantly impacted the natural phosphorus cycle, leading to ecological imbalances and environmental concerns. Agricultural practices, such as the extensive use of phosphorus-rich fertilizers, contribute to nutrient runoff into water bodies. This excess phosphorus can lead to eutrophication, where algal blooms deplete oxygen in aquatic systems, resulting in fish kills and loss of biodiversity. Moreover, as phosphorus accumulates in sediments, the long-term impacts can disrupt entire ecosystems and compromise water quality. Such feedback mechanisms illustrate how alterations in nutrient cycles can escalate issues within the biosphere, indicating the need for sustainable land-use practices and nutrient management to mitigate these adverse effects.

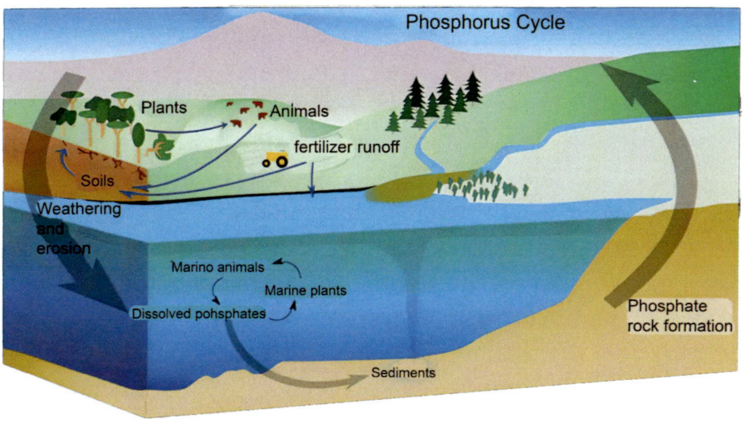

Figure 3.11. *The phosphorus cycle.*

Source: By BonniemfIncorporates work by NASA Earth Science Enterprise – Reworked by Bonniemf from the public domain file File:Carbon cycle-cute diagram.svg, CC BY-SA 3.0, https://commons.wikimedia.org/w/index.php?curid=29839469.

As noted in research, understanding the interactions between physical and chemical entities within ecosystems is paramount to managing the phosphorus cycle effectively (Goody et al.(n.d)). Addressing the implications of phosphorus cycling in an anthropogenic context highlights the interplay between ecosystems and broader environmental changes. The feedback between phosphorus cycle disruptions and climate change illustrates a complex relationship impacting global biodiversity and habitat sustainability. Adjustments in ecosystem productivity due to phosphorus loading can exacerbate climate change impacts by altering carbon cycling processes and greenhouse gas emissions. Moreover, the interdependence of phosphorus cycling with other elemental cycles underscores the necessity of integrative approaches to environmental management. Future research must prioritize these interconnected cycles to predict potential long-term changes affecting habitability on Earth (Drégelyi-Kiss et al., 2008). Understanding the phosphorus cycle is crucial for safeguarding ecosystems and ensuring the resilience of the biosphere against the pressures of human activity and climate variability (Figure 3.11).

3.7. BIODIVERSITY AND ECOSYSTEM HEALTH

The intricate relationship between biodiversity and ecosystem health is paramount in understanding environmental sustainability. Biodiversity, comprising the variety of life found in ecosystems, serves as a critical indicator of ecosystem resilience and functionality. A diverse range of species can enhance an ecosystem's ability to withstand disturbances such as climate change, pollution, and habitat degradation. When biodiversity is compromised, as seen in numerous freshwater ecosystems due to anthropogenic impacts (Ekins et al., 2020), the repercussions often lead to a decline in ecosystem services vital for human survival, such as clean water and food production. Therefore, preserving biodiversity is not just an ethical imperative but a necessary foundation for maintaining ecosystem health, which is essential for both ecological balance and human well-being. Moreover, the interplay between natural capital and social development highlights the significance of maintaining ecosystem health through effective biodiversity management. Natural capital encompasses the environmental assets that contribute to our quality of life, including clean air, water, and fertile land. As outlined in the Natural Capital Indicator Framework (NCIF), understanding, and monitoring the state of natural capital is crucial in aiding decision-making related to economic development (Ekins et al., 2020).

Countries can implement policies aimed at preserving biodiversity while managing their resources sustainably. This integration of natural capital accounting with biodiversity initiatives promises improved outcomes in ecosystem health, ensuring that the benefits derived from nature are safeguarded for future generations. By leveraging robust indicators, policymakers can navigate the complexities of human-nature relationships, fostering an environment where both biodiversity and human communities thrive. The rapid decline of various ecosystems worldwide exemplifies the urgent need to prioritize biodiversity for ecosystem health. Freshwater bodies, in particular, face multiple threats, including pollution, habitat destruction, and invasive species, which lead to significant biodiversity loss (Goethals et al., 2019). Such losses diminish the ability of ecosystems to provide essential services, creating a cascading effect on agriculture, fisheries, and water availability. This decline emphasizes the need for a holistic approach towards the sustainable development of these ecosystems, where conservation measures align with socioeconomic goals. By understanding and addressing the challenges linked to biodiversity loss, stakeholders can work together to establish sustainable practices that not only restore ecosystems but also enhance their resilience against future threats. The preservation of biodiversity is therefore a crucial component in achieving overall ecosystem health, which ultimately benefits both nature and humanity (Figure 3.12).

Figure 3.12. *The importance of biodiversity for healthy ecosystems and human life.*

Source: By Collection created by The Emirr – Collection of images Photo by LiquidGhoul. License: Creative Commons Attribution-Share Alike 3.0 Unported-Photo by Diliff. License: Creative Commons Attribution-Share Alike 3.0 Unported-Photo by Julius Rückert. License: Creative Commons Attribution-Share Alike 3.0 Unported(rotated) Photo by Hans Hillewaert (Lycaon). License: Creative Commons Attribution-Share Alike 2.5Photo by AngMoKio. License: Creative Commons Attribution-Share Alike 2.5Photo by Branko Kannenberg. License: License: Creative Commons Attribution-Share Alike 2.5Photo by Alexandre Buisse (Nattfodd). License: Creative Commons Attribution-Share Alike 3.0 UnportedPhoto by Malene Thyssen. License: Creative Commons Attribution-Share Alike 3.0 UnportedPhoto by Hans Hillewaert (Lycaon). License: Creative Commons Attribution-Share Alike 2.5Photo by Hans Hillewaert (Lycaon). License: Creative Commons Attribution-

Share Alike 2.5Photo by Richard Bartz. License: Creative Commons Attribution-Share Alike 2.5Photo by Hans Hillewaert (Lycaon). License: Creative Commons Attribution-Share Alike 2.5, CC BY-SA 3.0, https://commons.wikimedia.org/w/index.php?curid=20426804.

3.7.1. Definition of Biodiversity

Biodiversity encompasses the variety of life on Earth, reflecting the myriad of species, genetic differences, and ecosystems that exist across various environments. This diversity is crucial for sustaining ecosystem functions, promoting resilience, and facilitating processes that benefit human society, such as food production and climate regulation. As ecosystems evolve over time, their biological diversity contributes significantly to ecological stability, making it essential for conservation efforts. Effective management of biodiversity is increasingly recognized as vital to combatting environmental challenges, such as habitat destruction and climate change, thereby ensuring that ecosystems can continue to thrive and provide necessary services. Consequently, understanding, and defining biodiversity is pivotal in fostering awareness of its importance in sustaining both natural systems and human livelihoods. At the core of biodiversity is the understanding that variation exists not only among species but also within them, as seen through genetic diversity. This genetic variation is a foundation for adaptability and resilience in changing environments and is critical in promoting species survival in the face of threats. For instance, in ecosystems rich in genetic and species diversity, there are greater opportunities for a wide range of ecosystem services, highlighting how interconnected biodiversity and ecosystem functionality are. Understanding this definition of biodiversity also involves recognizing its implications on conservation strategies, such as the establishment of protected areas like the Natura 2000 network. These initiatives aim to maintain and restore favorable conservation status for species and habitats, as indicated in the project aimed at integrated management of these areas (Apeldoorn et al., 2009).

Thus, preserving biodiversity is not only an environmental necessity but also a critical factor in ensuring sustainable development. Local stakeholders often exhibit varying perceptions of biodiversity, which can profoundly influence conservation efforts and policy-making. A study assessing social preferences for ecosystem services in Chile highlighted the stark contrasts in how different groups value these services associated with biodiversity, such as drinking water and climate regulation. These values inform stakeholders' engagement with conservation measures and their advocacy for specific ecosystem services over others (Bidegain et al., 2019). Awareness and education about the benefits of biodiversity can promote a more unified approach to conservation, aligning the interests of diverse groups. By fostering discussions that integrate scientific knowledge and local perspectives, policymakers can devise strategies that effectively address both ecological needs and community values. In sum, the definition of biodiversity extends beyond mere species counts; it encompasses complex interrelations that require collaborative efforts for sustainable management and conservation.

3.7.2. Importance of Biodiversity for Ecosystem Stability

Ecosystem stability relies heavily on the intricate web of interactions among species within a given habitat. Greater biodiversity often leads to greater resilience, allowing ecosystems to adapt to changes such as environmental stressors, climate events, and biological invasions. Diverse ecological communities possess a wider range of functions, which enhances their ability to maintain processes such as nutrient cycling and energy flow. For instance, in a biodiverse ecosystem, various species may perform similar roles, ensuring that if one species declines, others can fulfill its function. This redundancy is critical; as highlighted by developments within Gaia theory, ecosystems operate as interdependent systems where both biotic and abiotic factors coalesce to support life (Rushton, 2008). Thus, protecting biodiversity not only sustains individual species but fortifies ecosystem dynamics essential for overall stability. Moreover, the relationship between biodiversity and ecosystem stability underscores the importance of managing our natural resources sustainably. It becomes increasingly evident that biodiversity loss directly impacts ecosystem services, which are vital for human well-being. The concept of a safe and just operating space reveals the trade-offs between social equity and environmental health, suggesting that excessive exploitation of natural resources can undermine the very ecosystems we depend upon for survival (Brown et al., 2014). Regions suffering from reduced biodiversity may experience diminished ecosystem functions, leading to challenges such as soil degradation, water quality issues, and diminished agricultural productivity. This ripple effect reinforces the argument that preserving biodiversity is not just an environmental concern but a vital socioeconomic necessity—ensuring that our ecosystems remain healthy and capable of supporting our growing population. In conclusion, the significance of biodiversity for ecosystem stability cannot be overstated. The interconnections between organisms create a complex structure that sustains life, demonstrating that fewer species often lead to a fragile ecosystem that is more susceptible to disturbances. Recognizing this complexity is essential for conservation efforts and sustainable resource management. It is critical to acknowledge that protecting biodiversity goes beyond preserving individual species; it is about maintaining the entire network of life that supports ecological integrity. As we confront global challenges such as climate change and resource depletion, understanding, and prioritizing biodiversity becomes a fundamental tenet of ecological resilience and sustainability, ensuring that ecosystems continue to thrive for future generations.

3.7.3. Threats to Biodiversity

The loss of biodiversity poses a significant challenge to the ecological balance and the sustainability of ecosystems worldwide. Factors such as habitat destruction, pollution, and climate change contribute to this alarming decline. Habitat destruction, particularly due to urbanization and agricultural expansion, has led to the degradation of crucial ecosystems, pushing numerous species toward extinction. For instance, wetland drainage has severely impacted aquatic life, diminishing biodiversity and disrupting food webs. Furthermore, pollution from industrial runoff and waste has introduced toxic substances into ecosystems, affecting the health of both terrestrial and aquatic organisms. To effectively address these pressing threats, a concerted effort involving policy reform and sustainable development practices is essential. Climate change emerges as a critical factor exacerbating the threats to biodiversity. As temperatures rise and weather patterns shift, many species find it increasingly challenging to adapt to their changing environments. For example, marine ecosystems are experiencing coral bleaching due to elevated sea temperatures, resulting in the loss of biodiversity in these vital habitats. The Millennium Ecosystem Assessment (2014) suggests that the ongoing trajectories of global change may drive human-environment systems toward highly degraded states, making it difficult for ecosystems to recover. Additionally, the implications of climate change extend beyond individual species; they affect entire ecosystems, leading to diminished services such as pollination, water purification, and carbon storage. If unaddressed, these interconnected issues threaten not only wildlife but also human well-being, emphasizing the urgency of collective action. Furthermore, invasive species significantly endanger native biodiversity by introducing competition, predation, and disease. These non-native species often thrive in altered ecosystems, where they can outcompete indigenous species for resources, leading to population declines or extinctions. (Goethals et al., 2019) highlights that freshwater resources like lakes and reservoirs are increasingly susceptible to such threats, with invasive species disrupting aquatic food webs and compromising water quality. As a result, the resilience of ecosystems is further compromised, leading to broader declines in biodiversity. Addressing the threat of invasive species necessitates stringent regulations on species introduction and enhanced public awareness. This proactive approach, coupled with habitat protection and restoration initiatives, can foster an environment conducive to the preservation of biodiversity, ensuring the stability of ecosystems and the services they provide to human societies (Figure 3.13).

Figure 3.13. *Deforestation.*

Source: By Jami Dwyer – https://www.flickr.com/photos/74281168@N00/173937750/, Public Domain, https://commons.wikimedia.org/w/index.php?curid=1363210.

3.8. HUMAN IMPACT ON ECOSYSTEMS

Human activities have fundamentally altered ecosystems, impacting biodiversity, carbon storage, and overall ecological balance. Land use change, particularly through agricultural expansion, has led to significant habitat loss, disrupting the intricate relationships within ecosystems. For instance, the establishment of croplands has historically resulted in carbon release from terrestrial ecosystems, as highlighted in the assessment that recognizes cropland establishment as a primary cause of carbon emissions up until 1958. The increased conversion of forests and grasslands into agricultural land not only diminishes plant and animal habitats but also undermines the ecosystem services that contribute to carbon sequestration. The long-term consequences of these alterations can degrade the resilience of ecosystems, hindering their ability to respond to environmental changes such as climate variability. The interplay between climate change and human activities further complicates these negative impacts on ecosystems. Rising atmospheric CO_2 levels, a byproduct of industrial activities and deforestation, impact plant growth and the capacity of ecosystems to sequester carbon. Models have demonstrated that while climate variability has minor long-term effects on carbon storage relative to increased atmospheric CO_2, it does influence interannual variability in ecosystem productivity (Hummel et al., 2011). This demonstrates that changes in climatic patterns, exacerbated by human actions, can lead to unpredictable responses within ecosystems, further indicating a significant human footprint in shaping ecological dynamics. The implications of these changes emphasize the urgency for sustainable practices, as human impact continues to extend beyond immediate alterations to address long-term global consequences. Addressing human impact on ecosystems requires a comprehensive understanding of ecological systems and the adoption of effective management strategies. Research initiatives that blend social and ecological perspectives can provide insights into sustainable adaptations to climate-induced changes in ecosystem function (Hummel et al., 2011). Prioritizing the conservation and restoration of ecosystem services is crucial for maintaining biodiversity and ensuring the stability of natural systems. Furthermore, understanding societal values and economic utilization of these services can empower communities to engage in biodiversity conservation actively. A transdisciplinary approach that incorporates various scientific, social, and economic factors can lead to informed decision-making in biodiversity and resource management, mitigating the enduring impacts of human activities on the biosphere. Such integrative strategies are vital for preserving ecological integrity in an era of rapid change (Figure 3.14).

Figure 3.14. *Human impacts on the environment.*

Source: By Anaskamal – Own work, CC BY-SA 4.0, https://commons.wikimedia.org/w/index.php?curid=64235736.

3.8.1. Pollution and Its Effects

The intricate relationship between pollution and its effects on the environment represents a pressing challenge for contemporary society. Pollution, in various forms such as air, water, and soil contaminants, disrupts the delicate balance of ecosystems, leading to detrimental consequences for biodiversity. For instance, chemical runoff from agricultural activities can introduce harmful substances into local water systems, causing eutrophication that depletes oxygen levels and threatens aquatic life. This disruption doesn't only harm individual species but also affects entire food webs and habitats, illustrating the cascading impact of pollution on ecological health.

The loss of species diversity, as a result of habitat destruction and toxicity, ultimately undermines ecosystem resilience, making it harder for nature to adapt to further environmental changes. Furthermore, pollution implications extend far beyond ecological degradation, impacting human health and societal structures. Airborne pollutants, primarily from fossil fuel combustion, contribute significantly to respiratory diseases and other health issues for populations in urban areas. As pollution persists, it can foster socio-economic disparities, as marginalized communities often bear the brunt of environmental hazards due to their proximity to industrial sites. The Northern Eurasia Future Initiative (NEFI) emphasizes the need for science-driven strategies to combat such challenges, highlighting the interconnectedness of human health and environmental stability (Albrechtova et al., 2017).

Figure 3.15. *Pollution and its effects.*

Source: By Moancho – Ihe onyonyo na-egosi ndị nọ ọrụ na mpaghara "Geo-physical Investigation of Dumpsite Leachate Plumes, CC BY-SA 4.0, https://commons.wikimedia.org/w/index.php?curid=64278511.

3.8.2. Deforestation and Habitat Loss

The rapid rate of deforestation poses significant threats to global biodiversity and the intricate ecosystems that sustain it. Areas such as the Amazon rainforest are particularly vulnerable, experiencing large-scale clearing for agriculture and urban development. This habitat destruction not only diminishes the variety of species inhabiting these regions but also disrupts critical ecological processes, such as nutrient cycling and carbon storage. As highlighted in current projections, global change may lead to regime shifts that degrade ecosystems, resulting in a loss of biodiversity and vital ecosystem services (Millennium Ecosystem Assessment, 2014). The consequences of this degradation extend beyond environmental impacts; they can jeopardize human well-being by reducing access to natural resources essential for survival. Furthermore, the push for biofuels has contributed to escalating deforestation, as land is cleared to plant energy crops. While biofuels are advocated as a climate change solution, their widespread production often sacrifices natural landscapes, leading to significant environmental repercussions. Advanced technologies that exploit cellulosic feedstocks could alleviate some pressures, yet simulations indicate that even these methods could result in substantial habitat loss (Cronin et al., 2009).

Figure 3.16. *Deforestation and habitat loss.*

Source: By Matt Zimmerman – Slash and burn agriculture in the Amazon, CC BY 2.0, https://commons.wikimedia.org/w/index.php?curid=33978259.

The tension between fulfilling energy needs and conserving biodiversity creates a complex dilemma. To address this, sustainable land management practices must be prioritized to prevent irreversible ecological damage while simultaneously advancing biofuel initiatives. Addressing deforestation and habitat loss is not merely an environmental concern; it is a pressing issue intertwined with socio-economic factors. Rural communities often rely on forests for their livelihoods, and the loss of habitat can lead to diminished resources and increased poverty. Policy-makers must implement strategies that balance development with conservation, promoting agroforestry and sustainable land use to ensure that both human populations and ecosystems thrive. Engaging local communities in conservation efforts can enhance the effectiveness of these strategies. Ultimately, combating deforestation will require a global commitment to develop policies that protect critical habitats while recognizing the needs of those who depend on them, creating a more sustainable and equitable future for all.

3.8.3. Climate Change and Its Consequences

Increasingly, climate change emerges as a primary driver of alteration across global ecosystems, challenging the delicate balance that sustains biodiversity. As human demands for resources escalate—from food production to energy consumption—ecosystems face compounding stresses that exacerbate these changes. The interplay of habitat destruction, pollution, and species invasions highlights the multifaceted nature of climate impacts. For instance, when habitats are degraded, not only do flora and fauna lose their homes, but the overall ecosystem functions that support life also falter. Ultimately, this puts extraordinary pressure on species to adapt or face extinction, as evident in the heightened vulnerabilities caused by anthropogenic activities that manipulate natural processes (Botella et al., 2016). Moreover, the feedback mechanisms triggered by climate change can lead to profound, often unexpected consequences for the biosphere. While climate change typically garners significant attention, it operates synergistically with other factors, creating a compounding effect that leads to "state shifts" in ecosystems (Botella et al., 2016). Such shifts can result in irreversible changes to ecological balance, making ecosystems less resilient to further disturbances. For instance, rising temperatures can exacerbate eutrophication in aquatic systems, depleting oxygen levels and leading to dead zones that threaten marine biodiversity. Understanding these interconnected processes emphasizes the importance of adopting a holistic approach to environmental policy and ecosystem management. Addressing climate change requires an integrated strategy that not only mitigates its effects but also enhances ecosystem resilience. This involves recognizing the complex interplay between human activity and natural systems while promoting sustainable practices. Actions such as restoring degraded habitats, reducing greenhouse gas emissions, and conserving biodiversity can create a more adaptive biosphere capable of withstanding the pressures of climate change. Acknowledging that the consequences of climate change extend well beyond mere temperature rises—that they encompass profound shifts in ecological dynamics—underscores the urgency for collaborative, global action aimed at preserving the integrity of our planet's ecosystems (Botella et al., 2016). Such efforts are essential not only for ecological well-being but also for ensuring a livable future for generations to come (Figure 3.17).

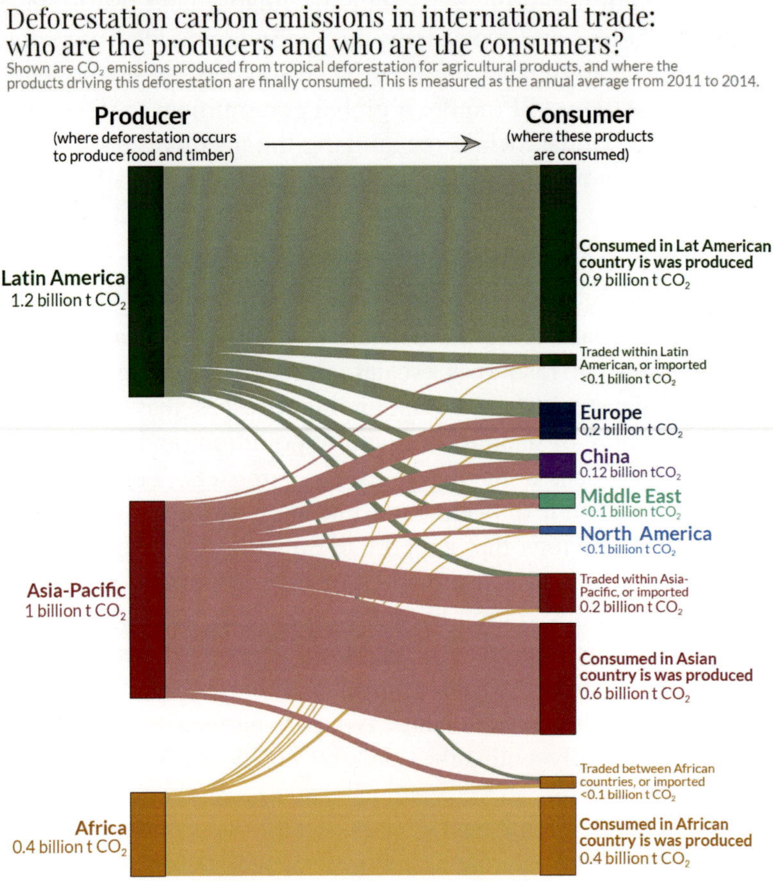

Figure 3.17. *Climate change and its consequences.*

Source: By Hannah Ritchie and Max Roser – https://ourworldindata.org/deforestation, CC BY-SA 4.0, https://commons.wikimedia.org/w/index.php?curid=113361464.

3.9. CONCLUSION

In understanding the intricate relationship between ecosystems and the biosphere, it becomes clear that our natural environments are not only places of beauty but also essential for human well-being. The protection of ecosystems such as coral reefs, mangroves, and beaches yields numerous benefits, illustrating the synergistic relationship between environmental health and human interests. Specifically, these ecosystems provide vital services, including safeguarding against beach erosion and supporting local fisheries, which ultimately enhance the potential for sustainable tourism growth. As indicated in the findings on coralline beaches and reefs, the careful management of these natural resources is imperative for both ecological balance and economic development (Cooper et al., 2010). Thus, our conclusion emphasizes that preserving these ecosystems is not just an environmental responsibility but also a pragmatic approach to ensuring human prosperity. Moreover, as climate change progresses, the interconnectedness of ecosystems and the biosphere presents significant challenges we must confront. The feedback loops between ecosystems and climatic changes can exacerbate the consequences of human-induced greenhouse gas emissions, leading to a range of ecologically detrimental effects. According to (Drégelyi-Kiss et al., 2008), higher global temperatures are likely to increase greenhouse gas levels, suggesting a cycle of worsening climatic conditions that can destabilize ecosystems and compromise their integrity. This feedback mechanism highlights the urgency for global cooperation and action aimed at mitigating climate change impacts, especially in vulnerable regions where ecosystems play a crucial role in sustaining livelihoods. In conclusion, the health of our ecosystems is inextricably linked to the stability of the biosphere, necessitating collective efforts to safeguard these vital natural resources. The implications of neglecting ecological balance are profound and far-reaching, affecting not only biodiversity but also economic and social structures. Recognizing the benefits derived from ecosystems, such as tourism potential and habitat protection, underscores the importance of adopting sustainable practices. As we move forward, we must integrate a deeper understanding of ecological dynamics into policy-making and community efforts, ensuring that both nature and humanity can thrive in harmony. The evidence clearly supports that proactive measures can yield substantial dividends in terms of both environmental sustainability and human advancement.

3.9.1. Summary of Key Points

In understanding ecosystems, it is crucial to consider the multifaceted roles that different components play in maintaining the balance of the biosphere. For example, coralline beaches, reefs, and mangroves provide vital ecosystem services that benefit both marine life and human communities. These habitats act as natural barriers, protecting coastlines against erosion while simultaneously nurturing diverse fish populations, which are essential for local fisheries. Furthermore, the potential for tourism growth in protected marine areas underscores the economic value of preserving these ecosystems. Such benefits highlight the interconnectedness of ecological health and human welfare, emphasizing the need for conservation efforts to ensure the sustainability of these vital resources (Cooper et al., 2010). Equally important is the understanding that Earth's habitability is influenced by intricate interactions among its physical, chemical, and biological systems. Research programs aimed at elucidating these interactions are crucial for assessing changes that may impact the planet's future. For instance, the distribution of essential nutrients such as carbon, nitrogen, phosphorous, sulfur, and water plays a significant role in ecosystem functionality. These nutrients are integral not only to sustaining life but also in understanding potential responses to anthropogenic influences, such as climate change and pollution. Thus, the examination of planetary life-support systems must incorporate a comprehensive framework that considers both natural and human-induced changes (Michael McElroy, 1983). The synthesis of knowledge regarding ecosystems and the biosphere ultimately reveals a pressing call to action for environmental stewardship. The recognized benefits of biodiverse habitats, combined with the insights gained from research on planetary life-support mechanisms, highlight the urgency of maintaining ecological integrity in the face of ongoing challenges. As we navigate the complexities of our relationship with the environment, it becomes increasingly essential to prioritize conservation and sustainable practices. By fostering an awareness of the intricate balance between ecosystems and human activities, we can work towards a more resilient and sustainable future that honors the delicate tapestry of life on Earth.

3.9.2. The Importance of Preserving Ecosystems

Ecosystems serve as the backbone of biodiversity, providing essential services that sustain life on Earth. The intricate web of interactions between species and their habitats creates a dynamic balance that promotes ecological stability. When ecosystems are preserved, they continue to offer vital resources such as clean air, freshwater, and fertile soil. Moreover, the preservation of these systems safeguards the genetic diversity necessary for species adaptation and resilience in the face of environmental changes. As highlighted in the context of conservation in Europe, maintaining healthy ecosystems aligns with directives aimed at protecting species and habitats under frameworks like the Birds and Habitats Directives, underscoring the interconnectedness of biodiversity and ecosystem management (Apeldoorn et al., 2009). Furthermore, ecosystems play a crucial role in regulating climate and supporting economic prosperity. They act as natural buffers against climate extremes, reducing the impacts of floods, droughts, and wildfires. For instance, wetlands, and forests are recognized for their ability to sequester carbon, thus mitigating climate change effects. By preserving these ecosystems, communities not only protect their natural heritage but also ensure long-term economic benefits through sustainable practices. This relationship emphasizes that conservation is not merely a moral imperative but a strategic approach to secure resources for future generations, as the decline of ecosystems often results in dire economic consequences for human societies (Sinha, 2011). Incorporating an integrated management approach is essential for the effective preservation of ecosystems. Such management not only addresses the ecological needs of species and habitats but also considers the social dimensions of conservation. Engaging stakeholders in the dialogue about ecosystem management fosters a sense of shared responsibility and investment in ecological outcomes. This collaborative approach can enhance local governance and build capacity for sustainable practices, ensuring that the benefits of biodiversity are recognized and protected. Ultimately, the importance of preserving ecosystems lies in their multifaceted contributions to life on Earth; failure to act restricts the future of both our natural environment and human well-being (Apeldoorn et al., 2009; Sinha, 2011).

3.9.3. Future Directions for Ecosystem Research and Conservation

A significant transformation in ecosystem research and conservation is underscored by the integration of advanced technologies such as remote sensing, artificial intelligence, and big data analytics. These innovations offer unprecedented capabilities to monitor ecological changes in real-time, allowing researchers to track the health of ecosystems and the organisms within them more efficiently. For instance, satellite imagery can reveal deforestation trends or habitat degradation over large areas, while drones equipped with sensors can assess biodiversity in hard-to-reach locations. Moreover, predictive modeling through AI can help identify potential ecological outcomes based on various conservation strategies. By adopting these technologies, scientists can not only enhance their understanding of ecosystems but also implement more effective conservation measures, ensuring a proactive rather than reactive approach to environmental management. Addressing the challenges posed by climate change is essential for the future of ecosystem research and conservation. The shifting climate presents unique pressures on species distributions, migration patterns, and ecological interactions. As temperatures rise and precipitation patterns change, many species are forced to adapt, migrate, or face extinction. Research efforts must prioritize understanding these dynamics by focusing on resilience strategies that can help ecosystems adapt to fluctuating climatic conditions. Additionally, conservation initiatives must integrate climate adaptation into their frameworks, such as through the establishment of wildlife corridors that facilitate species movement and genetic exchange. Engaging in collaborative research initiatives that cross disciplinary boundaries can yield valuable insights into how ecosystems respond to climate change, which in turn can inform policies and practices aimed at fostering both biodiversity and ecosystem vitality. Furthermore, the future of ecosystem conservation hinges on fostering public awareness and involvement in research efforts. An informed and engaged populace can drive positive change by advocating for sustainable practices and supporting conservation initiatives. This requires a concerted effort to educate communities about the importance of healthy ecosystems, creating opportunities for citizen science projects where individuals can participate in data collection and monitoring. By leveraging local knowledge and fostering connections between researchers and community members, conservation strategies can be tailored to meet localized needs simultaneously. Ultimately, the collaboration between scientists, policymakers, and the public will create a more holistic approach to ecosystem management, ensuring that conservation efforts are not only scientifically sound but also socially supported and culturally relevant.

REFERENCES

1. Aguayo-González, F., Lama-Ruiz, J. R., Martín-Gómez, A. M., & Ávila-Gutiérrez, M. J. (2020). Eco-Holonic 4.0 circular business model to conceptualize sustainable value chain towards digital transition. *MDPI AG*. https://core.ac.uk/download/299807283.pdf (accessed on 5 December 2024).
2. Alberti, M., Frank, A., & Kleidon, A. (2017). Earth as a hybrid planet: The Anthropocene in an evolutionary astrobiological context. *arXiv*. http://arxiv.org/abs/1708.08121 (accessed on 5 December 2024).
3. Albrechtova, J., Bergen, K., Chen, J., Groisman, P. Y., Gulev, S., Gutman, G., et al., (2017). Northern Eurasia Future Initiative (NEFI): Facing the challenges and pathways of global change in the twenty-first century. *University of New Hampshire Scholars' Repository*. https://core.ac.uk/download/215538897.pdf (accessed on 5 December 2024).
4. Ali, H., Khan, E., & Ilahi, I. (2019). Environmental chemistry and ecotoxicology of hazardous heavy metals: Environmental persistence, toxicity, and bioaccumulation. *Environmental Toxicology and Pharmacology, 68*, 1–14. https://doi.org/10.1155/2019/6730305.
5. Apeldoorn, R. C. van, Blust, G. de, Bouwma, I. M., Kruk, R. W., & Sier, A. R. J. (2009). Information and communication on the designation and management of Natura2000 sites. Main Report 3: Towards integrated management. *Alterra*. https://core.ac.uk/download/56183.pdf (accessed on 5 December 2024).
6. Berg, G., Rybakova, D., Fischer, D., Cernava, T., Champomier Vergès, M. C., Charles, T. C., Chen, X., Cocolin, L., Eversole, K., Herrero Corral, G., Kázou, M., Kinkel, L. L., Lange, L., Lima, N., Loy, A., Macklin, J., Maguin, E., Mauchline, T. H., McClure, R., Mitter, B., Ryan, M. J., Sarand, I., Smidt, H., Schelkle, B., Roume, H., Seghal Kiran, G., Selvin, J., Souza, R. S. C. de, van Overbeek, L. S., Singh, B. K., Wagner, M., Walsh, A. M., Sessitsch, A., & Schloter, M. (2020). Microbiome definition re-visited: Old concepts and new challenges. *Microorganisms, 8*, 875. https://doi.org/10.1186/s40168-020-00875-0.
7. Bidegain, I., Catalán, E., Cerda, C., López-Santiago, C., & Tironi, A. (2019). Social preferences for ecosystem services in a biodiversity hotspot in South America. *Public Library of Science (PLoS)*. https://repositorio.uam.es/bitstream/10486/690687/1/social_bidegain_po_2019.pdf (accessed on 5 December 2024).
8. Botella, F., Carrete, M., Clavero, M., DeVault, T. L., Donázar, J. A., Hermoso, V., Ibáñez, C., Losada, M. A., Polo, M. J., Pérez-García, J. M., Sánchez-Navarro, S., & Sánchez-Zapata, J. A. (2016). Effects of renewable

energy production and infrastructure on wildlife. *DigitalCommons@University of Nebraska–Lincoln*. https://core.ac.uk/download/188125059.pdf (accessed on 5 December 2024).

9. Brasoveanu, F., & Lisievici Brezeanu, A. P. (2010). The sustainable protection and exploitation of the living aquatic resources from the Danube Delta Biosphere Reserve. *Curentul Juridic, The Juridical Current*, *43*, 153–162.
10. Brown, S., Carstensen, J., Cole, M. J., Cornell, S. E., Dawson, T. P., Dearing, J. A., Doncaster, C. P., Dyke, J. G., Eigenbrod, F., Flörke, M., Haberl, H., Hossain, M. S., Jeffers, E., Langdon, P. G., Lenton, T. M., Mackay, A. W., Nykvist, B., Poppy, G. M., Raworth, K., Wang, R., & Zhang, K. (2014). Safe and just operating spaces for regional social-ecological systems. https://core.ac.uk/download/30662973.pdf (accessed on 5 December 2024).
11. Conti, M., English, S., & Martinez, V. (2012). Sustainable car life cycle design, taking inspiration from natural systems and thermodynamics. https://core.ac.uk/download/9340980.pdf (accessed on 5 December 2024).
12. Cooper, E., Wielgus, J., Burke, L., & Torres, R. (2010). *Coastal Capital: Dominican Republic: Case Studies on the Economic Value of Coastal Ecosystems in the Dominican Republic*. World Resources Institute (WRI). https://core.ac.uk/download/71350194.pdf (accessed on 5 December 2024).
13. Cronin, T. W., Felzer, B. S., Gurgel, A. C., Kicklighter, D. W., Melillo, J. M., Paltsev, S., Reilly, J. M., Schlosser, C. A., Sokolov, A. P., & Wang, X. (2009). Unintended environmental consequences of a global biofuels program. *MIT Joint Program on the Science and Policy of Global Change*. https://core.ac.uk/download/4409143.pdf (accessed on 5 December 2024).
14. Drégelyi-Kiss, Á., Drégelyi-Kiss, G., & Hufnagel, L. (2008). Ecosystems as climate controllers–Biotic feedbacks. *Applied Ecology and Environmental Research, 6*, 111–134. https://doi.org/10.15666/aeer/0602_111134.
15. Drégelyi-Kiss, G., Drégelyi-Kiss, Á., & Hufnagel, L. (2008). Ecosystems as climate controllers–Biotic feedbacks: A review. *Penkala Bt*. https://core.ac.uk/download/18426065.pdf (accessed on 5 December 2024).
16. Dyke, J. G., & Weaver, I. S. (2013). The emergence of environmental homeostasis in complex ecosystems. *PLoS Computational Biology, 9*(5), e1003050. https://doi.org/10.1371/journal.pcbi.1003050.
17. Ekins, P., Fairbrass, A., Mace, G., & Milligan, B. (2020). The Natural Capital Indicator Framework (NCIF): A framework of indicators for national natural capital reporting. *Elsevier BV*. https://core.ac.uk/download/326707296.pdf (accessed on 5 December 2024).

18. FAO, Gorshkov, G., Li, Lotka, M., Makarieva, M., Stern, T., Tutin, V., Van de Koppel, J., & Whittaker, R. (2010). Comprehending environmental and economic sustainability: Comparative analysis of stability principles in the biosphere and free market economy. *Wiley*. http://arxiv.org/abs/1011.5978 (accessed on 5 December 2024).

19. Frank, D., Reichstein, M., Bahn, M., Thonicke, K., Mahecha, M. D., Smith, P., van der Velde, M., Vicca, S., Babst, F., Beer, C., Buchmann, N., Canadell, J. G., Ciais, P., Cramer, W., Ibrom, A., Miglietta, F., Poulter, B., Rammig, A., Seneviratne, S. I., Walz, A., Wattenbach, M., Zavala, M. A., & Zscheischler, J. (2015). Effects of climate extremes on the terrestrial carbon cycle: Concepts, processes, and potential future impacts. *Global Change Biology, 21*(8), 2861–2880. https://doi.org/10.1111/gcb.12916.

20. Goethals, P., & Ho, L. T. (2019). Opportunities and challenges for the sustainability of lakes and reservoirs in relation to the sustainable development goals (SDGs). *MDPI AG*. https://core.ac.uk/download/225020142.pdf (accessed on 5 December 2024).

21. Goody, R. (n.d.). Global changes: Impacts on habitability. A scientific basis for assessment. https://core.ac.uk/download/pdf/42855210.pdf (accessed on 5 December 2024).

22. Ho, L. T., & Goethals, P. L. M. (2019). Opportunities and challenges for the sustainability of lakes and reservoirs in relation to the sustainable development goals (SDGs). *Water, 11*(7), 1462. https://doi.org/10.3390/w11071462.

23. Hummel, D., Jahn, T., & Schramm, E. (2011). Social-ecological analysis of climate-induced changes in biodiversity: Outline of a research concept. https://core.ac.uk/download/14522198.pdf (accessed on 5 December 2024).

24. Ineson, P., Lukac, M., & Milcu, A. (2011). The role of closed ecological systems in carbon cycle modelling. *Springer Science and Business Media LLC*. https://core.ac.uk/download/107822.pdf (accessed on 5 December 2024).

25. Kauffman, S., Longo, G., & Montévil, M. (2012). No entailing laws, but enablement in the evolution of the biosphere. https://core.ac.uk/download/275570919.pdf (accessed on 5 December 2024).

26. Ladanai, S. (2008). *Nutrient Relations in Coniferous Forests* (Acta Universitatis Agriculturae Sueciae, 2008:94). Department of Ecology, Swedish University of Agricultural Sciences. ISBN: 9789186195274.

27. McElroy, M. (1983). Global change: A biogeochemical perspective. https://core.ac.uk/download/pdf/42850184.pdf (accessed on 5 December 2024).

28. McGuire, A. D., et al. (2001). Carbon balance of the terrestrial biosphere in the twentieth century: Analyses of CO_2, climate, and land use effects with four process-based ecosystem models. *Global Biogeochemical Cycles, 15*(1), 183–206. https://doi.org/10.1029/2000GB001298.

29. McGuire, A. D., Melillo, J. M., Joyce, L. A., Kicklighter, D. W., Grace, A. L., Moore III, B., & Vorosmarty, C. J. (1992). Interactions between carbon and nitrogen dynamics in estimating net primary productivity for potential vegetation in North America. *Global Biogeochemical Cycles, 6*(2), 101–124. https://doi.org/10.1029/92GB00219.

30. Mendes, I. (2020). The circular economy: An ancient term that became polysemic. *ISEG–Departamento de Economia.* https://core.ac.uk/download/373459275.pdf (accessed on 5 December 2024).

31. Millennium Ecosystem Assessment [MA], United Nations Environment Programme [UNEP], Revenga, C., Mbow, C., Cooper, D., Bruley, E., Gilman, E., Foresight, G., Hurtt, G., Pereira, H. M., Figueiredo, J., Fernández-Manjarrés, J., Scharlemann, J. P. W., Stafford Smith, M., Walpole, M., Leadley, P., Biggs, R., Alkemade, R., Scholes, R., Guénette, S., Oberdorff, T., Sumaila, U. R., Proença, V., & Cheung, W. (2014). Interacting regional-scale regime shifts for biodiversity and ecosystem services. *Oxford University Press (OUP).* https://core.ac.uk/download/30608409.pdf (accessed on 5 December 2024).

32. Oleksińska, Z. (2015). Plants on duty – Phytotechnologies and phytoremediation at a glance. *Acta Universitatis Lodziensis. Folia Biologica et Oecologica, 11*, 23–29. https://doi.org/10.1515/fobio-2015-0004.

33. Patiño, D., Cabello Piñar, F. J., Casas Jiménez, J. J., Castro Martínez, A., Céspedes Duarte, L., Exposito Granados, M., Fernández Prados, J. S., Gallardo Pino, M. L., Giagnocavo, C. L., Giménez Moolhuyzen, M., Lacalle Marcos, A., López Rodríguez, M. D., Murillo López, B. E., Parra Gómez, S., Pumares Fernández, P., Rodríguez Díaz, F., Rodríguez, E., Ucles Aguilera, D., Valera Martínez, D. L., & Vásquez, L. M. (2019). Six collective challenges for the sustainability of Almería greenhouse horticulture. *MDPI AG.* https://core.ac.uk/download/286590515.pdf (accessed on 5 December 2024).

34. Ramos, Y., & Wang, D. (2011). Ecology. *Digital Commons @ Trinity.* https://core.ac.uk/download/216385245.pdf (accessed on 5 December 2024).

35. Rushton, E. A. C. (2008). An analysis of the impact of the Gaia theory on ecology and evolutionary theory. *Oxford Brookes University*. https://core.ac.uk/download/33469275.pdf (accessed on 5 December 2024).

36. Sinha, A. K. (2011). Conservation package for modal ecorace. https://

core.ac.uk/download/pdf/290162.pdf (accessed on 5 December 2024).

37. Veldhuis, M. P., Berg, M. P., Loreau, M., & Olff, H. (2018). Ecological autocatalysis: A central principle in ecosystem organization? *Ecological Monographs, 88*(3), 304–319. https://doi.org/10.1002/ecm.1292.
38. Zerbi, A., Aliaume, C., Lasserre, G., Garcia, S. M., & Do Chi, T. (2003). The ecosystem approach to fisheries: Issues, terminology, principles, institutional foundations, implementation and outlook. *Food and Agriculture Organization of the United Nations (FAO)*. https://core.ac.uk/download/75777389.pdf (accessed on 5 December 2024).

CHAPTER 4

COMMUNITY AND POPULATION ECOLOGY

Contents

INTRODUCTION

At the core of ecological studies is the intricate interplay of organisms within their environments, forming complex networks of interactions that define communities and populations. These interactions encompass various relationships, including predation, competition, and symbiosis, which together shape the dynamics of life in a given area. The thriving diversity within communities enhances ecosystem resilience and contributes to many ecosystem services beneficial to humans. By examining the interconnections, ecologists can assess how species composition and abundance influence environmental stability and health, thereby laying the groundwork for species conservation, resource management, and understanding the impacts of environmental stressors such as climate change. Population ecology delves deeper into the characteristics and dynamics of species populations within these communities. It focuses on aspects such as population size, density, distribution, and age structure, all essential for understanding the growth patterns of species over time. These patterns are influenced by birth and death rates, immigration, and emigration, and environmental pressures, which can either enhance or diminish population viability. By employing mathematical models and statistical analyses, ecologists can predict future population trends and evaluate the long-term sustainability of species. This understanding proves invaluable in devising strategies for wildlife management, addressing human-wildlife conflicts, and preserving biodiversity in increasingly fragmented habitats. Integral to both community and population ecology is the recognition of the impact that human activities have on these natural systems.

As urban development, agriculture, and industrialization expand, they disrupt traditional ecological balances, leading to species decline and habitat loss. Community ecologists, therefore, seek to understand how these anthropogenic effects alter community composition and interactions, while population ecologists assess the repercussions for individual species within changing environments. Collaborative efforts between ecologists and policymakers become crucial in mitigating these impacts through sustainable practices and conservation initiatives. By understanding the fundamental principles of community and population ecology, society can make informed decisions that promote coexistence between humanity and the natural world, ensuring healthier ecosystems for future generations.

4.1. DEFINITION OF COMMUNITY ECOLOGY

At the heart of the field of community ecology lies the intricate study of interactions among various species within a defined geographical area. This discipline goes beyond merely cataloging organisms; it examines the relationships, dynamics, and functions that underpin these communities. By exploring factors such as competition, predation, mutualism, and parasitism, ecologists can gain insights into community structure and stability. A pivotal component of this study is understanding how environmental changes affect these interactions. This understanding is vital, especially in light of pressing issues like climate change, where shifting temperatures can disrupt established relationships among species.

As these dynamics evolve, the viability of entire ecosystems is put at risk, emphasizing the necessity for community ecology to inform conservation strategies. Furthermore, the concept of community ecology extends to its application in managing and preserving biodiversity. Communities are not static; they constantly adapt in response to various biotic and abiotic factors. The importance of sustainable practices becomes evident when considering real-world implications; for example, as changing climates impact species distributions, community ecologists must determine the critical functions that species provide within their ecosystems.

Knowledge about these relationships is especially crucial for agricultural enterprises, as efficient land resource utilization can significantly impact species coexistence and productivity. Understanding the harmony and conflict among species within a community lends itself to better land and resource management practices that can mitigate impacts on biodiversity, ultimately striving for sustainable ecological solutions that align with community needs. Finally, the ongoing exploration of community ecology involves integrating scientific knowledge into broader environmental policy and management frameworks. Effective governance requires a clear understanding of the ecosystems in question, which can be accomplished through careful research and analysis. For instance, identifying key species that play a vital role in maintaining ecological balance allows policymakers to prioritize conservation efforts.

These lead to healthier ecosystems and also promotes resilience against human-induced pressures, such as overfishing and habitat destruction. By applying community ecological principles, we can foster environments that support diverse species, ensuring that the interconnectedness of life thrives amidst a changing world. As exemplified in studies of Arctic ecosystems, understanding these relationships will be crucial in developing strategies that protect vital species like Arctic cod, underscoring the far-reaching implications of community ecology (Figure 4.1).

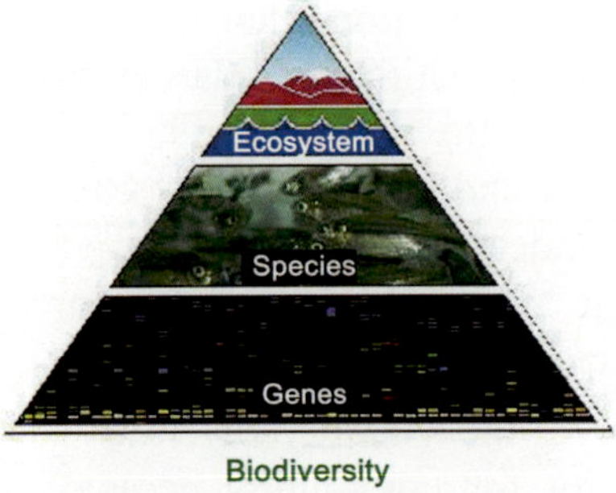

Figure 4.1. *Community in ecology.*

Source: https://www.coastalwiki.org/wiki/File:Driehoek.jpg.

4.2. DEFINITION OF POPULATION ECOLOGY

Understanding population ecology is essential as it examines the structures, patterns, and dynamics of species populations within defined environments. This branch of ecology focuses on the relationships between populations and their environment, analyzing factors such as birth rates, mortality rates, migration patterns, and interactions among species. By studying these components, population ecology aims to uncover how populations grow, shrink, or stabilize over time, providing insights into the stability and health of ecosystems. The impacts of invasive species, as discussed in invasion science, highlight the importance of population dynamics, as non-native species can drastically alter local ecosystems, threatening native populations and leading to significant ecological shifts (Angeler et al., 2016). At the heart of population ecology lies the balance of biotic and abiotic factors that influence population sizes and distributions. For example, resource availability, predation, and environmental conditions are crucial elements determining how a population interacts with its ecosystem. As evidenced by the concept of resilience, which describes how ecological systems can withstand disturbances (Angeler et al., 2016), the ability of a population to adapt to changes in its environment is essential for its long-term persistence. Population ecologists utilize various models, such as the logistic growth model, to illustrate how populations can approach carrying capacities, thus demonstrating the interplay between environmental limitations and biological potential. The study of population ecology also delves into the interactions between species within communities and the broader implications for biodiversity. Understanding these dynamics is vital for conservation efforts, as shifts in the population of one species can set off cascading effects throughout the ecosystem. For instance, predation dynamics can illustrate the concept of biotic resistance, where native species may limit the establishment and impact of non-native species (Richardson et al., 2013). Consequently, population ecology contributes to our understanding of individual species and aids in forming strategies for resource management and ecological restoration. Ultimately, the insights gleaned from population ecology facilitate better-informed decisions regarding biodiversity conservation and the sustainability of ecosystems (Figure 4.2).

Figure 4.2. *Population ecology.*

Source: By Sasata – Own work, CC BY-SA 3.0, https://commons.wikimedia.org/w/index.php?curid=5370799.

4.2.1. Importance of Studying These Ecological Fields

Understanding community and population ecology is crucial for effective biodiversity conservation and wildlife management. The intricate relationships between species and their environments determine the health of ecosystems. For instance, as illustrated by Fischer et al. (2015), urbanization profoundly impacts wildlife interactions and behaviors, demonstrating the necessity of studying how animals adapt to changing habitats. This knowledge is vital for conserving existing ecosystems and for designing effective conservation strategies that account for human expansion. By analyzing species interactions and dynamics, ecologists can identify ecological traps and critical habitats that require protection, ensuring that conservation efforts are prioritized in a manner that fosters resilience among populations. Moreover, exploring ecological connectivity sheds light on how populations interact across fragmented landscapes, which is increasingly relevant in an era of rapid habitat loss. Circuit theory, as discussed by McRae et al. (2008), provides a novel approach to modeling connectivity, offering insights into movement patterns and gene flow among populations. In practical terms, this knowledge aids in identifying wildlife corridors that facilitate safe passage between habitats, which is essential for maintaining genetic diversity and demographic stability.

The application of such models in landscape planning allows for a proactive approach to conservation, directly linking ecological research to actionable strategies that mitigate the adverse effects of urbanization and habitat fragmentation. Furthermore, the impact of ecological research extends beyond conservation, influencing policy, urban planning, and community engagement. As demonstrated by studies noting the interplay between species diversity and urban land use (Fischer et al., 2015), understanding these relationships informs decision-making processes that can promote sustainable development. Integrating ecological principles into urban planning allows for the creation of green spaces that enhance biodiversity while satisfying human needs. By prioritizing ecological understanding in policy frameworks, communities can foster coexistence between human activities and natural ecosystems, ensuring that future generations inherit landscapes rich in biodiversity and ecological integrity. Through continuous investigation in these fields, we can craft solutions that reflect the complexity and interdependence of life on Earth.

4.3. KEY CONCEPTS IN POPULATION ECOLOGY

Understanding the intricate dynamics of community ecology begins with the concept of species interactions, which shape the structure and composition of ecological communities. Interspecies interactions include predation, competition, mutualism, and commensalism, each influencing population sizes and distributions. For instance, competitive exclusion can lead to the dominance of certain species over others, potentially decreasing biodiversity. Such interactions may also create a competitive landscape where coexisting species evolve distinct niches to mitigate direct competition. This complexity enhances the resilience of communities, enabling them to adapt to environmental changes. Furthermore, demographic, and environmental stochasticity play crucial roles in shaping these dynamics. Stochastic processes introduce unpredictable variations that can significantly impact species interactions and community stability (Pol Capdevila et al., 2020). Hence, a comprehensive understanding of species interactions is fundamental to grasping the broader implications of community ecology. Another pivotal concept is the idea of community structure, which refers to how populations are arranged within an ecosystem.

Community structure is influenced by abiotic factors —such as temperature, light availability, and nutrient levels— in conjunction with biotic factors, like species interactions and dynamics. For example, the abundance of resources can dictate the number of individuals that an ecosystem can support, thereby affecting the overall diversity within the community. This relationship emphasizes the importance of trophic levels and food webs, wherein energy flow occurs from producers to various levels of consumers. Moreover, disturbances, whether natural or anthropogenic, disrupt these structures and can create cascading effects throughout the ecosystem. Resilience stems from the ability of communities to absorb disturbances while maintaining essential functions. Effective community management necessitates an understanding of these structural elements, as highlighted in various ecological studies (Pol Capdevila et al., 2020). The role of biodiversity within communities cannot be understated, as it contributes to ecosystem stability and resilience. Higher biodiversity often correlates with increased productivity and the provision of ecosystem services, such as pollination and nutrient cycling. Diverse communities are generally better equipped to withstand and recover from disturbances due to the presence of functional redundancies amongst species—where different species can fulfill similar ecological roles. This resilience framework highlights the importance of demographic resilience, wherein populations can sustain their structure and function despite external pressures. Recent research underscores the need for integrating various perspectives on stochasticity and resilience to better comprehend community dynamics. By understanding the intertwined roles of biodiversity, species interactions, and community structure, ecologists can develop effective conservation strategies to protect and restore ecological communities in the face of ongoing environmental change (Shoemaker et al., 2019).

4.3.1. Species Interactions (Predation, Competition, Mutualism)

The intricate web of interactions among species within ecosystems is fundamental to understanding community and population ecology. Predation, competition, and mutualism play pivotal roles in shaping the dynamics of these biological communities. Predation regulates prey populations, influencing both species richness and community structure. For instance, as Forbes (1887) suggests, predatory relationships can lead to the emergence of complex equilibria in ecosystems, maintaining life at sustainable levels even amidst destruction and conflict. This dynamic exemplifies the necessity of predatory species in preventing overpopulation of prey, ultimately resulting in healthier and more diverse ecosystems. Competition, on the other hand, typically manifests as either resource competition or interference among species (Figure 4.3).

Figure 4.3. *Species interactions.*

Source: https://slideplayer.com/10282035/35/images/slide_1.jpg.

As ecologists have documented, such interactions often dictate the distribution and abundance of species within a community, highlighting the notion that not all species can coexist indefinitely. For instance, when resources are limited, dominant species may outcompete subordinates, leading to decreased diversity over time. Recent analyses underscore this bias, as negative interactions such as competition tend to receive more research attention than positive interactions like mutualism (Jani Raerinne, 2020). This imbalance reflects historical trends in ecological research that favor competitive dynamics, perhaps due to their more noticeable impact on community structure. In contrast, mutualism fosters coexistence and interdependence among species, enhancing community stability and resilience. These beneficial interactions, where both species involved derive advantages, contribute not only to individual fitness but also to overall ecosystem function. For example, plant-pollinator interactions document how mutualism can bolster reproductive success and food security across numerous plant species. As research advances, insights into microbial consortia reveal that synthetic biology can engineer positive interactions that enhance biotechnological applications (McCarty et al., 2018). Thus, while competition and predation have historically dominated ecological frameworks, there is a growing recognition of the significance of mutualism in promoting biodiversity and ecological health, underscoring the need for a more balanced understanding of species interactions in community ecology.

4.3.2. Community Structure and Composition

The intricate dynamics of community structure and composition reveal how species interact within their environment, shaping ecosystems at multiple levels. The relationships among organisms—including competition, predation, and symbiosis—directly influence community composition. For instance, specific microbial interactions in marine environments demonstrate how surface colonization by microorganisms establishes a foundation for broader ecological interactions. These surface-associated communities exhibit unique structural characteristics and engage in significant biogeochemical processes that sustain ecosystem functions (Dang et al., 2015). By examining these interactions, ecologists can better understand how community composition influences and is influenced by environmental changes, providing insight into resilience and vulnerability in the face of ecological disturbances. Furthermore, distinct environmental conditions can lead to pronounced variations in microbial diversity and community structure, as seen in the case of acid mine drainage (AMD) systems. The harsh low-pH conditions present in AMD environments tend to restrict microbial diversity, fostering specialized communities adapted to extreme conditions. The limited diversity drives unique metabolic processes and element cycling, which are crucial for the ecosystem's functionality (Méndez García et al., 2015).

This reduction in biodiversity is not merely a consequence of environmental harshness; it reflects a complex interplay of selection pressures that shape community composition over time. By analyzing the potential impacts of such alterations on community structure, researchers can uncover patterns that also apply to other extreme or low-diversity ecosystems. Understanding community structure and composition requires a holistic approach that considers both biotic and abiotic factors. Integrating insights from microbial ecology and broader ecological frameworks allows for a more nuanced appreciation of how communities are assembled and maintained. The dynamics of community composition are influenced by multiple layers of interaction among species, including their responses to changing environmental conditions and anthropogenic influences. By synthesizing findings from studies across different ecosystems—such as those focusing on marine biofilms and AMD systems—we can develop a comprehensive understanding of the underlying principles governing community ecology. Such an integrative perspective not only advances theoretical knowledge but also informs practical conservation efforts aimed at preserving biodiversity and ecosystem health.

4.3.3. Ecological Niches and Habitat Diversity

Diverse ecological niches play a crucial role in enhancing biodiversity within various habitats. Each niche represents a specialized role that a species occupies in an ecosystem, including its interactions with other organisms and its environmental needs. Different species often exploit unique resources and perform specific functions, leading to a rich tapestry of life within an ecosystem. For instance, in a forest ecosystem, birds, mammals, insects, and plants interact in a complex web, supporting various ecological processes such as pollination and nutrient cycling. This diversity aids in stabilizing ecosystems and fostering resilience against environmental changes, as species with distinct functions can compensate for one another's decline. Understanding these niches, as exemplified in research highlighting the roles species play in their habitats, underlines the importance of habitat conservation in maintaining ecological balance (Estrada Peña et al., 2014). The relationship between habitat diversity and ecological niches is integral to understanding community dynamics. When habitats vary, they provide a range of conditions that can support different niches, potentially leading to greater species richness. For example, wetlands, forests, and grasslands each harbor unique conditions that allow for the establishment of specialized species. Such habitats function as a mosaic, where abiotic factors such as temperature, moisture, and substrate type create specific niches for various organisms. The creation of protected areas that incorporate such varied habitats is essential for conserving species that rely on specific ecological niches to thrive. Furthermore, anthropogenic pressures often lead to habitat destruction, thus diminishing these niches and threatening biodiversity. Research indicates that changes in habitat can disrupt the delicate balance necessary for certain species survival, emphasizing the need for holistic conservation strategies (Hompson et al., 2012) (Figure 4.4).

Figure 4.4. *The flightless dung beetle occupies an ecological niche: exploiting animal droppings as a food source.*

Source: https://en.wikipedia.org/wiki/Ecological_niche#/media/File:Flightless_Dung_Beetle_Circellium_Bachuss,_Addo_Elephant_National_Park,_South_Africa.JPG.

In examining the implications of habitat diversity on ecological niches, one must consider the potential consequences of species loss. The disruption of interconnected food webs due to a decline in specific species can lead to unforeseen ecological cascades. For instance, when top predators within a habitat diminish, it can result in an overpopulation of herbivores, which may subsequently overgraze vegetation and alter habitat structure. This evidence demonstrates the importance of maintaining both the diversity of habitats and the niches within those habitats. The survival of entire communities relies on the preservation of these intricate relationships. Moreover, understanding the vulnerabilities of specific niches to environmental changes further stresses the importance of targeted conservation efforts. By recognizing the interconnectedness of ecological niches and habitat diversity, we can better appreciate the necessity for proactive stewardship of our natural landscapes (Estrada Peña et al., 2014; Thompson, 2012).

4.4. KEY CONCEPTS IN POPULATION ECOLOGY

Understanding key concepts in population ecology is essential for analyzing how species interact with their environment and each other. One primary concept is the population dynamics, which refers to the changes in population size and composition over time due to births, deaths, immigration, and emigration. This dynamic interplay can significantly influence community structures and biodiversity. For example, a surge in prey species can lead to an increase in predator populations, demonstrating the intricate balance that exists within ecosystems. Furthermore, external factors like climate change and habitat destruction can disrupt these dynamics, leading to population declines or growth. Consequently, recognizing these interactions provides insights into the management of wildlife populations and the conservation of ecosystems. Another critical concept in this field is the carrying capacity, which denotes the maximum number of individuals an environment can sustainably support. This principle is foundational for understanding resource limitations and ecological balance. Factors such as food availability, water supply, and habitat space play a crucial role in determining carrying capacity. When a population exceeds its carrying capacity, it can lead to overexploitation of resources, habitat degradation, and ultimately, population crashes. This phenomenon highlights the importance of maintaining an equilibrium within ecosystems, as demonstrated by ongoing research in animal shelters where understanding population dynamics helps to ensure the health and welfare of both sheltered animals and the surrounding community (Horecka et al., 2022).

Ensuring that populations remain within sustainable limits is vital for long-term ecological stability. Species interactions within populations also bring forth the concepts of competition, predation, and symbiosis. These interactions shape community structures and influence evolutionary processes. Competition occurs when species vie for limited resources, leading to a natural selection process that fosters adaptability and survival. On the other hand, predation can regulate both prey and predator populations, maintaining the balance of ecosystems. Moreover, symbiotic relationships, including mutualism and commensalism, illustrate intricate connections between species where both or one benefits. These interactions can manifest in significant population changes and community compositions over time. Understanding these dynamics can better inform biodiversity conservation efforts and ecological restoration projects, highlighting the intricate web of life that sustains our natural environment, as seen in contemporary models for promoting active longevity within communities (Ivanova et al., 2024).

4.4.1. Population Dynamics and Growth Models

Understanding population dynamics is essential for predicting how species interact within ecosystems and respond to environmental changes. Population dynamics refers to the patterns and processes that influence the size and structure of populations over time. Factors such as birth rates, death rates, immigration, and emigration significantly impact population growth. Models such as the logistic growth model illustrate how populations grow exponentially until resources become limited, resulting in a leveling off, or carrying capacity. By employing these models, ecologists can simulate various scenarios, which can be crucial for managing species populations in the face of habitat alteration or climate change, thereby mitigating the risk of extinction among vulnerable populations (Figure 4.5).

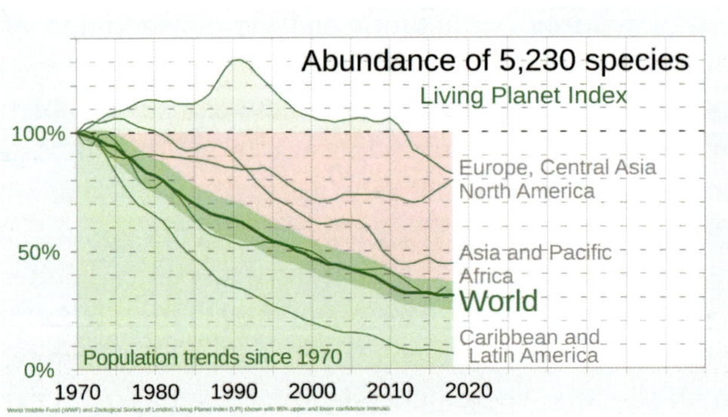

Figure 4.5. *Population dynamics.*

Source: By RCraig09 – Own work, CC BY-SA 4.0, https://commons.wikimedia.org/w/index.php?curid=139112554.

The integration of demographic data into mathematical models allows for a comprehensive understanding of population trends and behaviors. Matrix population models, for example, summarize critical life-history traits such as survival, growth, and reproduction, thus providing insight into population viability. As noted, the Compadre Plant Matrix Database permits researchers to analyze and compare demographic schedules across various species, enhancing the understanding of how ecological factors shape life-history traits (Salguero Gómez et al., 2014). By observing these traits in a structured way, scientists can make informed predictions about population responses to environmental pressures and identify effective conservation strategies. Furthermore, the information derived from these models can inform decisions regarding the management of endangered species and invasive populations, ensuring that ecological balance is maintained. Population growth models not only inform management practices but also contribute to a deeper understanding of community ecology. The interactions among different species within a community can significantly influence individual population dynamics. For instance, the presence of a dominant predator may regulate the population of its prey, while competition among similar species can lead to shifts in population size and distribution. Understanding these interactions is critical for predicting changes in community structure, especially in urbanized landscapes where human activity alters natural habitats. As seen in urban environments, increased interaction between wildlife and human populations can lead to emerging diseases, affecting both wildlife and public health (Hassell et al., 2016). Thus, models, and theories of population dynamics are vital tools for ecologists, helping them to make predictions that can guide effective conservation efforts and improve human-wildlife coexistence.

4.4.2. Factors Affecting Population Size (Biotic and Abiotic)

Understanding the population size of any species involves examining both abiotic and biotic factors. Abiotic factors, such as climate conditions, resource availability, and habitat stability, significantly influence population dynamics. For instance, temperature fluctuations and rainfall patterns can determine the growth rates of various species, affecting reproductive success and survival. Certain populations thrive under stable conditions, while those subjected to extreme weather may experience dramatic declines. This highlights the need for populations to have resilience against climate variability. Additionally, habitat characteristics, including soil quality and water availability, play crucial roles in supporting different life forms. Recognizing these abiotic elements is essential for comprehending how they directly correlate with population size and health. Biotic interactions, on the other hand, represent another layer of complexity in understanding population dynamics. Competition for resources among species can limit population sizes, ultimately leading to fewer individuals surviving to reproductive age. Predation also plays a crucial role; the presence of predators can drastically reduce prey populations, subsequently impacting the broader community.

The balance of these interactions determines population structures and can lead to different outcomes in community composition. For example, the study of diatoms highlights how local biotic relationships among species contribute to community assembly, revealing how these interactions weave into larger metacommunity dynamics (Álvarez-Cobelas et al., 2022). Understanding these dynamics helps predict population changes when species interact, establishing a framework for studying population ecology in various ecosystems. Furthermore, both abiotic and biotic factors are interconnected and can influence one another, further complicating population size estimations. A significant example is the potential effects of pesticide application, which can diminish beneficial biotic interactions while allowing certain pest populations to thrive (Gintoron et al., 2023). This interaction demonstrates how human activities can inadvertently alter population balances within ecosystems, emphasizing the need for careful management practices. The health and diversity of a species' environment, including aspects like connectivity to natural habitats, can also directly affect individual fitness and reproductive success. As such, the interplay between abiotic factors like habitat availability and biotic relationships such as predation creates a dynamic framework in which population sizes can fluctuate. Understanding these relationships offers vital insights into managing and conserving populations in a changing world (Figure 4.6).

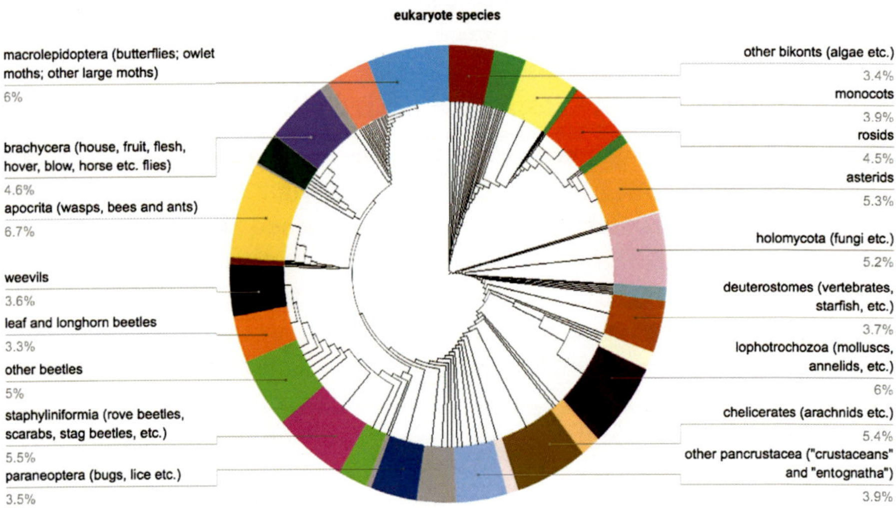

Figure 4.6. *Population size.*

Source: By Adam Dent – Own work, CC BY-SA 4.0, https://commons.wikimedia.org/w/index.php?curid=59928896.

4.4.3. Life History Strategies and Reproductive Strategies

Life history strategies are integral to understanding how organisms allocate resources throughout their lifetimes, impacting both survival and reproduction. These strategies encompass a range of traits that influence growth, development, and reproduction timing, governed by evolutionary pressures. For instance, species may adopt an r-selected strategy, characterized by early maturity, high reproduction rates, and lower parental investment. Contrarily, K-selected species typically mature later, invest more in offspring, and produce fewer young. Such distinctions in life history strategies can dictate community dynamics and population structure. Understanding these differing approaches allows ecologists to forecast population trends and changes within ecosystems, shedding light on the interactions between species and their environments. Reproductive strategies further refine our comprehension of life history tactics, focusing specifically on how organisms maximize reproductive success under variable ecological conditions. Different species exhibit diverse approaches based on environmental pressures, leading to various mating systems such as monogamy, polygyny, or polyandry. For instance, cooperative breeding, as detailed in studies on human communities, illustrates how maternal investment is supplemented through non-maternal support systems, enhancing the survival of the young and contributing to the fitness of the community as a whole (Kramer, 2005). This signifies that reproductive choices extend beyond individual survival; they encompass broader ecological interactions and stigmas, affirming the idea that reproductive strategies are as much about community dynamics as they are about individual lineage.

The intersections of life history and reproductive strategies emphasize the complexities of ecosystem functioning and sustainability. As freshwater biodiversity continues to face serious threats, understanding these strategies becomes crucial (Dudgeon et al., 2005). Human activities such as pollution, habitat degradation, and water extraction have profound implications on life history traits across species. For example, the alteration in the reproductive success of fish due to habitat modifications illustrates the extensive repercussions of anthropogenic disruptions (Karen Kramer, 2005). Therefore, researchers advocate for the application of reconciliation ecology, which fosters a balance between conservation efforts and human needs, highlighting the necessity for adaptive management strategies that integrate these historical and reproductive insights. This integrated approach is essential for promoting resilience within populations and maintaining biodiversity in increasingly challenged ecosystems.

4.5. THE ROLE OF BIODIVERSITY IN ECOSYSTEMS

Diversity within ecosystems serves as a foundation for their stability and resilience. Ecosystem productivity is closely linked to the variety and abundance of species present, with each organism playing a unique role in nutrient cycling and energy flow. For instance, diverse plant communities can optimize resource use by employing different strategies for acquiring sunlight, water, and nutrients.

This multifaceted approach allows ecosystems to better withstand environmental stresses such as drought or disease. In contrast, monocultures often lead to vulnerabilities; a single disturbance can decimate the entire population, triggering cascading effects throughout the ecosystem. Consequently, maintaining biodiversity is essential for immediate ecological responses and for long-term sustainability and adaptability. The role of biodiversity transcends mere species richness; it encompasses the complex interactions among organisms that shape community dynamics. Herbivores, for example, can exert significant influence on plant populations, regulating growth through their feeding activities.

Research has shown that these interactions may enhance diversity by preventing any single species from dominating the landscape, which fosters heterogeneity among plant communities (Jia et al., 2018). This dynamic interplay between species highlights the importance of both top-down and bottom-up processes in maintaining ecosystem equilibrium. Furthermore, ecosystems that support a variety of species are often more productive and efficient in nutrient cycling, enhancing their overall health and longevity. Moreover, the decline of biodiversity can have far-reaching implications for human health and well-being. Healthy ecosystems provide essential services, such as food production, clean water, and disease regulation. A diverse array of species is critical to sustaining food systems, which face challenges related to nutrition and environmental sustainability (Willett et al., 2019).

As human populations grow and climate change intensifies, the pressure on ecosystems increases, exacerbating the risk of biodiversity loss. Thus, protecting, and nurturing biodiversity is paramount not only for ecological integrity but also for supporting human health and robust food systems. Recognizing this interconnectedness is vital for developing effective conservation strategies that prioritize the preservation of biodiversity in the face of ongoing environmental challenges (Figure 4.7).

Figure 4.7. *Ecosystem biodiversity.*

Source: https://img.freepik.com/premium-photo/creative-collage-biodiversity-shape-sphere-ecosystem-protection_454018-1044.jpg.

4.5.1. Importance of Biodiversity for Ecosystem Stability

The interplay between biodiversity and ecosystem stability is a fundamental principle in ecology. Ecosystems with high biodiversity are often more resilient to disturbances such as climate change, pollution, and habitat destruction. This resilience stems from the variety of species that perform different roles within the ecosystem. A diverse community can ensure that if one species declines due to environmental pressures, others can fulfill critical ecological functions, maintaining productivity and stability. Additionally, high biodiversity contributes to the genetic variety within species, which further enhances adaptive capabilities in response to changing environmental pressures. This adaptability is crucial, particularly in light of increasing human-induced changes to ecosystems, as it supports overall ecological balance and function. When ecosystems lose biodiversity, their ability to recover from disturbances diminishes, leading to potential collapse. The reliance on a few dominant species can create vulnerability within ecosystems, which may result in reduced food web complexity and weakened ecosystem services. For instance, studies have shown that ecosystems characterized by reduced species variety often experience declines in nutrient cycling and energy flow. This erosion of function can have cascading effects on both local wildlife populations and human communities that rely on these ecosystems. By acknowledging that biodiversity loss can directly threaten ecosystem integrity, conservation efforts can be aimed at preserving species richness and habitat diversity, which are essential for maintaining stability and resilience in ecological communities (Willett et al., 2019). Moreover, the intricate relationship between microbial diversity and broader ecological health emphasizes the importance of preserving various life forms. Microbial communities, often overlooked, play significant roles in nutrient cycling, decomposition, and the maintenance of soil health, forming the foundation of terrestrial ecosystems. Research indicates that a diverse microbial populace contributes to the stability of ecosystems by supporting plant health and growth through nutrient availability and disease resistance (Kostic et al., 2014). As biodiversity contributes to the structural and functional complexity of ecosystems, it enhances their ability to withstand and recover from disturbances. Thus, safeguarding biodiversity is not only a matter of conserving individual species but is inherently tied to sustaining the ecological resilience necessary for maintaining a healthy planet.

4.5.2. Biodiversity and Ecosystem Services

The intricately linked concepts of biodiversity and ecosystem services are crucial for understanding the health of both ecosystems and the human communities that depend on them. Biodiversity encompasses the variety of life within a given ecosystem, including species diversity, genetic diversity, and ecological diversity. These elements work synergistically to form complex interactions that enhance ecosystem resilience, productivity, and adaptability. A rich diversity of species ensures a wider range of functions and services, such as pollination, nutrient cycling, and climate regulation, which are vital for sustaining life. As urbanization increases, it is essential to recognize how local biodiversity contributes to the ecosystem services essential for the well-being of urban residents, especially as up to 75% of the global population is projected to live in cities by 2050 (Ziter, 2015). Ecosystem services are the benefits that humans derive from healthy ecosystems, and they can be classified into four main categories: provisioning, regulating cultural, and supporting services. Provisioning services include the production of food, fresh water, and raw materials while regulating services encompass climate regulation, water purification, and pollination. Cultural services provide non-material benefits such as recreational opportunities and spiritual enrichment. Supporting services, like nutrient cycling and habitat provision, underpin the other services by maintaining ecosystem functionality. The loss of biodiversity threatens these services, potentially leading to significant ecological and economic repercussions.

Research has consistently shown that non-native species can disrupt native biodiversity and ecosystem services, illustrating how invasions serve as a critical factor affecting ecosystem resilience and functioning (Richardson et al., 2013). In light of ongoing environmental changes, including climate change and habitat loss, prioritizing biodiversity conservation is essential for maintaining ecosystem services. As indicated in the findings concerning urban ecosystem services (Ziter, 2015), the current research landscape tends to focus heavily on specific ecological functions in isolated contexts, often neglecting the broader implications of biodiversity loss across different ecosystems. To combat these challenges, multidisciplinary approaches that combine ecological research with policy and community involvement are necessary. Expanding the scope of biodiversity assessments and integrating them with ecosystem service evaluations will allow for more effective conservation strategies. These efforts are crucial not only for preserving biodiversity but also for ensuring the continued provision of ecosystem services that support human societies and foster ecological balance.

4.5.3. Threats to Biodiversity and Conservation Efforts

The loss of biodiversity represents a critical challenge for ecological stability and resilience. Various factors contribute to this decline, including habitat destruction, climate change, pollution, and the introduction of invasive species. Habitat destruction, often driven by human activities such as urban development and agriculture, results in the fragmentation of ecosystems, making it difficult for many species to thrive. In the case of the European mink (*Mustela lutreola*), habitat loss has reduced its populations and exacerbated competition with the more adaptable American mink (*Neovison vison*) (Haage et al., 2016). These competitive dynamics illustrate the broader theme of community ecology, where species interactions can significantly impact survival rates and biodiversity. Thus, understanding the multifaceted threats to biodiversity is essential for developing targeted conservation efforts that emphasize habitat preservation and restoration. Conservation initiatives must also address the behavioral aspects of species affected by declining populations. Research has shown that individual behavioral differences, such as personality traits, can significantly influence survival rates in various species, including those subjected to reintroduction programs (Haage et al., 2016). In the case of the European mink, personality traits like boldness and exploratory behavior were found to correlate with survival rates, suggesting that not all individuals are equally suited for success in the wild (Haage et al., 2016). Conservation efforts that account for these behavioral differences can enhance the effectiveness of programs intended to reestablish populations of endangered species. Additionally, understanding how these traits adaptively evolve provides insights that can guide future interventions and improve the success rates of conservation efforts in community and population ecology. The aftermath of invasive species introduction poses a particularly complex threat to biodiversity, as illustrated by the American minks and their impact on native ecosystems. Managing invasive populations is crucial for the conservation of endangered species like the European mink. Data indicate that native predators, such as red foxes and otters, can potentially suppress American mink populations, thereby providing a natural mechanism for restoring ecological balance (Haage et al., 2016). However, the effectiveness of such biological control methods requires careful ecological assessments to understand predator-prey dynamics in specific regions. Without these evaluations, conservation strategies may inadvertently disrupt existing ecosystems. Ultimately, integrating knowledge about species interactions and predator dynamics into conservation planning is vital for sustaining biodiversity and protecting vulnerable populations in increasingly fragmented habitats.

4.6. HUMAN IMPACT ON COMMUNITY AND POPULATION ECOLOGY

Human activities profoundly shape community and population ecology, as they introduce various stresses that can alter the balance of ecosystems. Urbanization, industrialization, and agricultural practices lead to habitat destruction, diminishing biodiversity and disrupting essential ecological processes. The research examining the relationship between economic variables and environmental quality reveals that demographic and ecological factors often have a more significant influence on pollution indices than economic growth itself, particularly in Emerging Market countries (Sutikno et al., 2024). This underscores the need for sustainable practices that prioritize ecological health over mere economic expansion. As urban areas continue to expand, the consequences of these pressures are evident; increased pollution and resource depletion affect local environments and have cascading effects on global ecological stability. Furthermore, the significance of understanding human ignorance towards environmental challenges cannot be overstated. A model analyzing the spread of diseases such as dengue illustrates how ignorance within populations can exacerbate ecological issues, leading to health crises that reflect broader ecological disruptions. This model found that without adequate public awareness and interventions, such as vector control and disease prevention measures, the risk of outbreaks could persist despite seemingly favorable ecological conditions (Aldila et al., 2023).

Effective communication and education about environmental impacts, public health, and sustainable practices are essential. By fostering a more informed citizenry, communities can mitigate negative ecological repercussions and promote a healthier interaction with their surrounding environments. The interplay between human behavior and community ecology emphasizes that sustainable development is not just a policy choice but a necessity for ecological resilience. As highlighted in the studies, the effectiveness of interventions hinges on both policy frameworks and community engagement. Incorporating community input into environmental decision-making encourages collective responsibility and sustainable practices. As ecological challenges grow increasingly complex, both individuals and governments must prioritize education and awareness as vital tools for fostering sustainable behaviors. Ultimately, the future of our ecosystems depends on the recognition of our interconnectedness with nature and the proactive steps taken to preserve it, ensuring that economic growth does not come at the expense of environmental health (Figure 4.8).

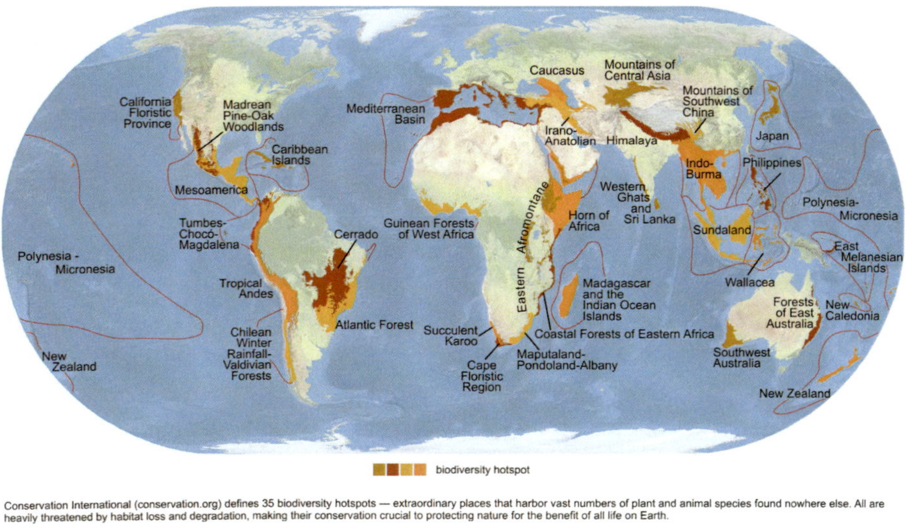

Figure 4.8. *Human impact on the community.*

Source: By Conservation International – Own work, CC BY-SA 4.0, https://commons.wikimedia.org/w/index.php?curid=34669870.

4.6.1. Urbanization and Habitat Fragmentation

Rapid urbanization is a defining characteristic of the 21st century, significantly reshaping natural habitats and ecological networks. The process leads to habitat fragmentation, where large continuous habitats are broken into smaller, isolated patches. This phenomenon alters ecosystems by reducing biodiversity, as species reliant on expansive areas for foraging, breeding, and migration struggle to cope with the diminished landscapes. Furthermore, urban growth often favors generalist species that can adapt to human-dominated environments while rendering specialized species vulnerable to extinction. The decline in species richness due to urban expansion is a pressing concern, given its implications for ecosystem services and stability, which are essential for human well-being. Habitat fragmentation also facilitates interactions between wildlife, livestock, and humans, increasing the risk of zoonotic disease transmission. As cities expand into natural areas, diverse wildlife populations come into closer contact with domesticated species and human communities. This increased interaction heightens the potential for pathogens to spill over from wild animals to domestic ones and subsequently to humans. The complex dynamics at these urban interfaces reveal how urbanization can amplify public health risks, with research indicating that roughly 60 to 80% of newly emerged infections have zoonotic origins, showcasing the interconnectedness between ecosystem health and human health outcomes (Hassell et al., 2016). Therefore, addressing urbanization's impacts requires a multifaceted approach that balances development with conservation. Urban planning and conservation strategies play vital roles in mitigating the effects of habitat fragmentation. Effective measures include creating wildlife corridors that connect isolated habitats, thereby facilitating species movement and promoting genetic diversity. Furthermore, incorporating green spaces in urban designs can enhance biodiversity and provide vital ecosystem services, such as air purification and climate regulation. These strategies not only protect vulnerable species but also foster human well-being by creating recreational spaces and improving urban aesthetics. Emphasizing the importance of sustainable urban development aligns with the ongoing recognition that healthy ecosystems are crucial for community resilience and the prevention of zoonotic diseases, ensuring ecosystems remain robust amid inevitable urban growth (Willett et al., 2019). This integrated approach can help reconcile the needs of human populations with the preservation of ecological integrity, reinforcing the importance of thoughtful interaction between communities and the environments they inhabit.

4.6.2. Climate Change and Its Effects on Populations

The impacts of climate change on populations are evident in the shifting thermal niches that many species must navigate. As temperatures rise, organisms face the choice of adapting to new conditions or migrating to more suitable habitats. For alpine plants, for instance, the ability to remain *in situ* hinges on their genetic adaptability and tolerance to extreme weather. As highlighted in the theoretical framework presented, populations in more topographically complex landscapes may exhibit greater resilience than those in simpler, flat terrains.

This complexity provides microclimatic refuges that can sustain biodiversity and potentially allow many species to persist despite changing climate conditions. Consequently, understanding how landscape topology affects species distribution and community dynamics becomes critical in predicting future ecological outcomes under climate stressors (Graae et al., 2017). Urban ecosystems undergo transformations that exacerbate the effects of climate change and influence the dynamics of both wildlife and human populations. Urbanization, representing a significant anthropogenic pressure, leads to habitat fragmentation, which can alter the movement and interactions of species. For instance, wildlife adaptations to urban environments, such as changes in behavior and community composition, play a role in the emergence of zoonotic diseases. As various species become more concentrated in urban settings, the potential for cross-species transmission increases, emphasizing the interconnectedness of human health and ecological stability. The complex dynamics within these urban-wildlife interfaces highlight that climate change affects populations through physical alterations and the emergent health risks and ecological consequences dependent on these changing environments (Hassell et al., 2016). Ultimately, the consequences of climate change on populations extend beyond individual species, impacting entire ecosystems and their resilience. The shifting conditions necessitate a re-evaluation of conservation strategies, emphasizing the importance of biodiversity preservation and ecosystem management to mitigate these effects. Measures such as protecting critical habitats and ensuring connectivity between fragmented landscapes can enhance species' capabilities to adapt to changing climates. This holistic approach is vital for maintaining ecological integrity and bolstering community-level resilience. By understanding the multifaceted implications of climate change on both populations and communities, stakeholders can make informed decisions that promote sustainability, foster adaptation, and support the health of ecosystems under dire environmental pressures (Graae et al., 2017; Hassell et al., 2016).

4.6.3. Pollution and Its Impact on Community Health

The association between pollution and community health is increasingly evident as industrial activities escalate, often resulting in significant ecological degradation. Rising levels of air and water contaminants, which are primarily the byproducts of industrial processes, pose immediate risks to public health. Exposure to pollutants such as nitrogen oxides and sulfur dioxide can lead to respiratory diseases and cardiovascular issues, particularly in vulnerable populations such as children and the elderly. Communities situated near industrial sites often express heightened concerns over health risks and environmental safety, indicating a critical intersection of ecological damage and community well-being. Ultimately, the persistence of such pollutants in the environment requires a thorough examination of their impact on health outcomes, spotlighting the need for improved regulatory frameworks to safeguard vulnerable populations. In regions heavily impacted by industrial pollution, public consciousness about contaminated sites has surged since the early 2000s, leading to increased advocacy for environmental justice and health protections. Studies reveal that the worst effects of industrial pollution manifest in aquatic and soil ecosystems, which in turn compromises food sources and water quality essential for community health (Figure 4.9).

Figure 4.9. *Pollution impacts community health.*

Source: https://en.wikipedia.org/wiki/File:Obvious_water_pollution.jpeg.

The degradation of these ecosystems can directly correlate with increased instances of health disparities, as those living in polluted areas face heightened risks of disease and adverse health outcomes. As highlighted by (Yuan et al., 2020), the urgency of evaluating ecological risks and their implications for human health is paramount in addressing the legacy of pollution. This growing concern emphasizes the importance of community-driven remedial efforts, fostering an awareness that can lead to significant improvements in both environmental and public health policies. Addressing pollution and its harmful effects on health requires a holistic approach integrating ecological risk management with public health initiatives. By recognizing the interdependence of healthy ecosystems and community well-being, policymakers can implement sustainable solutions that mitigate pollution and promote resilience in affected populations. The call for society-wide solutions, as emphasized in (Giudice et al., 2021), underlines the importance of addressing root causes of pollution, such as fossil fuel production, while fostering community engagement. There is an urgent need for collaboration between various stakeholders, including government agencies, local communities, and environmental organizations, to co-create strategies that prioritize health equity. Through this comprehensive framework, communities can work collaboratively to reclaim their environments and improve public health outcomes, ultimately achieving a holistic form of ecological sustainability and community empowerment.

4.7. CASE STUDIES IN COMMUNITY ECOLOGY

Examining case studies in community ecology reveals the intricate interdependencies and dynamics that shape ecological systems. The relationships between species, including predator-prey interactions and competition for resources, can significantly influence ecosystem stability. For instance, research indicates that urbanization alters host communities and pathogen dynamics, often leading to increased disease emergence, particularly at wildlife-livestock-human interfaces (Dell'Angelo et al., 2016). Analyzing specific scenarios, such as the effects of agricultural practices on local biodiversity, highlights how human activities directly impact ecological relationships. Particularly in developing countries, changes in land use have profound effects on wildlife populations, which can drive emerging infectious diseases into human populations. This interconnectedness implies that understanding species relationships within a community necessitates a holistic approach to both ecology and conservation.

Case studies frequently illustrate the concept of "commons grabbing," where customary land-use rights in communities are threatened by large-scale land acquisitions. This trend disrupts local agricultural practices and undermines community stability and resource management (Dell'Angelo et al., 2016). Such situations illuminate how community structures are challenged by external economic forces, often leading to conflicts over resources essential for survival. For example, the displacement of local farmers can lead to a decline in biodiversity as traditional land management practices are abandoned. As an instance, studies show that as rural residents lose their access to land, the delicate balance between wildlife and agricultural land is jeopardized, leading to increased human-wildlife conflicts and heightened disease risks, illustrating how community ecology is deeply intertwined with socio-economic contexts. Moreover, the impact of urbanization on infectious disease dynamics provides another crucial dimension. In cities, heightened interactions between humans and wildlife can catalyze the emergence of zoonotic diseases, particularly as diverse wildlife populations adapt to urban environments (Hassell et al., 2016). Understanding the factors that contribute to these human-animal interfaces is essential for public health and conservation efforts. This relationship underscores the importance of interdisciplinary approaches in community ecology—leveraging knowledge from urban planning, epidemiology, and ecology to develop strategies that mitigate health risks while conserving biodiversity. By integrating these various perspectives, researchers, and policymakers can formulate more effective interventions aimed at balancing the needs of human populations with the preservation of ecological integrity, thus fostering resilient communities capable of adapting to rapid environmental changes.

4.7.1. Coral Reef Ecosystems and Their Community Dynamics

Coral reefs are among the most biologically diverse ecosystems on Earth, serving as vital habitats for many marine species. These complex environments are structured by symbiotic relationships, particularly between corals and zooxanthellae, which provide essential nutrients through photosynthesis. The interconnectedness of coral species with other marine organisms establishes a dynamic community where each species plays a critical role. For instance, herbivorous fish maintain algal growth, which can otherwise smother coral, while predatory species help regulate the abundance of herbivores, thereby maintaining ecosystem balance. The intricate interactions among these diverse species create a resilient system capable of withstanding environmental changes, although increasing human pressures pose significant threats to their stability. However, understanding the dynamics of coral reef communities requires comprehensive demographic studies, as these assessments depict how various species adapt and flourish in response to changing environmental conditions. For example, research has indicated a diversity of species life strategies within coral ecosystems, showcasing varying investments in growth, survival, and reproduction.

Certain species exhibit adaptations that allow them to thrive in less-than-ideal conditions, illustrating a range of demographic paths that lead to ecological success. Such studies, as pointed out in the analysis of coral communities in New Caledonia, enhance our understanding of species responses under stress and can inform conservation strategies to protect these vital ecosystems. The explicit examination of coral demographics presents an opportunity to glean insights into the health and resilience of reef systems. Despite their resilience, coral reef ecosystems face substantial threats, including climate change, pollution, and unsustainable fishing practices. As evidenced by the impact of mining activities on local coral reefs in regions like Raja Ampat, human intervention can lead to significant damage, including soil erosion and marine pollution (Ulat et al., 2024). These disturbances not only disrupt local biodiversity but can also interfere with the established community dynamics, leading to shifts in species composition and function. To address these challenges, we must adopt sustainable management practices that prioritize both economic development and ecological integrity. Engaging local communities and stakeholders in conservation efforts can enhance ecological resilience and foster sustainable practices, ensuring the longevity and health of coral reef ecosystems. As we navigate the complexities of community and population ecology, recognizing the interconnectedness of species and their environments becomes essential for sustaining these invaluable marine ecosystems.

4.7.2. Forest Ecosystems and Species Interactions

In forest ecosystems, interactions between species play a critical role in shaping community dynamics and influencing population structures. These complex relationships encompass various types of species interactions, including predation, competition, mutualism, and parasitism. For example, predator-prey dynamics can significantly impact the population sizes of both groups involved, leading to fluctuations that can cascade through the ecosystem. Furthermore, plants, and herbivores engage in a constant battle, with plants evolving defensive mechanisms while herbivores adapt to overcome these defenses. The intricate web of interactions among species creates a balance that maintains ecological integrity and biodiversity, crucial for the resilience of forest ecosystems amidst environmental changes. Research on tropical butterflies highlights the importance of understanding species interactions in forest ecosystems. Tropical regions harbor immense biodiversity, with species interactions often more complex than in temperate systems. This complexity enables ecological processes such as mimicry, where certain butterfly species evolve to resemble toxic or unpalatable species, thereby enhancing their survival. Such interactions underscore the evolutionary consequences that arise from community dynamics, emphasizing the interconnectedness of trophic levels and the various roles species play within these ecosystems. By studying these interactions, ecologists can better comprehend how species diversity influences ecosystem stability and function, ultimately guiding conservation efforts. The conservation of forest ecosystems is increasingly complicated by habitat destruction and climate change, which directly impact species interactions. Understanding how these interactions shift in response to anthropogenic pressures is vital for effective management strategies. For instance, forest fragmentation can disrupt mutualistic relationships, such as those between plants and their pollinators, leading to declines in both species. Additionally, as highlighted in the research on tropical ecosystems, varying patterns of butterfly population dynamics demonstrate how climate change can alter reproductive strategies and adult lifespans among species. These changes can have profound implications not just for individual species but also for the broader ecological community. Therefore, a focus on understanding and preserving the intricate networks of species interactions is essential for sustaining forest ecosystems and their myriad ecological functions (Bonebrake et al., 2010; Thompson et al., 2012) (Figure 4.10).

Figure 4.10. *Forest ecosystems.*

Source: By Lubasi – Catedral Verde – Floresta Amazonica, CC BY-SA 2.0, https://commons.wikimedia.org/w/index.php?curid=18572902.

4.7.3. Grassland Ecosystems and Population Fluctuations

Life in grassland ecosystems is profoundly influenced by population fluctuations of both plants and animals. These fluctuations are driven by a variety of factors, including resource availability, climatic conditions, and interspecies interactions. In these dynamic environments, both native and invasive species interact to shape community structures and ecological outcomes. For instance, research suggests that diverse plant assemblages can resist invasion better than less diverse ones, highlighting the importance of native biodiversity in maintaining ecosystem stability under fluctuating populations (Maron et al., 2007). As populations of primary producers, such as grasses, change with seasons and environmental pressures, they create a cascading effect on herbivore populations, affecting their availability and abundance. This interdependence underscores the delicate balance within grassland ecosystems, wherein the health of plant populations is intrinsically linked to the survival and reproduction of herbivores and, subsequently, higher trophic levels. The role of disturbances, both natural and anthropogenic, cannot be overlooked when discussing population dynamics in grasslands. Fires, droughts, and grazing are among the factors that can reset or significantly alter the ecological landscape. These disturbances open up habitats and create opportunities for both established and invading species to flourish, influencing the composition and density of populations inhabiting these areas.

Recent studies have shown that after disturbances, plant communities may experience an initial surge of growth and diversity, followed by selection processes that favor certain species, effectively reshaping the ecosystem (Fridley et al., 2007). This temporal pattern highlights the cyclical nature of population fluctuations, underscored by the recovery and succession processes that follow disturbances. Understanding how these systems recover and adapt is crucial for conservation strategies and for predicting future changes in response to ongoing environmental pressures. Population fluctuations in grassland ecosystems also reflect broader ecological patterns that can provide insight into community dynamics and resilience. For instance, species such as herbivores may experience boom-bust cycles influenced by both climate and food availability, providing vital indicators of ecosystem health. The interaction between herbivore populations and plant communities is complex and often leads to feedback mechanisms where declines in plant diversity may similarly affect herbivore diversity and abundance. Importantly, invasion resistance may remain robust even in high resource availability scenarios, suggesting that maintenance of native diversity is key for ecological resilience (Maron et al., 2007). These findings emphasize that managing grassland ecosystems requires a holistic understanding of population dynamics, including the critical roles of disturbances and species interactions, to sustain biodiversity and ecosystem functions essential for human and wildlife communities alike.

4.8. CASE STUDIES IN POPULATION ECOLOGY

Research in population ecology often utilizes case studies to understand the dynamics between species and their environment. One significant example focuses on the interactions between wolves and moose on Isle Royale, where a vector-autoregressive model was employed to capture the varying effects over time. This research, presented through the dsem package, demonstrated improved analytical precision, particularly in dealing with missing data, thus allowing a clearer insight into complex community dynamics. Such studies are vital as they provide empirical evidence that supports theoretical frameworks within ecology, allowing for a more nuanced understanding of how species interact in fluctuating environmental conditions (Thorson et al., 2024). By engaging with these case studies, researchers can affirm existing ecological theories and refine predictive models that factor in the time-lagged effects of interspecies relationships. Another compelling investigation involves the California Current ecosystem, where researchers have identified significant trophic cascades linked to declining sea otter populations. The dsem analysis revealed how the decrease in sunflower starfish correlates with an uptick in urchin populations, resulting in detrimental effects on kelp density. Such findings underscore the intricate connections within marine ecosystems and highlight how population dynamics can lead to unexpected ecological consequences. The methodical application of dynamic structural equation modeling allowed scientists to parse out simultaneous and lagged interactions among multiple species, revealing complex feedback loops that stimulate or inhibit population growth. These insights are critical not only for understanding existing ecological relationships but also for informing conservation strategies aimed at restoring balance within these ecosystems (Thorson et al., 2024).

Exploring more broadly, the case of Alaska pollock exemplifies how changing environmental conditions, specifically declining sea ice, can disrupt population dynamics and impact species survival. The study employs generalized Lotka-Volterra equations to analyze how reduced cold-water habitats harm copepod populations, which are essential for the early-life survival of pollock. This research illustrates the necessity of adapting theoretical frameworks to incorporate not just mean interactions but also variances and higher-order moments of interactions among species. By considering non-Gaussian dynamics, researchers begin to uncover the instability in population distributions that empirical data often reflect. Thus, case studies in population ecology serve as critical tools in evaluating how species interact amidst changing environmental factors, ultimately contributing to more effective management and conservation efforts within affected ecosystems.

4.8.1. The Dynamics of Predator-Prey Relationships

The intricate dance of predator-prey relationships significantly shapes ecosystem dynamics and stability. These interactions can be described as a balance, with predators regulating prey populations, thereby preventing any one species from overwhelming the resources in an ecosystem. This balance, however, is dynamic and can be influenced by various factors such as habitat availability, food resources, and environmental changes. For instance, as predators become more efficacious in hunting, they can push their prey populations to lower densities. This reduction, in turn, can lead to a decline in the predators as their food source diminishes. The cycles of abundance and scarcity demonstrate the profound interconnectedness within ecological communities, emphasizing that each species plays a critical role in sustaining the health of its habitat. In analyzing predator-prey dynamics, researchers often employ mathematical models to gain insights into population fluctuations and interaction patterns. For example, the Rosenzweig-MacArthur model captures the intrinsic growth rates of prey populations and the functional response of predators. Such models elucidate how the spatial dynamics of prey dispersal can affect overall population stability. Importantly, movement patterns play a role; in situations where predators are present, prey may exhibit altered dispersal behaviors, further complicating these dynamics. For instance, Barman et al (2021) highlight that incorporating factors like time delays associated with prey movement can provide a more nuanced understanding of stability within these relationships. The model demonstrates that prey dispersal, under different densities, can lead to either extinction or stabilization of populations, reinforcing the idea that predator-prey dynamics are more than mere interactions—they are complex systems that evolve over time. Moreover, empirical studies contribute valuable data to our understanding of these interactions, reinforcing theoretical findings. For instance, recent observations of martial eagles have revealed insights into their predation behavior, offering concrete estimates of kill rates and prey biomass recycling. Such findings indicate that these apex predators play a vital role in shaping their ecosystems by influencing prey population dynamics and biodiversity. Hatfield et al (2024) discuss these relationships by emphasizing that the health of prey populations directly impacts the entire food web, demonstrating the cascading effects of predator activity. Thus, the dynamics of predator-prey relationships highlight the interplay of survival strategies and underscore their importance in maintaining ecological balance and species diversity. Understanding these interactions helps ecologists develop better conservation strategies and predict the ecological impacts of environmental changes.

4.8.2. Invasive Species and Their Population Impacts

The introduction of invasive species into new ecosystems often leads to significant population impacts among native species, disrupting the established balance of those environments. These non-native organisms can outcompete established residents for resources such as food, light, and space, leading to declines in native populations. For example, invasive plants often adapt quickly and proliferate rapidly due to their lack of natural predators in their new habitat. This dynamic not only hinders the growth and reproduction of native flora but may also alter soil composition and nutrient availability, further exacerbating the challenges faced by the indigenous species. Moreover, some invasive species can evolve advantageous traits that allow them to exploit resources more effectively, ultimately leading to a diminished variety of native plants and animals. The consequences of these invasions underscore the fragile interdependencies within ecosystems and the cascading effects that occur when external pressures disrupt that balance.

In addition to direct competition, invasive species can also indirectly affect resident populations by altering habitat and ecosystem functions. For instance, the spread of invasive species can change fire regimes, nutrient cycling, and water availability within an ecosystem, impacting all organisms that rely on those resources. This phenomenon is illustrated by how invasive grasses may alter fire dynamics, making areas more prone to wildfires that native species are less adapted to survive. Furthermore, as invasive species establish themselves, they can draw attention away from conservation efforts aimed at protecting vulnerable native populations. The increasing dominance of invasive organisms often leads to an overwhelmed management response, diverting resources and focus away from preserving biodiversity. Thus, invasive species are not just competitors; they represent multifaceted threats that shape the biological and ecological landscape, compounding the urgency of addressing their spread in conservation strategies. The legacy of plant invasions extends beyond immediate ecological impacts, influencing the long-term trajectory of affected communities. As invasive species alter the soil seed bank—the reservoir of seeds in the soil that can give rise to new plants—they contribute to a shift in the reproductive dynamics of resident plant communities. Changes within the seed bank can hinder the establishment of native species, creating a vicious cycle that promotes further invasions, as expounded in (Gioria et al., 2015). This transition often leads ecosystems toward a homogenized state dominated by a few invasive species, diminishing overall biodiversity. Furthermore, research emphasizes that successful management of invasive populations can facilitate the recovery of native communities by enhancing their regenerative potential through improved seed bank viability. Thus, understanding the interplay between invasive species and their population impacts is essential for developing effective restoration and conservation efforts that prioritize the resilience of native ecosystems.

4.8.3. Conservation Efforts for Endangered Species

Efforts to conserve endangered species have taken on unprecedented urgency due to the accelerating rates of habitat loss, climate change, and poaching. Effective conservation requires a multifaceted approach that combines habitat protection, restoration, and laws that govern wildlife trade. Policy frameworks, such as the Endangered Species Act (ESA) in the United States, play a critical role in regulating activities that threaten biodiversity. For instance, successful case studies of species recovery highlight the importance of partnerships between governmental agencies, non-profits, and local communities. These collaborations often hinge on public education and engagement, ensuring that stakeholders understand the intricate balance within ecosystems. As Costanza et al (2017) emphasize, addressing human impacts on habitats is essential for resilient species populations and highlights that conservation strategies must adapt to ongoing environmental changes to foster a sustainable future. In exploring the social dimensions of conservation, it becomes clear that local communities are integral to successful outcomes. Empowering these communities through involvement in conservation initiatives can lead to better stewardship of natural resources and biodiversity. Engaging indigenous and local populations often results in their traditional ecological knowledge being harnessed, which is invaluable in managing species and their habitats effectively. Additionally, community-driven conservation projects have demonstrated that local perspectives can foster innovative practices for wildlife protection and habitat restoration. According to Beever et al (2015), understanding the adaptive capacities allows conservationists to create tailored approaches that account for local conditions, which further enhances the resilience of both species and ecosystems. This understanding underscores the need for conservation strategies to be as inclusive as possible, integrating both top-down and bottom-up approaches to ecological management. Ultimately, the success of conservation efforts depends on the integration of scientific research, community stakeholder engagement, and adaptive management practices. A significant take-home message from ongoing studies and conservation planning is that dynamic ecosystems require flexible strategies that can respond to changing environmental conditions. For instance, methodologies that assess the adaptive capacity of species can inform adaptive management practices that mitigate the impacts of climate change and habitat loss. Acknowledging and incorporating comprehensive ecological data into planning decisions is crucial for creating effective conservation strategies. As the conversation around conservation evolves, it is clear that a collaborative effort involving scientists, policymakers, and local communities will yield more effective outcomes in preserving endangered species for future generations.

4.9. CONCLUSION

The exploration of community and population ecology reveals the profound connections between social structures and health outcomes in various species, including humans and insects. Understanding these relationships is vital, particularly in the context of mental health. For instance, research shows that individuals with a positive sense of community exhibit significantly lower symptoms of depression, anxiety, and stress compared to those lacking such a connection (Park et al., 2023). This finding illustrates how social capital can be leveraged to promote better health outcomes, highlighting the essential role that the community plays in shaping both individual and collective well-being. Here, the implications extend beyond human populations; they resonate across ecological interactions by underscoring how symbiosis within communities, such as between aphids and their bacterial symbionts, can enhance fitness and reproductive success, thereby influencing population dynamics. Moreover, the intricate balance between social relationships and ecological interactions emphasizes the necessity of robust community structures. In insect populations, such as *S. miscanthi*, the presence of beneficial symbiotic bacteria significantly enhances survival, longevity, and fecundity (Li et al., 2022). This dynamic serves as a parallel to human communities, where strong social ties can improve mental health and cognitive resilience.

The essence of these relationships paints a broader picture of community ecology, where mutual support systems foster resilience against environmental stressors. As evidenced by both human psychological studies and ecological research, nurturing these community ties is crucial in mitigating adverse effects associated with isolation and stress, which can degrade individual health and destabilize populations. In conclusion, the interplay of community and population ecology underscores the importance of social and biological interdependencies. As seen in the beneficial effects of a positive community sense on mental health (Park et al., 2023) and the enhancement of reproductive success in aphids due to symbiotic bacteria (Li et al., 2022), fostering strong connections is critical for sustaining both human and ecological health. Thus, the future of community ecology research must focus on developing strategies that enhance social cohesion while promoting awareness of the ecological interactions that contribute to overall well-being. By recognizing the symbiotic relationship between social structures and population dynamics, researchers can advocate for holistic approaches that nurture both community bonds and ecological resilience, paving the way for healthier societies and environments.

4.9.1. Summary of Key Points Discussed

Understanding the dynamics of tree ontogeny is critical for advancing approaches in sustainable forest management, especially in northern European hemiboreal forests. The research highlights unique growth patterns and regeneration strategies among tree species, emphasizing the concept of multiple niche positions within forest dynamics. Recognizing the genetic trajectories involved in tree development allows foresters to adopt management strategies that echo natural ecological processes. A key takeaway is that closer-to-nature management prioritizes ecological integrity and forest resilience, which can significantly improve the vitality of tree communities, thereby enhancing overall forest health (Petrokas et al., 2024). This perspective underscores the importance of incorporating ecological principles into forestry practices. Similarly, the genetic architecture of adaptive traits in species, such as Darwin's finches, reveals the intricate interplay between evolution and environmental pressures. The identification of six loci significantly influencing beak size illustrates how genetic variation can facilitate rapid evolutionary changes in response to natural selection. The profound impact of hybridization and natural selection illustrates that adaptive traits can arise swiftly due to shifts in environmental conditions, like droughts, affecting food availability (Enbody et al., 2023). This correlation between genetics and ecological adaptation underscores the importance of understanding population dynamics within communities, thereby enriching the future research on community and population ecology. In summary, a nuanced understanding of both tree ontogeny and genetic architecture informs and enhances ecological research and practical applications. By examining the distinct strategies of tree species in hemiboreal forests alongside the adaptive traits of finches, researchers can uncover underlying principles that govern community dynamics. The integration of genetic knowledge with ecological practices fosters more effective management strategies that consider evolutionary processes, ultimately leading to healthier, more resilient ecosystems. These insights reinforce the essential connection between genetic factors and ecological resilience, which is vital for sustainable environmental stewardship (Enbody et al., 2023; Petrokas et al., 2024).

4.9.2. The Future of Community and Population Ecology Research

One promising avenue for the future of community and population ecology research is the integration of advanced technological tools and methodologies. As urbanization increases and more people inhabit cities, understanding the dynamics of urban ecosystems becomes crucial. Current studies are often geographically limited, focusing primarily on western, developed regions while neglecting diverse urban landscapes globally. By utilizing geographic information systems (GIS), remote sensing, and big data analytics, researchers can explore ecological patterns and processes across different cities and environments more effectively. This kind of analysis is essential to uncovering the complex relationships between biodiversity and ecosystem services. For instance, understanding how these relationships function can inform urban planning and conservation efforts that prioritize sustainable ecological practices. Future research can significantly enhance our comprehension of community dynamics within increasingly urbanized settings by expanding the geographical scope and methodological approaches (Han et al., 2016). The evolving challenges posed by climate change, habitat loss, and species extinction also demand a recalibration of research priorities in community and population ecology. As the interactions between organisms and their environments become more intricate and unpredictable, interdisciplinary collaboration will be vital. Integrating insights from sociocultural studies and environmental sciences could lead to a more comprehensive understanding of how ecological communities adapt and respond to external pressures. Studies focusing on the impacts of urbanization on species interactions, as well as biodiversity–ecosystem service relationships, will illuminate the synergies and tradeoffs inherent in urban settings. Such insights are crucial not only for understanding ecological dynamics but also for informing policymakers and urban planners. By pursuing these interdisciplinary approaches, future research can enhance the resilience of communities in the face of rapid environmental change (Han et al., 2016). Lastly, the future of community and population ecology research must emphasize the role of public engagement and citizen science in shaping ecological outcomes. The rise of technology has enabled broader participation in ecological monitoring and data collection, allowing citizens to contribute valuable information about local ecosystems. This trend fosters greater awareness and understanding of ecological issues, facilitating a public discourse that bridges the gap between scientific research and community needs. To further enhance this engagement, researchers should focus on translating complex ecological concepts into accessible formats for diverse audiences. This could involve educational initiatives, workshops, and collaborative projects that empower communities to take an active role in the stewardship of their environments. By cultivating a strong connection between research and community involvement, the future of community and population ecology can drive meaningful change, ensuring that ecological health remains a priority for all stakeholders (Ziter, 2015).

4.9.3. The Importance of Ecological Awareness and Action

Understanding the intricate relationships between living organisms and their environments is essential for building a sustainable future. Ecological awareness fosters a deep comprehension of how human actions impact biodiversity and ecosystem balance. Recognizing the interdependencies among species allows us to appreciate the crucial role each organism plays, from the smallest microorganisms to the largest mammals. This awareness enhances our sense of stewardship over the planet and drives individuals and communities to engage in conservation efforts. By promoting ecological literacy, we equip future generations with the knowledge needed to tackle pressing environmental challenges, such as habitat destruction and climate change, thus ensuring that the natural world continues to thrive alongside human development. Proactive engagement and action stemming from ecological awareness can lead to significant changes at both local and global levels. Community initiatives, such as local clean-up drives or tree-planting events, serve as powerful catalysts for environmental restoration and education. When individuals come together to address ecological issues, they cultivate a sense of shared responsibility and empowerment that can influence broader societal norms. Additionally, such participatory actions can spur policy changes, as communities that prioritize ecological health may advocate for sustainable regulations or funding for renewable resources. By bridging the gap between awareness and action, communities can address the root causes of ecological decline and inspire a collective commitment to sustainability. Ultimately, fostering ecological awareness and translating it into meaningful action can enhance the resilience of both human populations and natural ecosystems. As communities embrace environmentally sustainable practices, they reduce their ecological footprint and enhance their quality of life. For instance, urban areas that prioritize green spaces and biodiversity often report improved mental health and community cohesion. Furthermore, the integration of sustainable practices into daily life can lead to economic benefits, such as the development of green jobs in renewable energy or conservation sectors. By uniting ecological awareness with actionable strategies, society can pave the way toward a more balanced coexistence with the natural world, ensuring a thriving planet for generations to come.

REFERENCES

1. Aldila, D., Puspadani, A., & Rusin, R. (2023). Mathematical analysis of the impact of community ignorance on the population dynamics of dengue. *Frontiers in Applied Mathematics and Statistics, 9*, 1094971. https://doi.org/10.3389/fams.2023.1094971.
2. Álvarez-Cobelas, M., & Rojo, C. (2022). Linking the community and metacommunity perspectives: Biotic relationships are key in benthic diatom ecology. *Water*, *14*(23), 3805. https://doi.org/10.3390/w14233805.
3. Angeler, D. G., & Allen, C. R. (2016). Quantifying resilience. *Journal of Applied Ecology, 53*(5), 617–624. https://doi.org/10.1111/1365-2664.12649.
4. Azaele, S., & Maritan, A. (2023). Large system population dynamics with non-Gaussian interactions. *Semantics Scholar*. https://www.semanticscholar.org/paper/3b4a52164b1bf771c9846f4ef00f8cd244f64cfa (accessed on 5 December 2024).
5. Baker, P. J. (2009). Tropical forest community ecology. *Annals of Botany, 104*(1), undefined. https://doi.org/10.1093/aob/mcp110.
6. Barman, B., & Ghosh, B. (2021). Dynamics of a spatially coupled model with delayed prey dispersal. *Semantics Scholar*, *42*, 400–414. https://www.semanticscholar.org/paper/66ecdab9a10bf010ab81223ca0615d4ef5644ddc (accessed on 5 December 2024).
7. Beever, E. A., O'Leary, J. F., Mengelt, C., West, J. M., Julius, S., Green, N., Magness, D. R., Petes, L. E., Stein, B. A., Nicotra, A. B., Hellmann, J. J., Robertson, A., Staudinger, M. D., Rosenberg, A. A., Babij, E., Brennan, J., Schuurman, G. W., & Hofmann, G. E. (2015). Improving conservation outcomes with a new paradigm for understanding species' fundamental and realized adaptive capacity. *Conservation Letters, 9*(2), 131–137. https://doi.org/10.1111/conl.12190.
8. Bonebrake, T. C., Ponisio, L. C., Boggs, C. L., & Ehrlich, P. R. (2010). More than just indicators: A review of tropical butterfly ecology and conservation. *Biological Conservation, 143*(8), 1831–1841. https://doi.org/10.1016/j.biocon.2010.04.044.
9. Capdevila, P., Stott, I., Beger, M., & Salguero-Gómez, R. (2020). Towards a comparative framework of demographic resilience. *Trends in Ecology & Evolution, 35*(9), 776–786. https://doi.org/10.1016/j.tree.2020.05.001.
10. Costanza, R., de Groot, R. S., Braat, L., Kubiszewski, I., Fioramonti, L., Sutton, P. C., Farber, S., & Grasso, M. (2017). Twenty years of ecosystem services: How far have we come and how far do we still need to go? *Ecosystem Services, 28*, 1–16. https://doi.org/10.1016/j.ecoser.2017.09.008.

11. Dang, H., & Lovell, C. R. (2015). Microbial surface colonization and biofilm development in marine environments. *Microbiology and Molecular Biology Reviews*, *80*, 91–138. https://doi.org/10.1128/mmbr.00037-15.
12. Dell'Angelo, J., D'Odorico, P., Rulli, M. C., & Marchand, P. (2016). The tragedy of the grabbed commons: Coercion and dispossession in the global land rush. *World Development*, *92*, 1–12. https://doi.org/10.1016/j.worlddev.2016.11.005.
13. Dolzhykova, I. (2024). Methodological aspects of the organizational-economic mechanism of efficiency in the utilization of agricultural enterprises' land resources. *Semantics Scholar*. https://www.semanticscholar.org/paper/b951e5a8c50347cfc1e143281fe48be95b2208bd (accessed on 5 December 2024).
14. Dudgeon, D., Arthington, A. H., Gessner, M. O., Kawabata, Z., Knowler, D. J., Lévêque, C., Naiman, R. J., Prieur-Richard, A. H., Soto, D., Stiassny, M. L., & Sullivan, C. A. (2006). Freshwater biodiversity: Importance, threats, status and conservation challenges. *Biological Reviews*, *81*(2), 163–182. https://doi.org/10.1017/S1464793105006950.
15. Enbody, E. D., Sendell-Price, A. T., Sprehn, C. G., Rubin, C. J., Visscher, P. M., Grant, B. R., Grant, P. R., & Andersson, L. (2023). Community-wide genome sequencing reveals 30 years of Darwin's finch evolution. *Science*, *381*(6665), eadf6218. https://doi.org/10.1126/science.adf6218.
16. Estrada-Peña, A., Ostfeld, R. S., Peterson, A. T., Poulin, R., & de la Fuente, J. (2014). Effects of environmental change on zoonotic disease risk: An ecological primer. *Trends in Parasitology, 30*(4), 205–214. https://doi.org/10.1016/j.pt.2014.02.003.
17. Fischer, J. D., Schneider, S. C., Ahlers, A. A., & Miller, J. R. (2015). Categorizing wildlife responses to urbanization and conservation implications of terminology. *Conservation Biology, 29*(4), 1246–1248. https://doi.org/10.1111/cobi.12451.
18. Geoffroy, M., Bouchard, C., Flores, H., Robert, D., Gjøsæter, H., Hoover, C., Hop, H., Hussey, N. E., Nahrgang, J., Steiner, N., Bender, M., Berge, J., Castellani, G., Chernova, N., Copeman, L., David, C. L., Deary, A., Divoky, G., Dolgov, A. V., Duffy-Anderson, J., Dupont, N., Durant, J. M., Elliott, K., Gauthier, S., Goldstein, E. D., Gradinger, R., Hedges, K., Herbig, J., Laurel, B., Loseto, L., Maes, S., Mark, F. C., Mosbech, A., Pedro, S., Pettitt-Wade, H., Prokopchuk, I., Renaud, P. E., Schembri, S., Vestfals, C., & Walkusz, W. (2023). The circumpolar impacts of climate change and anthropogenic stressors on Arctic cod (Boreogadus saida) and its ecosystem. *Elementa: Science of the Anthropocene*, *11*(1), 00097. https://doi.org/10.1525/elementa.2022.00097.

19. Gintoron, C. S., Mohammed, M. A., Sazali, S., Deka, E. Q., Ong, K., Shamsi, I., & King, P. J. H. (2023). Factors affecting pollination and pollinators in oil palm plantations: A review with an emphasis on the Elaeidobius kamerunicus weevil (Coleoptera: Curculionidae). *Semantic Scholar*. https://www.semanticscholar.org/paper/4d41587ea85126e485ca28592c19cf8fd588b3b8 (accessed on 5 December 2024).
20. Gioria, M., & Pyšek, P. (2016). The legacy of plant invasions: Changes in the soil seed bank of invaded plant communities. *BioScience, 66*(1), 40–53. https://doi.org/10.1093/biosci/biv165.
21. Giudice, L. C., Llamas Clark, E. F., DeNicola, N. G., Pandipati, S., Zlatnik, M. G., Decena, D. C. D., Woodruff, T. J., & Conry, J. A. (2021). Climate change, women's health, and the role of obstetricians and gynecologists in leadership. *International Journal of Gynecology & Obstetrics*, *155*, 345–356. https://doi.org/10.1002/ijgo.13958.
22. Graae, B., Vandvik, V., Armbruster, W., Eiserhardt, W., Svenning, J. C., Hylander, K., Ehrlén, J., Speed, J., Klanderud, K., Bråthen, K., Milbau, A., Opedal, Ø. H., Alsos, I., Ejrnæs, R., Bruun, H., Birks, H., Westergaard, K., Birks, H., & Lenoir, J. (2017). Stay or go–how topographic complexity influences alpine plant population and community responses to climate change. *Perspectives in Plant Ecology, Evolution and Systematics*, *30*. https://doi.org/10.1016/j.ppees.2017.09.008.
23. Haage, M., Maran, T., Bergvall, U. A., Elmhagen, B., & Angerbjörn, A. (2016). Evolutionary maintenance of personality–a field experiment on survival and personality. *Semantics Scholar*. https://www.semanticscholar.org/paper/46992fafb71f321b267b708245f667fe0ed11d8c (accessed on 5 December 2024).
24. Haage, M., Pasanen Mortensen, M., Sidorovich, V., Angerbjörn, A., & Elmhagen, B. (2016). American mink in north-eastern Europe–abundances, population trends and dynamics in relation to the red fox and otter. *Semantics Scholar*. https://www.semanticscholar.org/paper/f67f53a0bf774870a936ab88798ea474a0356d61 (accessed on 5 December 2024).
25. Han, B. A., Kramer, A. M., & Drake, J. M. (2016). Global patterns of zoonotic disease in mammals. *Trends in Parasitology*, *32*, 565–577. https://doi.org/10.1016/j.pt.2016.04.007.
26. Hassell, J. M., Begon, M., Ward, M. J., & Fèvre, E. M. (2016). Urbanization and disease emergence: Dynamics at the wildlife–livestock–human interface. *Trends in Ecology & Evolution, 32*(1), 55–67. https://doi.org/10.1016/j.tree.2016.09.012.
27. Hatfield, R. S., Davis, A. G., Buij, R., Cox, J. J., Kapila, S., Parmuntoro,

L., Thomsett, S., Virani, M. Z., Njoroge, P., & van Langevelde, F. (2024). Africa's overlooked top predator: Towards a better understanding of martial eagle feeding ecology in the Maasai Mara, Kenya. *Wildlife Biology*. https://doi.org/10.1002/wlb3.01223.

28. Horecka, K., & Neal, S. (2022). Critical problems for research in animal sheltering, a conceptual analysis. *Semantics Scholar*, *9*. https://www.semanticscholar.org/paper/4ac84b1ce75f71462ae12f456a880a5a1fe2d658 (accessed on 5 December 2024).

29. Ivanova, A. E., & Vangorodskaya, S. A. (2024). Active longevity: Concept definition, key determinants and contradictions. *Semantics Scholar*. https://www.semanticscholar.org/paper/0eb26b1781e509c38e6dfdfe7b270f1357a3e525 (accessed on 5 December 2024).

30. Jia, S., Wang, X., Yuan, Z., Lin, F., Ye, J., Hao, Z., & Luskin, M. S. (2018). Global signal of top-down control of terrestrial plant communities by herbivores. *Proceedings of the National Academy of Sciences*, *115*, 6237–6242. https://doi.org/10.1073/pnas.1707984115.

31. Kayal, M., Mevrel, E., & Ballard, J. (2023). Coral demographic performances in New Caledonia, a video transect approach to operationalize imagery-based investigation of population and community dynamics. *bioRxiv*. https://doi.org/10.1101/2023.05.12.540552.

32. Kostic, A. D., Xavier, R. J., & Gevers, D. (2014). The microbiome in inflammatory bowel disease: Current status and the future ahead. *Gastroenterology*, *146*(6), 1489–1499. https://doi.org/10.1053/j.gastro.2014.02.009.

33. Kramer, K. L. (2005). Children's help and the pace of reproduction: Cooperative breeding in humans. *Evolutionary Anthropology*, *14*, 224–237. https://doi.org/10.1002/evan.20082.

34. Li, X., Sun, Y., Tian, X., Wang, C., Li, Q., Li, Q., Zhu, S., Lan, C., Zhang, Y., Li, X., Ding, R., & Zhu, X. (2023). Sitobion miscanthi L type symbiont enhances the fitness and feeding behavior of the host grain aphid. *Pest Management Science*, *79*(4), 1362–1371. https://doi.org/10.1002/ps.7308.

35. McCarty, N. S., & Ledesma Amaro, R. (2018). Synthetic biology tools to engineer microbial communities for biotechnology. *Trends in Biotechnology*, *37*, 181–197. https://doi.org/10.1016/j.tibtech.2018.11.002.

36. McRae, B. H., Dickson, B. G., Keitt, T. H., & Shah, V. B. (2008). Using circuit theory to model connectivity in ecology, evolution, and conservation. *Ecology*, *89*(10), 2712–2724. https://doi.org/10.1890/07-1861.1.

37. Méndez-García, C., Peláez, A. I., Mesa, V., Sánchez, J., Golyshina, O.

V., & Ferrer, M. (2015). Microbial diversity and metabolic networks in acid mine drainage habitats. *Frontiers in Microbiology, 6*. https://doi.org/10.3389/fmicb.2015.00475.

38. Park, E. Y., Oliver, T. R., Peppard, P. E., & Malecki, K. C. (2023). Sense of community and mental health: A cross-sectional analysis from a household survey in Wisconsin. *Family Medicine and Community Health, 11*(2), e001971. https://doi.org/10.1136/fmch-2022-001971.
39. Petrokas, R., Manton, M., & Kavaliauskas, D. (2024). Tree regeneration and ontogenetic strategies of northern European hemiboreal forests: transitioning towards closer-to-nature forest management. *Semantics Scholar*. https://www.semanticscholar.org/paper/e76e71f9d4e8559ff2a0bb5e96caca4660d0a0f4 (accessed on 5 December 2024).
40. Raerinne, J. (2020). Ghosts of competition and predation past: Why ecologists value negative over positive interactions. *Biological and Environmental Sciences, 101*. https://doi.org/10.1002/bes2.1766.
41. Richardson, D. M., & Ricciardi, A. (2013). Misleading criticisms of invasion science: A field guide. *Diversity and Distributions, 19*(12), 1461–1467. https://doi.org/10.1111/ddi.12150.
42. Salguero-Gómez, R., Jones, O. R., Archer, C. R., Buckley, Y. M., Che-Castaldo, J., Caswell, H., et al., (2015). The COMPADRE Plant Matrix Database: An open online repository for plant demography. *Journal of Ecology, 103*(2), 202–218. https://doi.org/10.1111/1365-2745.12334.
43. Shoemaker, L. G., Sullivan, L. L., Donohue, I., Cabral, J. S., Williams, R. J., Mayfield, M. M., Chase, J. M., Chu, C., Harpole, W. S., Huth, A., HilleRisLambers, J., James, A. R. M., Kraft, N. J. B., May, F., Muthukrishnan, R., Satterlee, S., Taubert, F., Wang, X., Wiegand, T., Yang, Q., & Abbott, K. C. (2020). Integrating the underlying structure of stochasticity into community ecology. *Ecology, 101*(2), e02922. https://doi.org/10.1002/ecy.2922.
44. Sutikno, A., Royali, S., Sholiha, F. U., & Suliswanto, M. S. W. (2024). Investigating the impact of economic growth on pollution index in emerging market countries. *Semantics Scholar*. https://www.semanticscholar.org/paper/4464b3c6e36a625d52ceab260b525d4e031c4ea7 (accessed on 5 December 2024).
45. Thompson, R. M., Brose, U., Dunne, J. A., Hall, R. O., Hladyz, S., Kitching, R. L., Martinez, N. D., Rantala, H. M., Romanuk, T. N., Stouffer, D. B., & Tylianakis, J. M. (2012). Food webs: Reconciling the structure and function of biodiversity. *Trends in Ecology & Evolution, 27*(12), 689–697. https://doi.org/10.1016/j.tree.2012.08.005.

46. Thorson, J. T., Andrews, A. G. III, Essington, T. E., & Large, S. I. (2024). Dynamic structural equation models synthesize ecosystem dynamics constrained by ecological mechanisms. *Methods in Ecology and Evolution*, *15*, 744–755. https://doi.org/10.1111/2041-210X.14289.

47. Ulat, M. A., Handayani, H., Mulya, A., Poltak, H., & Ismail, I. (2024). Analysis of the social, economic, and ecological impact of mining activities of PT. Gag Nickel on society and coral reef ecosystem in Gag Island, Raja Ampat District. *Semantic Scholar*. https://www.semanticscholar.org/paper/3278fcc566abbaf8f7b61b240521432c8f9e75fc (accessed on 5 December 2024).

48. Willett, W., Rockström, J., Loken, B., Springmann, M., Lang, T., Vermeulen, S., Garnett, T., Tilman, D., DeClerck, F., Wood, A., Jonell, M., Clark, M., Gordon, L. J., Fanzo, J., Hawkes, C., Zurayk, R., Rivera, J. A., De Vries, W., Majele Sibanda, L., Afshin, A., Chaudhary, A., Herrero, M., Agustina, R., Branca, F., Lartey, A., Fan, S., Crona, B., Fox, E., Bignet, V., Troell, M., Lindahl, T., Singh, S., Cornell, S. E., Srinath Reddy, K., Narain, S., Nishtar, S., & Murray, C. J. L. (2019). Food in the Anthropocene: The EAT-Lancet Commission on healthy diets from sustainable food systems. *The Lancet*, *393*(10170), 447–492. https://doi.org/10.1016/S0140-6736(18)31788-4.

49. Yuan, J., Lü, Y., Wang, C., Cao, X., Chen, C., Cui, H., Zhang, M., Wang, C., Li, X., Johnson, A. C., Sweetman, A. J., & Du, D. (2020). Ecology of industrial pollution in China. *Environmental Pollution*, *6*. https://doi.org/10.1080/20964129.2020.1779010.

50. Ziter, C. (2015). The biodiversity–ecosystem service relationship in urban areas: A quantitative review. *Oikos*, *125*, 761–768. https://doi.org/10.1111/oik.02883.

CHAPTER 5

CONSERVATION AND BIODIVERSITY

Contents

INTRODUCTION

Efforts to protect the natural world are increasingly urgent as the consequences of environmental degradation become more pronounced. This urgency is primarily driven by trends such as habitat loss, climate change, and pollution, which threaten not only countless species but also the ecosystems that sustain human life. Today, the crisis of biodiversity loss represents a significant challenge to achieving a sustainable future, as it undermines the resilience of natural systems. The intricate web of life on Earth is not only vital for the survival of flora and fauna but also plays a crucial role in human health, economic stability, and cultural identity.

To address these complex issues, a concerted global effort focused on conservation and biodiversity preservation is essential, as it can mitigate adverse effects and promote a more harmonious coexistence between humanity and nature. Achieving effective conservation requires a multifaceted approach that incorporates scientific research, community involvement, and policy frameworks. Scientific studies help to elucidate the intricate relationships within ecosystems and how various factors contribute to biodiversity loss, thereby informing strategies that can mitigate such impacts. Simultaneously, engaging local and indigenous communities is vital, as their knowledge and commitment can drive sustainable practices and enhance conservation efforts.

Moreover, government policies play a significant role in habitat protection and resource management; effective laws must prioritize environmental considerations alongside economic development. By creating a robust framework that encompasses science, community activism, and policy intervention, we can better safeguard the diverse life forms that enrich our planet and establish a sustainable future for generations to come.

The sacrifices made in the name of progress often come at the expense of natural ecosystems, leading to dire outcomes for both biodiversity and human welfare. It is imperative to recognize that every species plays a unique role in its environment, contributing to the overall health and functionality of ecosystems. The loss of a single species can lead to a cascade of consequences, disrupting food webs and ecological balance. Conversely, healthy ecosystems exhibit resilience and adaptability in the face of environmental change, underscoring the importance of safeguarding biodiversity. In light of these dynamics, an urgent call to action emerges; not only must we work toward conserving existing wildlife and habitats, but we must also restore areas that have been damaged by human activity. By prioritizing conservation and fostering biodiversity, we can nurture the ecosystems of the planet and, by extension, secure our own future on Earth.

5.1. DEFINITION OF CONSERVATION

The essence of conservation lies in the sustainable management of natural resources to ensure that biodiversity is preserved for future generations. Conservation is not a static process; rather, it demands a dynamic approach that adapts to changing environmental conditions and human needs. In this context, it involves implementing strategies that mitigate human-induced threats to ecosystems, such as habitat destruction, pollution, and climate change. The impact of these pressures on biodiversity is profound, as evidenced by the increasing extinction rates of plant and animal species worldwide. Recognizing these challenges underscores the importance of establishing effective political institutions and policies that prioritize ecological health, as stated in recent studies about the relationship between democratic governance and biodiversity efforts (Rydén et al., 2019). A comprehensive understanding of conservation also emphasizes the interconnectedness of ecological systems and the species that inhabit them. Ecosystems provide crucial services such as clean air, water filtration, and soil fertility, which are essential for human survival and well-being. This interdependence implies that conservation efforts must consider both ecological and human dimensions, fostering a sense of stewardship among communities. By promoting sustainable practices and raising awareness about the importance of biodiversity, conservation can help alleviate some of the pressures facing global food systems, which currently threaten both human health and environmental sustainability (Willett et al., 2019). Thus, conservation not only protects diverse species but also supports the resilience of ecosystems that people rely on for their livelihoods. Ultimately, effective conservation must be grounded in a shared commitment to ecological integrity and social equity. This involves engaging diverse stakeholders, from government entities to local communities, in the decision-making processes that shape conservation strategies. Collaborative approaches enable a more holistic perspective that takes into account varying cultural values and economic interests. Furthermore, empirical studies suggest that democratic institutions, characterized by transparency and active civic participation, are crucial for the success of biodiversity conservation initiatives (Rydén et al., 2019). As society grapples with the urgent need for sustainable practices, conservation efforts must be clearly defined, strategically implemented, and continuously evaluated to adapt to emerging challenges and foster a harmonious coexistence with nature.

5.2. THE IMPORTANCE OF BIODIVERSITY

The interconnectedness of ecosystems underscores how biodiversity enriches environmental resilience. Diverse biological systems contribute not only to the stability of ecosystems but also to their productivity. For instance, a variety of species can perform different ecological roles, which collectively enhance nutrient cycling and pollination, essential processes for healthy agriculture. The decline of biodiversity, as noted in recent analyses, greatly hampers these essential services, leading to reduced yields and heightened vulnerability to pests and diseases (Tscharntke et al., 2021). In essence, every species lost can compromise not only the health of an ecosystem but also the agricultural outputs that communities rely upon, highlighting the complexity and interdependence of life on Earth. Recognizing the crucial role that biodiversity plays in human health and food security is paramount. Food systems that rely on a limited number of species are less resilient to shocks such as climate change, disease, and other environmental stresses. The current global food production models often prioritize high-yield monoculture, which can exacerbate malnutrition and hunger. Over 820 million people are affected by insufficient food, and many suffer from diet-related diseases amplified by a lack of nutritional diversity (Willett et al., 2019).

Figure 5.1. *Importance of biodiversity.*

Source: By Dano – Biodiversity, CC BY 2.0, https://commons.wikimedia.org/w/index.php?curid=147567470.

A robust biodiversity ensures a variety of crops and livestock, which can enhance nutrient availability and dietary quality for populations worldwide. Ensuring sustainable food systems depends critically on maintaining this diversity, which contributes both to agricultural resilience and the overall health of populations. Conservation efforts aimed at preserving biodiversity are not merely ecological imperatives; they are also fundamental to sustainable development. As the global population continues to grow, the demand for food and natural resources increases, putting immense pressure on ecosystems. Strategies that focus on enhancing landscape diversity—such as through the integration of seminatural habitats and diversified cropping systems—can foster biodiversity while maintaining agricultural productivity. Research indicates that such ecological approaches can help mitigate the known yield losses associated with monoculture and conventional farming practices (Tscharntke et al., 2021). Emphasizing initiatives that balance human needs with ecological sustainability can lead to healthier ecosystems and vibrant economies. Ultimately, successful conservation practices are essential not just for protecting various species but also for ensuring a sustainable future for humanity and the planet (Figure 5.1).

5.2.1. Ecosystem Services Provided by Biodiversity

Biodiversity is fundamental to the sustainability of ecosystem services, which are the direct benefits that humans derive from healthy environments. These services include provisioning, regulating, cultural, and supporting benefits that are critical for human survival and well-being. For instance, diverse ecosystems enhance the resilience of agricultural systems, increasing crop yields and food security. The variety of species within these systems can also aid in pest control and pollination, further boosting agricultural productivity. By maintaining a pool of genetic diversity, ecosystems can adapt to changing conditions, ensuring the continued provision of essential resources for future generations. Without robust biodiversity, ecosystems become vulnerable to disturbances, leading to a decline in these vital services. The relationship between biodiversity and ecosystem functions is increasingly emphasized in research, particularly in understanding how different species contribute to ecosystem services. Studies indicate that diverse species play distinct roles in maintaining ecosystem health, supporting the concept of biodiversity–ecosystem functioning (B–EF). An increase in species richness can enhance the efficacy of ecosystem processes like nutrient cycling, water filtration, and carbon sequestration. Specifically, certain species exhibit functional traits that enable them to perform unique roles within their environments, thus optimizing ecosystem services. As noted in recent research, expanding protected areas and safeguarding natural habitats can substantially improve the regulation of crucial functions, such as nutrient cycling, which has far-reaching implications for agriculture and water quality management (Duncan et al., 2015). The implications of conserving biodiversity extend beyond immediate ecosystem services to encompass broader environmental goals, including climate change mitigation. By protecting vital habitats, we can not only conserve species but also optimize the capacity of the planet to regulate greenhouse gases and provide clean water. According to research, achieving global biodiversity targets could lead to substantial reductions in carbon emissions and enhance the ecological resilience of natural systems, thus contributing to climate adaptation strategies (Zeng et al., 2022). Furthermore, the preservation of biodiversity is essential for maintaining the complex interactions within ecosystems that underpin various ecosystem services. As we confront environmental challenges such as pollution, habitat loss, and climate change, prioritizing biodiversity conservation becomes imperative for ensuring the sustainability of essential ecosystem services that benefit both human populations and the natural world.

5.2.2. Role of Biodiversity in Food Security

Biodiversity plays a crucial role in ensuring food security by supporting the resilience of agricultural systems. Diverse ecosystems contribute to the stability of food production through various mechanisms, such as pest control and pollination. For example, a variety of species can help mitigate the risks of crop failure associated with disease outbreaks or climate fluctuations. Additionally, agroecological practices that promote biodiversity—such as crop rotation and the integration of perennial plants—can enhance soil health, leading to improved crop yields over time. Emphasizing biodiversity within farming practices is essential for reducing dependency on chemical fertilizers and pesticides, which can have detrimental effects on both ecosystems and human health. Moreover, the genetic diversity found within agricultural crops is vital for adapting to changing environmental conditions. Different crop varieties may possess unique traits that confer resistance to pests, diseases, or extreme weather events, thus safeguarding food supplies. The loss of genetic diversity can hinder the potential for developing these resilient crops, making food systems more vulnerable to threats. As highlighted in research, certified organic farming shows a significant increase in local species richness, though it may lead to lower yields compared to conventional practices (Tscharntke et al., 2021). This trade-off underscores the necessity for a balanced approach wherein biodiversity conservation aligns with agricultural productivity, reinforcing the idea that biodiversity is integral to maintaining food security. Integrating biodiversity into food security strategies fosters a holistic approach that addresses environmental sustainability and public health. Diverse agricultural practices encourage healthy ecosystems, which in turn provide essential services such as nutrient cycling and water filtration. The decline of biodiversity, often driven by industrial agriculture and habitat loss, presents a significant threat to food systems globally (Intergovernmental Panel on Climate Change, 2023). We can restore habitats and promote biodiversity, ultimately enhancing food production systems by adopting conservation-oriented measures, such as planting cover crops and implementing agroforestry systems. Therefore, prioritizing biodiversity strengthens food security and contributes to broader environmental conservation efforts, creating resilient systems capable of withstanding future challenges (Figure 5.2).

Figure 5.2. *Role of biodiversity in food security.*

Source: By I, Luc Viatour, CC BY-SA 3.0, https://commons.wikimedia.org/w/index.php?curid=998412.

5.2.3. Cultural Significance of Diverse Species

Biodiversity encompasses not just the variety of species in an ecosystem but also the cultural significance linked to those species. Various communities, particularly indigenous populations, have developed rich traditional knowledge systems that underpin their relationships with diverse flora and fauna. These local understandings are pivotal for the cultural identity and heritage of these communities, influencing practices related to medicine, agriculture, and sustainable resource management. For instance, ethnobotanical studies reveal that specific plant species are revered for their medicinal properties, informing the healing practices that have been passed down through generations. The Hamar community in Southwestern Ethiopia, for instance, relies heavily on traditional knowledge regarding over 145 plant species used to treat various ailments (Bekele et al., 2022). This deep-rooted connection showcases how biodiversity sustains cultural identities while also providing practical benefits. The conservation of diverse species is intrinsically linked to preserving cultural practices and identities. As anthropogenic pressures lead to habitat loss and the extinction of species, the traditional knowledge systems that depend on biodiversity face immense risk. The erosion of these systems not only threatens the plants and animals themselves but also the socio-cultural fabric of the communities that rely on them.

The phenomenon of "cultural keystone species" highlights the critical role certain species play in cultural identity and practices; their removal can induce social dislocation and loss of expert knowledge. As evident in the ongoing efforts recognized by the United Nations Decade on Ecosystem Restoration, conserving biodiversity is thus as much a cultural undertaking as it is an ecological one (Fischer et al., 2020). Strengthening the preservation of these species effectively supports community resilience and sustainable development. Furthermore, fostering a wider acknowledgment of the cultural significance of diverse species can lead to more inclusive conservation strategies. Acknowledging the intertwined relationship between biodiversity and human culture encourages collaborative efforts among conservationists, local communities, and policymakers. By integrating traditional ecological knowledge into modern conservation practices, as illustrated by examples from the field, a more holistic approach can emerge. This multifaceted strategy helps to ensure the preservation of species and ecosystems and strengthens the cultural identities that depend on them. The recognition of relational values in resource management emphasizes that human connections to nature extend beyond utilitarian views, fostering a stewardship ethic that values biodiversity for its cultural significance (Fischer et al., 2020). Ultimately, safeguarding diverse species is vital not only for ecological integrity but also for the richness of human culture itself.

5.3. THREATS TO BIODIVERSITY

Among the most pressing threats to biodiversity is habitat loss, which has become increasingly pervasive due to urbanization, agriculture, and deforestation. As natural habitats are transformed into agricultural fields or developed spaces, the intricate networks of life they support are disrupted, leading to a decline in species populations. This reduction not only decreases biodiversity but also diminishes the resilience of ecosystems, making them more vulnerable to environmental stressors. The consequences of habitat degradation are profound; numerous species face extinction, and the services these ecosystems provide—such as pollination, water purification, and carbon storage—are significantly diminished. According to (Kalpana et al., 2024), the decline in pollinator populations, for instance, has critical implications for plant reproduction and genetic diversity, underscoring the interconnectedness of species within their ecosystems. In addition to habitat destruction, pollution poses a substantial threat to biodiversity, affecting both terrestrial and aquatic ecosystems. Chemical pollutants, such as pesticides and heavy metals, can have devastating effects on flora and fauna. Toxic substances can disrupt reproductive systems, impair immune responses, and even lead to mortality in various species. Further exacerbating this crisis is the issue of climate change, which alters habitats and shifts species distributions, often beyond their ability to adapt or migrate. As highlighted in (Gunnam, 2024), the synergy of pollution and climate change can magnify the loss of biological diversity, creating a cascading effect that undermines ecosystem integrity. Thus, addressing pollution is not only necessary to protect individual species but also vital for preserving the ecological balance that supports all life forms. The introduction of invasive species represents another critical challenge to biodiversity, as these organisms can outcompete native species for resources, leading to local extinctions. In many cases, invasive species are introduced intentionally or accidentally through human activities, further emphasizing the role of human agency in this global crisis. The encroachment of invasive species disrupts established ecological relationships, altering food webs and nutrient cycling within ecosystems. Moreover, the threat posed by invasive species can be compounded by climate change, which may provide favorable conditions for these non-native species to flourish. It is essential to recognize that the loss of biodiversity not only stems from direct human actions but also from the broader interconnected systems in which these actions occur. Thus, the conservation of biodiversity requires a multifaceted approach that addresses habitat loss, pollution, and invasive species collaboratively, fostering resilience against ongoing environmental changes.

5.3.1. Habitat Destruction and Fragmentation

Habitat destruction and fragmentation pose significant threats to global biodiversity, leading to the alarming decline in species populations and overall ecosystem health. As ecosystems are altered through urbanization, agriculture, and deforestation, the natural habitats of countless species are disrupted, resulting in habitat loss. This fragmentation breaks ecosystems into isolated patches, making it increasingly challenging for wildlife to migrate, breed, and find food. According to recent research, freshwater ecosystems are particularly vulnerable, experiencing biodiversity loss at a rate more than double that of terrestrial or marine environments, as wetlands disappear rapidly from the landscape (Tickner et al., 2020). This depletion not only affects the species inhabiting these areas but also disrupts essential ecosystem services, such as water purification and carbon sequestration, which human communities rely on for their survival. As habitats become fragmented, the remaining patches often support fewer species, leading to diminished ecological resilience. This fragmentation reduces the capacity of ecosystems to recover from disturbances, such as climate change or natural disasters and directly impacts the populations of flora and fauna that depend on continuous and interconnected habitats. Additionally, isolated populations may face decreased genetic diversity, reducing their ability to adapt to changing environmental conditions and increasing the risk of extinction (Gann et al., 2019) (Figure 5.3).

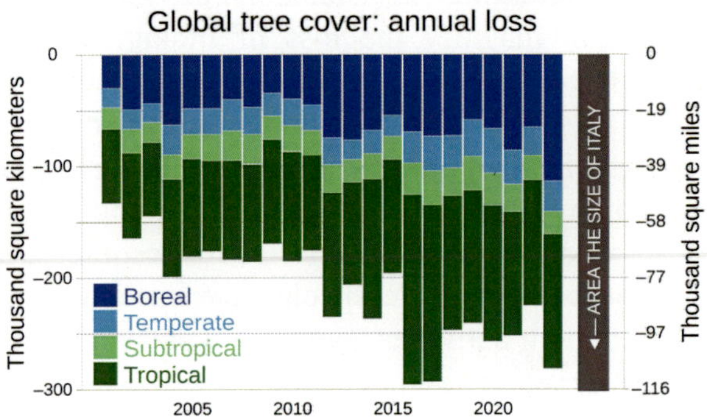

Figure 5.3. *Habitat destruction.*

Source: By RCraig09 – Own work, CC BY-SA 4.0, https://commons.wikimedia.org/w/index.php?curid=102770581.

A lack of genetic variability can leave species more susceptible to diseases and limit their evolutionary potential, effectively weakening the entire ecosystem that relies on their presence. Consequently, conservation strategies must prioritize not only the protection of existing habitats but also the restoration of connectivity between fragmented landscapes. Addressing habitat destruction and fragmentation requires coordinated efforts from various stakeholders, including governments, local communities, and conservation organizations. Effective management strategies are essential to mitigate these threats and promote biodiversity conservation. Such strategies may include creating wildlife corridors that facilitate species movement across fragmented landscapes, implementing sustainable land-use practices, and engaging in the restoration of degraded habitats. By fostering collaboration among diverse groups and incorporating traditional ecological knowledge, conservation initiatives can enhance the resilience of ecosystems and support the recovery of affected species. Ultimately, acknowledging the interconnectedness of habitats and implementing proactive conservation measures are crucial to reversing the trend of habitat destruction, ensuring that both biodiversity and human well-being can thrive in the face of ongoing environmental challenges (Gann et al., 2019; Tickner et al., 2020).

5.3.2. Climate Change Impacts

The relentless progression of climate change has far-reaching implications for biodiversity, particularly in fragile ecosystems like the Arctic. The rapid alteration of weather patterns threatens iconic species such as the polar bear, whose survival is intricately linked to melting sea ice and declining prey availability. As detailed in the findings from the conference summarizing the effects on polar bear habitats, these changes disrupt their natural behaviors, ranging from hunting to migration patterns. This ecological imbalance illustrates how climate change endangers specific species and weakens the entire ecosystem. As temperatures rise and habitats shift, the intricate relationships that define these environments are unraveled, leading to a gradual loss of biodiversity that could have cascading effects on broader ecological networks (Wang, 2024). In addition to direct effects on species, climate change complicates existing conservation efforts through the introduction of new challenges for management practices. The concept of marine protected areas (MPAs) is pivotal in preserving biodiversity; however, they are increasingly threatened by ocean warming and acidification. An analysis of MPA management plans revealed a stark underrepresentation of on-the-ground adaptation measures specifically focused on combating climate change impacts. With only 4.7% of documented actions aimed at enhancing biodiversity conservation, there exists a critical "biodiversity adaptation gap" that needs to be addressed (Corelli et al., 2024). This gap signals an urgent need for new strategies and collaborative efforts, enabling MPA managers to effectively address the climate-induced changes affecting marine ecosystems. As we continue to witness the profound impacts of climate change on biodiversity globally, the need for sophisticated and integrated conservation approaches becomes ever more apparent. Enhanced research into species adaptability and resilience, particularly among vulnerable populations like polar bears, is vital for formulating effective conservation policies. Similarly, strengthening the adaptive capacities of MPAs through increased monitoring and specific action plans will be essential to mitigate the adverse effects of climate change. By recognizing the interconnectedness of climate change and biodiversity, we can foster more effective conservation initiatives that aim to protect individual species and enhance the overall resilience of ecosystems in the face of inevitable environmental changes. The urgency to act is clear, as the health of our planet relies heavily on our ability to safeguard its rich biodiversity from the escalating threats posed by a warming world (Figure 5.4).

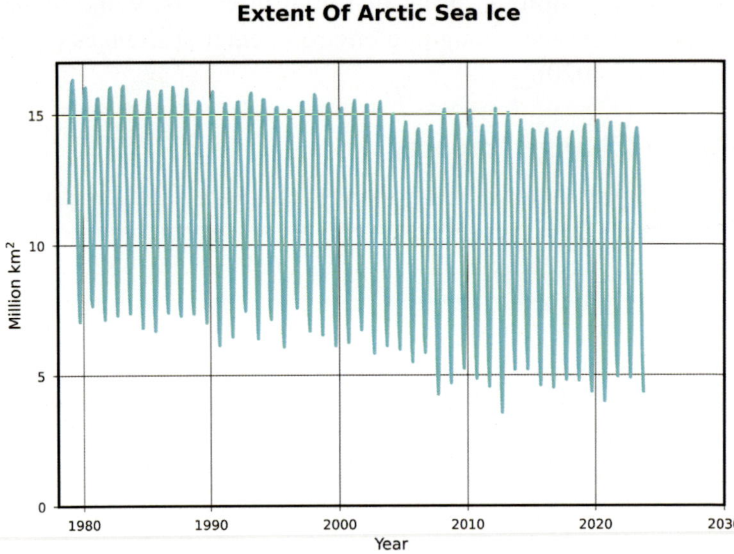

Figure 5.4. *The impact of climate change on biodiversity.*

Source: By Dominiklenne – Own work, Public Domain, https://commons.wikimedia.org/w/index.php?curid=11321458.

5.3.3. Invasive Species and Their Effects

The introduction of invasive species into new ecosystems poses significant threats to biodiversity and ecosystem stability. These non-native organisms often outcompete local flora and fauna for resources, leading to declines or extinctions of indigenous species. According to research, invasive species account for over 40% of the global extinction events documented since the 1600s. This disruption not only diminishes species richness but also alters habitat structures and ecosystem functions, which can lead to cascading effects throughout the trophic levels. In many cases, the invasive species lack natural predators in their new environments, allowing their populations to proliferate unchecked, further exacerbating the strain on local ecosystems. The expansive footprint of invasive species underscores the urgent need for effective management strategies and conservation efforts aimed at protecting vulnerable native populations.

Furthermore, the economic implications of invasive species are profound. Invasive organisms can inflict damage on agriculture, fisheries, and natural resources, costing economies hundreds of billions of dollars annually. For instance, species like the zebra mussel and the Burmese python have altered local waterways and ecosystems, affecting industries that rely on these environments. As local species dwindle and ecosystems are compromised, communities reliant on these resources for their livelihoods suffer. In addition, invasive species can introduce new diseases and parasites that threaten both wildlife and human health. This dual challenge of ecosystem preservation and economic stability highlights the need for a multifaceted approach to managing invasive species, as the failure to do so can lead to both ecological and economic crises. Addressing the issue of invasive species requires collaboration and a commitment to conservation initiatives. Effective management involves early detection and rapid response systems that can mitigate the introduction and spread of these organisms. Public education and awareness campaigns can also enhance community involvement in monitoring local ecosystems. Organizations that focus on biodiversity conservation and habitat restoration, as noted in the IPCC Sixth Assessment Report (Intergovernmental Panel on Climate Change, 2023), emphasize the importance of integrated strategies that combine ecological understanding with societal needs. The current trends of biodiversity loss, driven by factors including invasive species, call for urgent action. Only through comprehensive and collaborative efforts can we hope to safeguard ecosystems and ensure the viability of both natural and human communities for future generations (Sandra Díaz et al., 2019).

5.4. CONSERVATION STRATEGIES

Implementing effective conservation strategies is crucial in the ongoing fight against biodiversity loss. One approach is the expansion of protected areas to safeguard habitats and promote ecosystem resilience. Establishing these conservation zones can provide a buffer against both human exploitation and environmental changes induced by climate fluctuations. However, simply increasing the number of reserves is not enough. It requires a coordinated management plan that incorporates the ecological needs of threatened species while addressing the social dimensions of land use. Such balanced strategies ensure that conservation efforts are sustainable and widely supported by local communities, ultimately facilitating the long-term preservation of biodiversity. Collaboration between different stakeholders is vital in developing and implementing successful conservation strategies (Figure 5.5).

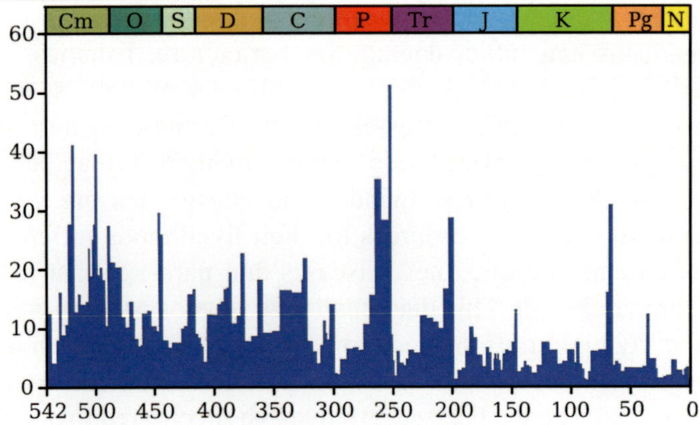

Figure 5.5. *Conservation is needed for biodiversity.*

Source: https://en.wikipedia.org/wiki/File:Extinction_intensity.svg.

Engagement with local communities, researchers, policymakers, and non-governmental organizations fosters a more inclusive approach to biodiversity management. For instance, the integration of indigenous knowledge into conservation planning often uncovers sustainable practices that have been effective for centuries. The importance of this collaborative effort is underscored in recent findings, suggesting that reversing declines in biodiversity requires an ambitious and coordinated effort across sectors, including agriculture and habitat management (Leclère et al., 2020). By forging alliances that unite various interests, conservation initiatives can tackle complex environmental issues more effectively, ensuring that protection efforts yield meaningful results. Lastly, adaptive management plays a critical role in the effectiveness of conservation strategies. As environmental conditions change due to factors like climate change, conservation practices must evolve accordingly. Monitoring biodiversity and ecosystem responses allows for timely adjustments to management plans, making them more responsive to emerging challenges. For example, the meta-analysis revealed that biodiversity enhances ecosystem functioning, particularly under stressful conditions brought on by global change (Hong et al., 2021). This adaptability is key to not only preserving biodiversity but also maintaining ecosystem services essential for human well-being. Thus, incorporating flexible strategies that can respond to new data and shifting ecological landscapes is fundamental for robust conservation efforts in the face of rapid environmental change.

5.4.1. Protected Areas and National Parks

Protected areas and national parks play a vital role in the conservation of ecosystems and biodiversity. These designated regions serve as sanctuaries for threatened species, enabling them to thrive in their natural environments without the pressures of urbanization and industrialization. They create a space for essential ecological processes, such as nutrient cycling and natural hydrology, which are critical for maintaining healthy ecosystems. Furthermore, protected areas often promote research and education, fostering a greater understanding of biodiversity and encouraging stewardship among local communities. The efficacy of these areas hinges on effective management strategies that address threats such as encroachment, habitat destruction, and climate change. Challenges surrounding the management of protected areas are increasingly complex, as evidenced by findings from Bhawal National Park in Bangladesh. This park faces significant threats, including proximity to urban development and industrial pressures, which exacerbate environmental degradation. The study highlights critical issues such as fuelwood collection, forest land encroachment, and agricultural expansion, all of which contribute to deforestation and biodiversity loss. The alarming reduction of plant and wildlife species, by roughly one-third and one-half respectively, underscores the urgent need for adaptive management strategies that can mitigate these pressures and support conservation efforts in protected areas.

Addressing these challenges requires a concerted effort involving local communities, policymakers, and conservation organizations to develop a cohesive framework for managing these vital ecosystems effectively. To ensure the longevity of biodiversity in protected areas, innovative approaches to conservation and stakeholder involvement are crucial. Creating partnerships with local communities can empower them to take active roles in protecting their environments while creating sustainable livelihoods. Furthermore, integrating scientific research into conservation strategies can yield valuable insights into species behaviors and ecosystem dynamics. By employing data-driven management practices, such as those suggested by the results of the Bhawal National Park study, conservationists can address specific threats more effectively. Ultimately, the future of national parks and protected areas hinges on our ability to blend traditional conservation methods with modern strategies, thereby creating resilient ecosystems that can withstand environmental pressures and thrive for generations to come.

5.4.2. Restoration Ecology Initiatives

A pivotal aspect of restoration ecology initiatives is their ability to enhance ecosystem health by promoting biodiversity recovery in degraded areas. In Mozambique, comprehensive biodiversity conservation strategies have yielded notable positive outcomes, particularly through the management of protected areas and community-centered initiatives. These efforts have led to significant species recovery and habitat preservation, demonstrating that collaborative approaches can foster resilience within ecosystems. The findings from the desk study underscore the importance of integrating community-based conservation efforts that empower local populations, thereby ensuring sustainable long-term stewardship of natural resources. Such measures restore ecological integrity and reinforce the connection between local communities and their environments, fostering a mutual commitment to conservation efforts (Ngwenya, 2024). The challenges encountered in restoration ecology highlight the need for adaptive management strategies that effectively address ongoing threats. Habitat fragmentation, pollution, and climate change can undermine restoration goals, necessitating innovative approaches that adapt to dynamic environmental conditions. For instance, prescribed burns in the Midwest serve as a restoration technique that opens up dense forest canopies, creating conducive conditions for certain bird populations to thrive. Research indicates that these practices have enhanced nesting success among various avian species connected to disturbed habitats compared to their undisturbed counterparts (Wilderness, 2007). Such evidence suggests that incorporating natural disturbance regimes into management plans can yield beneficial outcomes for both restoration and biodiversity conservation. By addressing ecological challenges through informed initiatives, restoration ecology can contribute meaningfully to the overall health and stability of ecosystems. Public policy plays a crucial role in the success of restoration ecology initiatives, requiring collaboration across multiple sectors. Integrating biodiversity conservation into national and local development strategies is essential for promoting resource sustainability and ecosystem resilience. As demonstrated in Mozambique, successful conservation initiatives necessitate alignment with broader social and economic policies, ensuring that the needs and rights of local communities are prioritized. Furthermore, adopting evidence-based practices, as seen in the reforms of oak savanna management, can optimize restoration outcomes while mitigating adverse ecological impacts (Wilderness, 2007; Ngwenya, 2024)). The effective intertwining of science, policy, and community engagement can create a robust framework for restoration ecology. Ultimately, fostering partnerships among stakeholders will enhance global efforts to combat biodiversity loss and restore vital ecosystems, making significant strides toward achieving long-term conservation goals.

5.4.3. Sustainable Resource Management Practices

The interconnectedness of global food systems and environmental sustainability underscores the importance of sustainable resource management practices. As the population climbs toward an estimated 9.1 billion by 2050, food production must increase without compromising ecological integrity. This can be achieved through innovative agricultural techniques that prioritize efficiency and conservation. Notably, integrating conservation agriculture can enhance resource use efficiency, ensuring that more food is produced with less environmental degradation. A multifaceted approach that emphasizes the restoration of marginal lands is crucial, as these areas often harbor untapped potential for sustainable agricultural practices. By applying specific crop management strategies tailored to local soil conditions, the agricultural sector can foster both productivity and resilience, thereby contributing to biodiversity conservation efforts within ecosystems impacted by intensive farming practices (Willett et al., 2019). Another vital aspect of sustainable resource management is the reduction of food waste, which directly ties into achieving food security while minimizing the environmental footprint. More than 820 million people currently face hunger, illustrating the disparity between food production and distribution (Willett et al., 2019) (Figure 5.6).

Figure 5.6. *Sustainable resource management.*

Source: https://img.freepik.com/premium-photo/sustainable-resource-management_810293-12724.jpg.

Implementing better management practices across the food supply chain can alleviate this crisis. Strategies such as improving supply chain logistics, increasing consumer awareness about sustainable consumption, and repurposing surplus food can play pivotal roles. Higher efficiency in food systems not only ensures that resources are utilized judiciously but also reduces the overall demand for additional agricultural land, thus preserving vital ecosystems. By confronting the issue of food waste head-on, sustainable management practices can create a ripple effect, fostering both economic stability and environmental preservation. Finally, the promotion of sustainable resource management requires a reorientation of policies that incentivize ecological stewardship. Governments and organizations must create frameworks that encourage practices such as agroecology, organic farming, and regenerative agriculture, all of which enhance biodiversity while sustaining yields. The transformation towards more sustainable agricultural systems is not merely a choice but rather a necessity to fulfill the dual objectives of food security and conservation. Policies should focus on collaborative efforts between stakeholders, integrating innovative technologies, and fostering community engagement in conservation initiatives. By adhering to these principles and emphasizing their applicability to both present and future agricultural practices, society can make significant strides toward a more sustainable coexistence with our natural environment (Schröder et al., 2017).

5.5. THE ROLE OF LEGISLATION IN CONSERVATION

Legislation plays a pivotal role in conservation efforts by establishing frameworks that govern the use and protection of natural resources. Effective laws can create protected areas, regulate resource extraction, and enforce penalties for ecological violations, thereby safeguarding biodiversity. Countries accountable under international agreements, such as the Convention on Biological Diversity, often enact domestic laws that reflect these commitments. These regulations aim to preserve endangered species and ensure the sustainable management of landscapes and ecosystems essential for the well-being of local communities. For instance, when conservation policies restrict the interaction of Indigenous peoples and local communities (IPLC) with their environments, there can be long-term consequences related to knowledge loss associated with resource management. Thus, legislation serves as both a protective mechanism and a guiding framework fostering responsible stewardship of natural resources. Moreover, the effectiveness of conservation legislation is often tested by real-world implementation challenges. Notably, laws can sometimes inadvertently disconnect IPLC from their traditional lands, as certain protective measures may limit access to vital resources. This can lead to cultural and knowledge erosion and affect biodiversity conservation, as IPLCs are typically stewards of their environments. Such policies, when overly restrictive, can warp local ecological knowledge, forcing communities to adapt to new, often less effective methods of resource management (Pier Paolo Gatta, 2022). A critical analysis of these frameworks suggests that successful legislation must account for the intrinsic relationships between communities and their environments, allowing for adaptive management that reflects the realities of social-ecological systems. Therefore, a balanced approach prioritizing community empowerment in conjunction with legislative measures can enhance the effectiveness of conservation efforts. Additionally, the evolving dynamics of climate change necessitate updating and strengthening conservation laws to address emerging challenges. As environmental conditions shift, the legal frameworks that once provided stability may require modification to incorporate adaptive strategies that reflect changing ecosystems. For instance, recognizing the rights and practices of Indigenous peoples in legislative provisions can lead to more resilient conservation outcomes, as traditional ecological knowledge often holds keys to sustainable management practices (Pier Paolo Gatta, 2022). The interplay between legislation and community practices is crucial for fostering an environment where conservation efforts can thrive amidst ongoing ecological pressures. Legislation that embraces flexibility and responsiveness can help mitigate the rigidity that leads to biocultural hysteresis, ensuring that both ecosystems and communities can adapt together. Ultimately, robust conservation legislation must continuously evolve to be relevant and effective, supporting both biodiversity and the cultural heritage of IPLC (Figure 5.7).

Figure 5.7. *Convention on biological diversity (CBD).*

Source: By Continentaleurope – Own work, CC BY-SA 4.0, https://commons.wikimedia.org/w/index.php?curid=53779386.

5.5.1. International Agreements and Treaties

The significance of international agreements and treaties has become increasingly evident in the context of environmental conservation and biodiversity protection. These legal frameworks offer vital platforms for nations to collaborate on pressing global issues, such as climate change, marine biodiversity, and the sustainable use of non-renewable resources. Treaties like the Convention on Biological Diversity (CBD) and the Paris Agreement exemplify how countries can align their efforts, establishing common goals and shared responsibilities. However, the effectiveness of these agreements often hinges on the participating nations' commitment to adhere to stipulated guidelines and implement necessary local policies. As negotiations continue, it is essential to assess not just the legal language but the political will behind these treaties, which ultimately determines their success in promoting conservation efforts globally.

The ongoing negotiations surrounding the new legally binding instrument for marine biodiversity beyond national jurisdiction (BBNJ) exemplify the complexities associated with international accords. With over one hundred governments involved, discussions cover critical areas such as marine genetic resources, area-based management tools, and environmental impact assessments (Arne Langlet et al., 2022). Such multifaceted deliberations highlight the need for a cohesive strategy that entails not just biodiversity conservation, but also fair access to marine resources and equitable sharing of benefits. The emergence of a BBNJ regime complex, comprised of numerous international organizations (IOs), will play a pivotal role in implementation, as these entities often possess unique expertise and protocols that can enhance treaty effectiveness (Langlet et al., 2022). Consequently, the interrelation between these organizations and the BBNJ will shape the future of marine conservation efforts.

As international treaties evolve, the importance of maintaining robust science-policy interfaces cannot be overstated. These connections facilitate the flow of research and evidence that inform treaty negotiations and strengthen regulatory frameworks. For instance, the systematic review of BBNJ-related literature underscores the necessity of addressing knowledge gaps in marine biodiversity conservation (-von Wysocki et al., 2020). Engaging scientists to provide data-driven insights allows policymakers to craft effective strategies based on empirical evidence. Moreover, emphasizing transformative change, highlighted in the existing academic discourse, reinforces the call for adaptive management approaches in marine environments. Ultimately, fostering collaboration between scientists, policymakers, and stakeholders is crucial, as it creates a dynamic framework that supports the overarching goal of preserving biodiversity through cooperative international agreements and treaties.

5.5.2. National Laws and Policies

The role of national laws and policies in conservation and biodiversity is paramount, as these frameworks set the stage for how a country manages its natural resources. Effective legislation can facilitate the protection of ecosystems and regulate activities that pose a threat to biodiversity. For instance, policies can be designed to enhance above- and below-ground species diversity, reduce synthetic inputs, and implement sustainable agricultural practices, all of which are crucial for achieving ecological intensification (Garibaldi et al., 2019). Administrative measures, such as the promotion of conservation easements or protected areas, can also serve to safeguard crucial habitats and maintain ecosystem functionality. Ultimately, without strong legal frameworks, efforts to conserve biodiversity risk becoming disorganized and ineffective, hence the imperative for nations to devise comprehensive policies that align with local and global conservation goals. Moreover, the implementation of national laws fosters a collective responsibility among citizens, businesses, and government agencies. By establishing legal standards, governments can incentivize sustainable practices, such as developing resilient food systems that prioritize nutrition and environmental health. As highlighted by many studies, globally integrated food policies can directly impact food security and public health, as well as minimize negative externalities, including environmental degradation and public health crises (Willett et al., 2019). Furthermore, engaging local communities in the creation and enforcement of these laws can lead to more effective conservation outcomes. This participatory approach ensures that policies address the unique ecological and socio-economic contexts of different regions, allowing for tailored solutions that resonate with local populations. Consequently, national laws not only set a regulatory framework but also promote a societal ethos of conservation stewardship. Finally, evaluating the effectiveness of national laws and policies in conservation and biodiversity is essential for continuous improvement and accountability. Regular assessments can identify gaps in legislation and highlight successful initiatives that can be replicated or scaled. As outlined in the concept of ecological intensification, policies should also consider multiple benefits, understanding that enhanced ecosystem services often lead to increased agricultural productivity and improved livelihoods (Garibaldi et al., 2019). Moreover, tracking the long-term impacts of conservation strategies can guide future legislative efforts, ensuring that they remain responsive to changing ecological conditions and societal needs. By fostering a culture of adaptive management and learning, nations can better address the dynamic challenges posed by biodiversity loss and environmental degradation, ultimately moving towards a sustainable future where both people and nature can thrive.

5.5.3. Local Governance and Community Involvement

Local governance plays a crucial role in promoting biodiversity conservation at the community level. By engaging local stakeholders in decision-making processes, governance structures can tailor conservation strategies to the unique ecological, social, and economic contexts of specific regions. This localized approach fosters a sense of ownership among community members, who are often more invested in preserving their surrounding environment. When local governments involve citizens in discussions around resource management, they can leverage indigenous knowledge and community insights, which can result in more effective and sustainable conservation practices. As outlined in the IPCC Sixth Assessment Report, integrating community perspectives into climate resilience strategies not only aids in biodiversity preservation but also supports the development of adaptive measures that can benefit local populations (Intergovernmental Panel on Climate Change, 2023). However, the effectiveness of local governance is often diminished by gaps in stakeholder engagement, particularly concerning environmental issues like microplastic pollution. Research highlights the inadequacy of current management strategies, which often overlook the importance of community participation and citizen science initiatives (Onyena et al., 2021). By failing to incorporate community input, local governments risk implementing policies that do not resonate with or address the actual challenges faced by their constituents. Furthermore, the lack of collaborative approaches can hinder the effectiveness of conservation efforts. Establishing partnerships among local governments, NGOs, and citizen groups is essential for harnessing collective action and fostering shared responsibility for biodiversity. Such alliances can lead to more impactful conservation outcomes while simultaneously empowering communities to take charge of their environmental stewardship. Ultimately, enhancing local governance through community involvement is vital for the success of conservation efforts. Initiatives that encourage active participation can bridge the gap between policy and practice, ensuring that conservation measures are grounded in local realities. Educational campaigns and strategic incentives can galvanize communities to reduce their environmental footprint and advocate for stronger protective measures. By prioritizing community engagement in conservation policies and practices, local governments can address pressing issues such as microplastic leakage and biodiversity loss more effectively. This integrative approach not only aids in restoring ecosystems but also fosters a culture of sustainability that benefits future generations (Intergovernmental Panel on Climate Change, 2023; Onyena et al., 2021). Thus, creating opportunities for active involvement can empower local citizens to become stewards of their natural resources, ultimately enhancing both conservation success and community resilience.

5.6. THE IMPACT OF HUMAN ACTIVITY

Human activities have significantly altered the natural landscape, resulting in profound consequences for biodiversity. Urbanization, deforestation, and agricultural expansion are primary drivers of habitat destruction, which in turn leads to the loss of species and ecosystems. Industrial practices often disrupt migratory patterns and breeding grounds, causing a decline in populations of various organisms. The processes within ecosystems become increasingly homogenized due to such human interventions, resulting in a loss of species richness. Furthermore, the increased fragmentation of habitats leads to isolated populations, which diminishes genetic diversity and adapting capacity over time.

This interconnectedness of ecosystems highlights the urgent need for sustainable practices that balance human development and conservation efforts. The increasing prevalence of invasive species is another critical consequence of human activity that compounds the challenges faced by conservationists. Global trade and travel have facilitated the movement of species beyond their native habitats, where they often outcompete indigenous organisms for resources. This biotic homogenization can lead to a decrease in resilience among ecosystems, as native species struggle against the pressures of competition and predation introduced by invasive species. The importance of understanding beta-diversity emphasizes the necessity of assessing species assemblages to gauge the impacts of these introductions (Corlett et al., 2020).

Effective management strategies must consider the intricate relationships between species, as well as the potential for invasive species to disrupt the ecological balance, further endangering regional biodiversity. Additionally, the repercussions of climate change, primarily driven by human activities, pose a looming threat to biodiversity. Rising temperatures, altered precipitation patterns, and extreme weather events create conditions that challenge species' survival and adaptive capability. Ecosystems that once thrived under stable conditions are now faced with unprecedented stressors, contributing to shifts in species distributions and increased extinction rates.

The COVID-19 pandemic, as highlighted in Corlett et al (2020), has also sparked a conversation about the links between human health, biodiversity loss, and emerging diseases. The connection between habitat destruction and zoonotic diseases underscores the importance of preserving biodiversity not just for ecological reasons, but also for public health. Overall, the impact of human activity extends far beyond immediate effects; it disrupts the delicate balance of ecosystems, necessitating urgent conservation strategies to mitigate these risks (Figure 5.8).

Figure 5.8. *Human impact on ecosystems.*

Source: By NASA, (originally uploaded to en: Wikipedia by http://en.wikipedia.org/wiki/User:Seth_Ilys) – http://eol.jsc.nasa.gov/scripts/sseop/photo.pl?mission=ISS002&roll=E&frame=5654, Public Domain, https://commons.wikimedia.org/w/index.php?curid=2897139.

5.6.1. Urbanization and Its Effects on Ecosystems

The rapid transformation of landscapes driven by urbanization presents significant challenges to local ecosystems. Urban environments often lead to habitat fragmentation, where natural habitats are divided into smaller, isolated patches. This can create barriers for species that require larger, continuous areas for migration, breeding, and foraging. As cities expand, the high density of built structures typically replaces vast areas of vegetation, resulting in a decline in biodiversity. The negative impact of urbanization on terrestrial arthropods, as shown in studies revealing diminished diversity and abundance in more urbanized landscapes, underscores the severity of this issue (María Silvina Fenoglio et al., 2020). With several taxonomic groups, particularly sensitive ones like Coleoptera and Lepidoptera, exhibiting the most significant declines, the ecological balance within urban ecosystems becomes precarious. Additional concerns arise from the changes in ecological interactions that accompany urban growth. Urbanization not only reduces habitat availability but often alters environmental conditions including temperature, moisture, and light, which can disrupt existing biotic relationships. For instance, the presence of pollutants and increased impervious surfaces in cities can create harsh conditions that some species cannot withstand.

As noted in the CityScapeLab Berlin, understanding the intricate dynamics between different taxa and their responses to these urban conditions is essential for effective biodiversity conservation (Moritz von der Lippe et al., 2020). This platform exemplifies how controlled studies can reveal the nuances of urban ecosystems, allowing researchers to isolate urbanization effects from specific habitat influences, thereby paving the way for informed conservation strategies. Ultimately, the ramifications of urbanization extend beyond mere species loss; they encompass broader ecological functions essential for maintaining ecosystem health. Urban areas that lose biodiversity often experience a decline in vital ecosystem services, such as pollination, soil health, and water filtration. These services are not only crucial for the surviving flora and fauna but also significantly impact human well-being. As evident from the ongoing biodiversity crisis, urban planners and conservationists must prioritize the integration of green spaces and biodiversity-friendly designs into city landscapes to mitigate the adverse impacts of urbanization. By fostering healthier urban ecosystems, it becomes possible to create a sustainable coexistence between urban growth and biodiversity, ensuring that cities can thrive while preserving vital ecological functions.

5.6.2. Agriculture and Biodiversity Loss

Agriculture, while essential for feeding the growing global population, has become a significant driver of biodiversity loss through practices that prioritize short-term yields over ecological balance. Intensive farming methods, such as monoculture and excessive use of chemical fertilizers and pesticides, not only diminish soil health but also disrupt the habitat of various species. As forests are cleared to make way for agricultural expansion, vital ecosystems are destroyed, leading to the extinction of numerous plant and animal species. According to research, land degradation in Africa, exacerbated by rising agricultural demands and urbanization, has already affected 65% of the continent, showcasing the broad impact of agricultural practices on biodiversity (Ayeni et al., 2024). This reduction in biodiversity undermines the resilience of ecosystems, making them more susceptible to diseases and climate change, ultimately threatening agricultural productivity itself. The interplay between agriculture and biodiversity is complex, underscoring the need for sustainable farming practices that can mitigate these adverse effects. One promising approach is the promotion of agroecological methods, which integrate traditional farming knowledge with modern science to enhance biodiversity while maintaining crop yields. Practices such as crop rotation, polyculture, and organic farming not only foster a healthier ecosystem but also improve soil fertility and reduce the reliance on chemical inputs. Ayeni et al (2024) emphasize that the adoption of sustainable practices helps in carbon sequestration and minimizes the carbon footprints associated with conventional agriculture. By focusing on techniques that support both agricultural productivity and biodiversity, farmers can create systems that are resilient, productive, and environmentally sound, thereby contributing to broader conservation goals. Ultimately, addressing the challenges between agriculture and biodiversity loss necessitates collaborative efforts among stakeholders, including government, farmers, and communities. Effective policy frameworks that incentivize sustainable agricultural practices alongside rigorous environmental regulations are crucial in this endeavor. Education and outreach programs can empower farmers with the knowledge needed to implement biodiversity-friendly practices, promoting conservation at the grassroots level. Additionally, Community-Based Natural Resource Management (CBNRM) initiatives can play a vital role in fostering collaboration among local populations to manage resources sustainably. By pursuing integrated approaches that prioritize biodiversity, we can not only safeguard ecosystems but also secure the agricultural systems that support human life, ultimately achieving a harmonious balance between nature and food production (Arya Arun, 2024) (Figure 5.9).

Figure 5.9. *Biodiversity loss.*

Source: By Schlegel – Own work, CC BY-SA 4.0, https://commons.wikimedia.org/w/index.php?curid=3176252s.

5.6.3. Pollution and Its Consequences

The pervasive issue of pollution presents significant threats to both biodiversity and human health. Air, water, and soil pollution contribute to the degradation of ecosystems, leading to a decline in plant and animal species. The introduction of harmful chemicals and waste into natural habitats disrupts intricate relationships within ecosystems, often resulting in the extinction of vulnerable species. Moreover, air pollution specifically has been linked with a range of serious health issues, including respiratory diseases and cardiovascular problems. This degradation of biodiversity not only compromises the resilience of ecosystems but also undermines the essential services they provide, such as clean air, water purification, and pollination. Additionally, the consequences of pollution extend beyond immediate environmental impacts to affect food systems globally. The Intergovernmental Panel on Climate Change highlights the potential of food systems to support both human health and environmental sustainability; however, they are placed under severe stress due to pollution (Willett et al., 2019). Contaminated water sources and soil can compromise agricultural productivity, affecting the quality of food produced and limiting access to nutritious diets. These challenges often leave communities vulnerable, resulting in significant health disparities, particularly within low-income populations. Addressing the repercussions of pollution in agriculture is crucial not only for ensuring food security but also for promoting ecological health and preventing further biodiversity loss. Ultimately, these multifaceted consequences necessitate an urgent and integrated approach to conservation and sustainability. As ecosystems suffer under the weight of pollutants, the interconnected nature of pollution, health, and biodiversity becomes increasingly clear. The IPCC asserts that climate-resilient development can mitigate some of these adverse effects through effective pollution control and adaptive strategies (Intergovernmental Panel on Climate Change, 2023). Creating policies that reduce pollutive practices, promote sustainable agriculture, and enhance conservation efforts is imperative. By prioritizing these actions, society can pave the way toward healthier environments, improved human health, and a more balanced coexistence between humanity and the biodiversity that sustains life on Earth.

5.7. CASE STUDIES IN CONSERVATION

Conservation efforts often draw from successful case studies to create effective strategies for protecting biodiversity. The myriad of challenges faced by ecosystems today necessitate adaptive management practices that are informed by previous successes and failures. For instance, the restoration project on the Elwha River in Washington State represents a landmark case in removing dams to revive natural river flows and fish habitats. This initiative resulted in the resurgence of native salmon populations, demonstrating that restoring ecological functions can lead to significant biodiversity recovery. By analyzing projects like this, conservationists can develop frameworks that promote similar restoration efforts across different ecosystems, thereby enhancing ecological resilience and maintaining biodiversity. The integration of sustainable practices into various industries can also serve as a model for conservation. A notable example is Longwood Gardens' commitment to sustainability, which is reflected in its innovative approach to water management and carbon footprint reduction (Turner Skoff et al., 2024). The Gardens utilize green infrastructure alongside traditional engineering methods to ensure water conservation and quality. Their comprehensive strategies highlight the potential for gardens and natural spaces to lead the way in promoting environmental stewardship and biodiversity. By focusing on material circularity and habitat preservation, such case studies provide practical insights for other organizations aiming to integrate conservation principles into their operational frameworks. Lastly, hydropower, while a crucial source of renewable energy, presents unique conservation challenges that need addressing to protect aquatic biodiversity. The Aswan High Dam exemplifies the complex balance between energy production and ecological health, as it altered the natural flow regime of the Nile River, impacting fish populations and habitat integrity (Yang, 2024). By examining these ecologically significant changes, researchers, and policymakers can identify effective management strategies that mitigate negative impacts, such as implementing fish passage facilities and habitat restoration efforts. Thus, the lessons drawn from such case studies reveal the importance of collaborative approaches in conserving biodiversity while meeting the energy needs of human populations.

5.7.1. Successful Conservation Programs

A fundamental aspect of successful conservation programs is the integration of local communities into the conservation process. Engaging local stakeholders ensures that conservation efforts are culturally relevant and economically beneficial. As evident from the rich biodiversity of Nigeria, conservation strategies must not only focus on ecological preservation but also on socio-economic factors. For instance, effective programs often involve community-based initiatives that educate and empower locals to participate actively in conservation efforts. This approach fosters a sense of ownership and responsibility towards natural resources, which is critical for long-term sustainability. By aligning conservation goals with the economic needs and cultural practices of local populations, programs can see improved compliance and engagement, ultimately leading to more significant positive outcomes in biodiversity preservation (Ugboju et al., 2023).

The effectiveness of conservation initiatives is often bolstered through strategic partnerships between various organizations. The North Carolina Zoo exemplifies this collaborative approach by partnering with local and state agencies, non-profit organizations, and academic institutions to strengthen its conservation programs. Such collaborations allow for the pooling of resources, sharing of expertise, and implementation of comprehensive strategies that can address diverse conservation challenges. By leveraging the strengths of multiple stakeholders, these programs can scale their efforts and increase their impact on local biodiversity. The success of these partnerships highlights the necessity of interdisciplinary cooperation in conservation efforts, where diverse organizational capabilities contribute to the shared goal of protecting wildlife and their habitats effectively (Roznik et al., 2023). Adaptive management and innovative practices are also essential components of successful conservation programs. With the ongoing challenges posed by climate change and habitat degradation, programs must be flexible and responsive to new environmental data and changing circumstances. The research conducted in Nigeria reveals that while many conservation strategies show promise, they also require continual assessment and adjustments to remain effective amid evolving threats (Ugboju et al., 2023). Incorporating advanced technologies, such as remote sensing and genomic tools, can enhance conservation strategies by providing accurate data for decision-making. Furthermore, integrating technological advancements with community education initiatives amplifies the effectiveness of conservation efforts, fostering a culture of sustainability. Such a multifaceted approach not only improves immediate conservation outcomes but also strengthens resilience against future ecological challenges.

5.7.2. Lessons Learned from Failed Initiatives

Experiences from historical conservation initiatives reveal critical insights into what can lead to their failure. A common theme among these unsuccessful efforts is the neglect of local context and stakeholder engagement. Initiatives that imposed top-down solutions often ignored the socio-economic realities of indigenous communities and local populations. Effective conservation requires collaboration with these stakeholders, who possess essential knowledge of their ecosystems. For example, without adequate involvement from local communities, conservation measures may unintentionally disrupt livelihoods and foster resistance. This is evident in cases where large-scale reforestation was pursued without considering the existing agricultural practices of the communities, leading to conflict and non-compliance. Therefore, ensuring that conservation strategies are inclusive and respectful of local customs is fundamental to their success. Another contributing factor to the failure of past conservation initiatives has been a narrow focus solely on specific ecological outcomes, often at the expense of broader environmental and socio-economic considerations. Many programs prioritized afforestation or reforestation as a primary goal while overlooking the importance of protecting existing ecosystems, as evidenced by the emphasis on tree planting without adequate protections for standing forests. Such an approach risks not only failing to achieve biodiversity goals but also harming local ecosystems by transitioning naturally biodiverse areas into monoculture plantations, which often lack ecological resilience. Reports suggest that these misguided efforts have led to unintended consequences, such as increased competition for land and exacerbated food insecurity (Seymour, 2020). A more holistic approach that recognizes the interdependence of ecosystems and local communities is crucial for sustainable biodiversity conservation. Finally, the evaluation of failed initiatives underscores the importance of adaptive management and ongoing assessment in conservation strategies. Successful conservation is not a static process but requires continual monitoring, flexibility, and responsiveness to changing environmental conditions and community needs. Past efforts have often been implemented with fixed goals and insufficient frameworks for assessment and adjustment. For instance, projects that did not account for environmental changes, such as climate impacts, faced higher risks of failure. This necessitates a shift toward adaptive management approaches that prioritize learning, experimentation, and feedback loops in the conservation process (Maxwell et al., 2020). By fostering an environment where strategies can evolve based on real-world experiences, conservation initiatives stand a better chance of achieving their biodiversity objectives while simultaneously addressing the needs of communities and ecosystems.

5.7.3. Innovative Approaches to Biodiversity Conservation

Addressing the accelerating decline in global biodiversity requires innovative approaches that transcend traditional conservation methods. One such approach is the integration of technology in monitoring and managing ecosystems. Utilizing satellite imagery, machine learning, and drone technology can significantly enhance our ability to track changes in land use and habitat degradation—critical factors contributing to biodiversity loss. For instance, these technologies allow for real-time data collection, enabling conservationists to assess and respond to environmental changes more rapidly. This aligns with the view expressed in Geist (2021), where the decline of ecosystems and species abundance is noted as a pressing concern. By leveraging technology, conservation efforts can become more efficient and targeted, fostering resilience in ecosystems under threat from human activities. Another promising avenue lies in community-based conservation, which emphasizes local stewardship and engagement in biodiversity initiatives. Empowering local communities to manage natural resources ensures that conservation practices are culturally relevant and economically viable. Such participatory approaches not only increase the effectiveness of conservation strategies but also help to distribute the socioeconomic benefits of biodiversity sustainably. This is particularly important in regions facing pressures from infrastructure development and agricultural expansion (Geist, 2021). The success of these initiatives often hinges on building trust and collaboration between stakeholders, leading to more sustainable management practices that recognize both human and ecological needs. Lastly, embracing adaptive management frameworks can enhance the resilience of conservation strategies in the face of rapid environmental changes. These frameworks involve continual monitoring, evaluation, and adjustment of conservation practices based on new scientific findings and ecological shifts. By adopting an iterative process, conservationists can learn from both successes and failures, making informed decisions that improve biodiversity outcomes. This adaptive approach acknowledges the complexities and uncertainties inherent in managing ecosystems (Díaz et al., 2019). Ultimately, innovative conservation strategies that combine technology, community involvement, and adaptive management can not only mitigate biodiversity loss but also foster a deeper connection between people and nature, ensuring that both thrive in the future.

5.8. CONCLUSION

The pressing nature of conservation and biodiversity emphasizes the importance of implementing sustainable practices across food systems and ecological projects. As highlighted by recent findings, while global food production has managed to keep pace with population growth, it has often failed to consider the nutritional quality of diets, resulting in significant health issues such as obesity and non-communicable diseases (Willett et al., 2019). This underscores the need for an integrated approach that not only addresses food security but also enhances ecosystem resilience. Ultimately, promoting healthy diets through sustainable agricultural methods can create a dual benefit: nourishing the population while preserving the environment. Therefore, redefining food production with biodiversity in mind should be a central element of conservation efforts, ensuring that human wellness and ecological health are not at odds. When examining constructed wetlands as a strategy for biodiversity enhancement, the insights drawn from existing literature reveal a mixed landscape of outcomes. Although constructed wetlands (CWs) have been recognized for their potential to support biodiversity, many studies lack rigorous methodological foundations or consistent monitoring protocols (Wiegleb et al., 2017). This variability in research approaches limits the ability to draw comprehensive conclusions regarding the effectiveness of CWs in improving biodiversity. However, certain case studies indicate that when best practices are applied, such as ensuring appropriate landscape connectivity and adopting disturbance regimes tailored to the site, CWs can serve as valuable ecosystems within a wider conservation framework. Therefore, future projects must incorporate robust monitoring and evaluation processes to refine their strategies and maximize ecological benefits. In summary, the intersection of conservation practices and biodiversity necessitates a holistic understanding of how human activities impact ecological systems. The challenges presented by food insecurity and health-related issues demand that conservationists consider the broader implications of food systems on biodiversity (Willett et al., 2019). Simultaneously, constructing effective wetlands and other restorative projects must prioritize evidence-based practices to ensure ecological gains (Wiegleb et al., 2017). Ultimately, the conclusion drawn from the current discourse is that sustainable development must harmonize human and environmental objectives to foster resilience against ongoing global challenges. By prioritizing integrated strategies that align nutrition, health, and biodiversity, society can cultivate a future where both people and ecosystems thrive.

5.8.1. Summary of Key Points

Addressing the challenges of biodiversity conservation requires a multifaceted approach that engages diverse stakeholders. Recognizing the interconnectedness of species within ecosystems is essential for effective conservation efforts. For instance, studies on rare salamanders in Chongqing highlight the need for tailored conservation strategies that account for specific habitat preferences and climate conditions affecting these species. As the projections indicate, suitable habitat areas for the Wushan Salamander and others are expected to diminish significantly by 2090 due to climate change impacts and human disturbances (Ma et al., 2024). This underscores the importance of mitigating anthropogenic effects while preserving natural habitats to ensure the survival of vulnerable species. Incorporating scientific findings into conservation policies can provide a framework for fostering biodiversity in rapidly changing environments. Moreover, the integration of technology, particularly artificial intelligence, into conservation decision-making processes is gaining traction as a means to enhance resource allocation and planning. The AI-Assisted Decision Making for Conservation workshop underscored significant research questions that encompass both ecological and methodological advances (Xu et al., 2023). Modern conservation efforts can benefit from AI applications that analyze data to predict ecological outcomes, optimize habitat restoration, and manage species populations more effectively. By leveraging technology in this manner, conservationists can identify potential threats and prioritize interventions that yield the highest ecological benefits. Such informed approaches can ensure that conservation measures are scientifically sound and responsive to the ever-evolving challenges presented by climate change and habitat degradation. Ultimately, collaborative efforts among scientists, policymakers, and the public play a crucial role in the successful conservation of biodiversity. Engaging communities fosters a sense of stewardship and enhances public awareness of conservation issues such as habitat loss and climate change impacts. The implications of the studies on salamander distributions illustrate the urgency for collective action in protecting vulnerable species amidst ongoing environmental changes (Ma et al., 2024). Additionally, as highlighted in discussions during the conservation workshop, fostering interdisciplinary partnerships can bridge gaps between ecological research, technology, and community engagement strategies. By synthesizing diverse perspectives and innovations, conservation initiatives can be more robust and effective, ultimately contributing to the sustainable management of biodiversity for future generations.

5.8.2. The Future of Conservation Efforts

As we look toward the future of conservation efforts, it is evident that a multifaceted approach is essential for addressing the complex challenges facing biodiversity. Climate change, habitat loss, and invasive species are critical threats, necessitating strategies that go beyond traditional conservation techniques. In particular, protecting biodiversity hotspots—regions of high species richness and endemism—should be prioritized. For instance, the high-altitude areas of the Cretan mountains demonstrate both significant biodiversity and evidence of being under severe threat due to the projected "escalator to extinction" phenomenon. Such areas must be preserved not only for their unique flora but also for the evolutionary heritage they represent (Kougioumoutzis et al., 2020). By establishing focused conservation initiatives in these regions, we can create resilient ecosystems that endure even amid changing climatic conditions. Equally important is the recognition of the vital role that mutualistic plant-pollinator networks play in maintaining ecological balance. As noted in recent research, the interactions between wild bees and their floral hosts are crucial for pollination and, by extension, for food production and ecosystem stability (Mathiasson et al., 2020). Unfortunately, these networks are increasingly vulnerable to anthropogenic pressures and climate change, which have already led to the decline of native bee species and the loss of critical plant-pollinator interactions. The need for effective conservation strategies, particularly those focused on habitat restoration and the conservation of both pollinators and plants, will be pivotal in ensuring the functionality of these networks. By incorporating ecological research into conservation planning, we can better understand these network dynamics and build resilience against future disruptions. Ultimately, the future of conservation efforts hinges on our ability to integrate scientific research into practical applications that emphasize adaptive management and collaborative governance. Addressing the challenges posed by climate change requires not only safeguarding existing protected areas but also expanding and linking them through ecological corridors to enhance species mobility and genetic diversity. Such strategies, informed by data on endemism and extinction risk highlight the crucial need for proactive planning (Kougioumoutzis et al., 2020). Moreover, public awareness and community involvement are essential in fostering a conservation-minded society. By cultivating partnerships between governments, NGOs, and local communities, we can implement sustainable practices that will protect biodiversity for future generations. Only through such collective efforts will we ensure the resilience and richness of our natural environments in the face of ongoing environmental changes.

5.8.3. Call to Action for Individuals and Communities

Every individual has the power to spark change within their communities by adopting sustainable practices that promote conservation and protect biodiversity. Simple actions, such as reducing waste, conserving water, and utilizing energy-efficient appliances, can collectively lead to significant environmental improvements. By choosing to support local farms and businesses, individuals help bolster sustainable agricultural practices that are kinder to ecosystems. Moreover, educating oneself and others about the importance of conserving habitats and protecting endangered species can foster a culture of environmental awareness and responsibility. These small but impactful steps can energize community members and initiate broader discussions on conservation strategies. Communities play a pivotal role in the collective movement toward environmental protection. By organizing events such as clean-up drives, tree planting initiatives, and biodiversity awareness workshops, residents can unite in their efforts to safeguard their local environments. Schools, local governments, and non-profits should collaborate to create programs that educate the younger generation about the significance of biodiversity and the intricate relationships within ecosystems. Through such initiatives, communities can cultivate a sense of stewardship, where individuals feel personally invested in the health of their surroundings. Moreover, grassroots efforts can put pressure on policymakers to enact legislation that prioritizes conservation and environmental sustainability. Ultimately, the call to action extends beyond individual behavior and community engagement. Collective advocacy is essential for driving systemic change that addresses the root causes of biodiversity loss and environmental degradation. Collaborating with environmental organizations, attending town hall meetings, and voicing concerns through public forums can amplify community efforts and influence policy decisions. Furthermore, communities should lobby for the protection of local habitats and support efforts to create wildlife corridors and protected areas. By fostering a network of passionate advocates, individuals can ensure that awareness translates into action, contributing to a more sustainable future. This holistic approach underscores the shared responsibility we all bear in preserving our planet's rich biodiversity for generations to come.

REFERENCES

1. Adikwu, E., Abatan, A., Obaedo, B., Balogun, O., & Oluwatoyin, A. (2023). Assessing the effectiveness of biodiversity conservation strategies in Nigeria: A comprehensive review. *International Journal of Applied Research in Social Sciences, 5*, 577–598. https://doi.org/10.51594/ijarss.v5i10.665.
2. Alauddin, M., Hossain, M. N., Islam, M. B., Islam, S., & Islam, K. (2020). Management strategies for sustainable forest biodiversity conservation in protected areas of Bangladesh: A study of Bhawal National Park, Gazipur. *Natural Resources, 3*, 56–72. https://doi.org/10.33002/nr2581.6853.03035.
3. Arun, A. (2024). Assessing the efficacy of CBNRM frameworks in enhancing agricultural sustainability and conservation efforts. https://www.semanticscholar.org/paper/f66fe00455d93ab525e4b944b3a564dda85efaeb (accessed on 5 December 2024).
4. Ayeni, O., & Olagoke-Komolafe, O. E. (2024). Environmental impact of modern agricultural practices: Strategies for reducing carbon footprint and promoting conservation. https://www.semanticscholar.org/paper/33d21d88b882b5347f45f1419a6cb574c8d9e781 (accessed on 5 December 2024).
5. Bekele, M., Woldeyes, F., Lulekal, E., Bekele, T., & Demissew, S. (2022). Ethnobotanical investigation of medicinal plants in Buska Mountain range, Hamar district, Southwestern Ethiopia. *BMC Complementary Medicine and Therapies, 18*. https://doi.org/10.1186/s13002-022-00558-0.
6. Chicago Wilderness. (2007). *To the Forest Preserve District of Cook County: Ecosystem Restoration and the Viability of Bird Populations in the Chicago Wilderness*. https://www.semanticscholar.org/paper/11b3fc0c9bc633c28c759643e00517dea0bda756 (accessed on 5 December 2024).
7. Collins, C. M., Audusseau, H., Hassall, C., Keyghobadi, N., Sinu, P., & Saunders, M. E. (2024). Insect ecology and conservation in urban areas: An overview of knowledge and needs. *Insects, 17*, 169–181.
8. Corelli, V., Boerder, K., Hunter, K. L., Lavoie, I., & Tittensor, D. P. (2024). The biodiversity adaptation gap: Management actions for marine protected areas in the face of climate change. *Semantics Scholar*. https://www.semanticscholar.org/paper/b12ab1662ca833a27f147dffac69098b9a905357 (accessed on 5 December 2024).
9. Corlett, R. T., Primack, R. B., Devictor, V., Maas, B., Goswami, V. R., Bates, A. E., Koh, L. P., Regan, T. J., Loyola, R., Pakeman, R. J., Cumming, G. S., Pidgeon, A. M., Johns, D., & Roth, R. (2020). Impacts of the coronavirus

pandemic on biodiversity conservation. *Biological Conservation, 246*, 108571. https://doi.org/10.1016/j.biocon.2020.108571.

10. Del Socorro, L. F. (2023). Guardians of the green: A chapter on the impacts of climate change on forest ecosystems and its mitigation. https://www.semanticscholar.org/paper/12cf27c865aff9dc0c66a021d884275f3ead60f0 (accessed on 5 December 2024).
11. Díaz, S., Settele, J., Brondízio, E. S., Ngo, H. T., Agard, J., Arneth, A., et al., (2019). Pervasive human-driven decline of life on Earth points to the need for transformative change. *Science, 366*(6471), eaax3100. https://doi.org/10.1126/science.aax3100.
12. Duncan, C., Thompson, J., & Pettorelli, N. (2015). The quest for a mechanistic understanding of biodiversity–ecosystem services relationships. *Semantics Scholar*. https://www.semanticscholar.org/paper/e7e1122c339ff352685a2594fefb1ca4b13cc081 (accessed on 5 December 2024).
13. Fenoglio, M. S., Rossetti, M. R., & Videla, M. (2020). Negative effects of urbanization on terrestrial arthropod communities: A meta-analysis. *Global Ecology and Biogeography, 29*, 1412–1429. https://doi.org/10.1111/geb.13107.
14. Fischer, J., Riechers, M., Loos, J., Martín López, B., & Temperton, V. M. (2020). Making the UN Decade on Ecosystem Restoration a social-ecological endeavour. *Trends in Ecology & Evolution, 36*, 20–28. https://doi.org/10.1016/j.tree.2020.08.018.
15. Gann, G. D., McDonald, T., Walder, B., Aronson, J., Nelson, C. R., Jonson, J., Hallett, J. G., Eisenberg, C., Guariguata, M. R., Liu, J., Hua, F., Echeverría, C., Gonzales, E. K., Shaw, N. L., Decleer, K., & Dixon, K. W. (2019). International principles and standards for the practice of ecological restoration (2nd ed.). *Restoration Ecology, 27*. https://doi.org/10.1111/rec.13035.
16. Garibaldi, L. A., Pérez Méndez, N., Garratt, M. P. D., Gemmill Herren, B., Miguez, F. E., & Dicks, L. V. (2019). Policies for ecological intensification of crop production. *Trends in Ecology & Evolution, 34*, 282–286. https://doi.org/10.1016/j.tree.2019.01.003.
17. Gatta, P. P. (2022). *The State of World Fisheries and Aquaculture 2022*. https://doi.org/10.4060/cc0461en.
18. Geist, J. (2021). Editorial: Green or red: Challenges for fish and freshwater biodiversity conservation related to hydropower. *Aquatic Conservation: Marine and Freshwater Ecosystems, 31*, 1551–1558. https://doi.org/10.1002/aqc.3597.

19. Gunnam, V. (2024). Combating biodiversity loss: Artificial intelligence solutions for sustainable ecological preservation. *Darpan International Research Analysis, 12*, 210–222. https://doi.org/10.36676/dira.v12.i2.150.
20. Hong, P., Schmid, B., De Laender, F., Eisenhauer, N., Zhang, X., Chen, H., Craven, D., De Boeck, H. J., Hautier, Y., Petchey, O. L., Reich, P. B., Steudel, B., Striebel, M., Thakur, M. P., & Wang, S. (2021). Biodiversity promotes ecosystem functioning despite environmental change. *Ecology Letters, 25*, 555–569. https://doi.org/10.1111/ele.13936.
21. https://www.semanticscholar.org/paper/99228ec654b1e6c6912ca41c6da83ed2469ee95b (accessed on 5 December 2024).
22. Intergovernmental Panel on Climate Change. (2023). *Climate Change 2022–Impacts, Adaptation and Vulnerability*. https://doi.org/10.1017/9781009325844.
23. Kalpana, K., Reddy, G. S. V., Chellem, S. R., Rajalakshmi, J., Indrani, K., Lavanya, P., & Navya, D. S. (2024). The role of pollinators in enhancing biodiversity and pollination mechanisms: A review. *Uttar Pradesh Journal of Zoology, 45*(13), 226–241. https://doi.org/10.56557/upjoz/2024/v45i134150.
24. Kougioumoutzis, K., Kokkoris, I., Panitsa, M., Trigas, P., Strid, A., & Dimopoulos, P. (2020). Plant diversity patterns and conservation implications under climate-change scenarios in the Mediterranean: The case of Crete (Aegean, Greece). *Diversity, 12*, 270. https://doi.org/10.3390/d12070270.
25. Langlet, A., & Vadrot, A. B. M. (2022). Not 'undermining' who? Unpacking the emerging BBNJ regime complex. *Marine Policy, 147*, 105372. https://doi.org/10.1016/j.marpol.2022.105372.
26. Leclère, D., Obersteiner, M., Barrett, M., Butchart, S. H. M., Chaudhary, A., De Palma, A., et al., (2020). Bending the curve of terrestrial biodiversity needs an integrated strategy. *Nature, 585*(7826), 551–556. https://doi.org/10.1038/s41586-020-2705-y.
27. Lyver, P. O'B., Timoti, P., Davis, T. J., & Tylianakis, J. M. (2019). Biocultural hysteresis inhibits adaptation to environmental change. *Trends in Ecology & Evolution, 34*, 771–780. https://doi.org/10.1016/j.tree.2019.04.002.
28. Ma, Q., Wan, L., Shi, S., & Wang, Z. (2024). Impact of climate change on the distribution of three rare salamanders (*Liua shihi*, *Pseudohynobius jinfo*, and *Tylototriton wenxianensis*) in Chongqing, China, and their conservation implications. *Animals, 14*(5), 672. https://doi.org/10.3390/ani14050672.

29. Mathiasson, M. E., & Rehan, S. M. (2020). Wild bee declines linked to plant-pollinator network changes and plant species introductions. *Insect Conservation and Diversity, 13*, 595–605. https://doi.org/10.1111/icad.12429.

30. Maxwell, S. L., Cazalis, V., Dudley, N., et al. (2020). Area-based conservation in the twenty-first century. *Nature, 586*, 217–227. https://doi.org/10.1038/s41586-020-2773-z.

31. Medellín, R. A., Wiederholt, R., & López Hoffman, L. (2017). Conservation relevance of bat caves for biodiversity and ecosystem services. *Biological Conservation, 211*, 45–50. https://doi.org/10.1016/j.biocon.2017.01.012.

32. Ngwenya, F. (2024). Biodiversity conservation strategies and their impact on ecosystem health in Mozambique. *International Journal of Natural Sciences, 4*(1), 32–45. https://doi.org/10.47604/ijns.2660.

33. Onyena, A. P., Aniche, D., Ogbolu, B., Rakib, M. R. J., Uddin, J., & Walker, T. R. (2021). Governance strategies for mitigating microplastic pollution in the marine environment: A review. *Microplastics, 1*, 15–46. https://doi.org/10.3390/microplastics1010003.

34. Reyes, G. P., Nandakumar, K., & Stevens, R. (2019). Biological conservation. In *Biological Conservation* (pp. 87–120). https://doi.org/10.4018/978-1-5225-7302-9.ch005.

35. Roznik, E. A., Buckanoff, H., Langston, R. W., Shupp, C. J., & Smith, D. (2023). Conservation through collaboration: Regional conservation programs of the North Carolina Zoo. *Journal of Zoological and Botanical Gardens, 4*(2), 292–311. https://doi.org/10.3390/jzbg4020025.

36. Rydén, O., Zizka, A., Jagers, S. C., Lindberg, S. I., & Antonelli, A. (2019). Linking democracy and biodiversity conservation: Empirical evidence and research gaps. *Ambio, 49*, 419–433. https://doi.org/10.1007/s13280-019-01210-0.

37. Schröder, P., Beckers, B., Daniels, S., Gnädinger, F., Maestri, E., Marmiroli, N., Mench, M., Millán, R., Obermeier, M. M., Oustrière, N., Persson, T., Poschenrieder, C., Rineau, F., Rutkowska, B., Schmid, T., Szulc, W., Witters, N., & Sæbø, A. (2017). Intensify production, transform biomass to energy and novel goods and protect soils in Europe—A vision how to mobilize marginal lands. *Science of The Total Environment, 616, 617*, 1101–1123. https://doi.org/10.1016/j.scitotenv.2017.10.209.

38. Seymour, F. (2020). Seeing the forests as well as the (trillion) trees in corporate climate strategies. *One Earth, 2*, 390–393. https://doi.org/10.1016/j.oneear.2020.05.006.

39. Socolar, J. B., Gilroy, J. J., Kunin, W. E., & Edwards, D. P. (2016). How should beta-diversity inform biodiversity conservation? *Trends in Ecology & Evolution, 31*(1), 67–80. https://doi.org/10.1016/j.tree.2015.11.005.
40. Tessnow-von Wysocki, I., & Vadrot, A. B. M. (2020). The voice of science on marine biodiversity negotiations: A systematic literature review. *Frontiers in Marine Science, 7*, 614282. https://doi.org/10.3389/fmars.2020.614282.
41. Tickner, D., Opperman, J. J., Abell, R., Acreman, M., Arthington, A. H., Bunn, S. E., Cooke, S. J., Dalton, J., Darwall, W., Edwards, G., Harrison, I., Hughes, K. A., Jones, T., Leclère, D., Lynch, A. J., Leonard, P., McClain, M. E., Muruven, D., Olden, J. D., Ormerod, S. J., Robinson, J. A., Tharme, R. E., Thieme, M., Tockner, K., Wright, M., & Young, L. (2020). Bending the curve of global freshwater biodiversity loss: An emergency recovery plan. *BioScience, 70*, 330–342. https://doi.org/10.1093/biosci/biaa002.
42. Tscharntke, T., Graß, I., Wanger, T. C., Westphal, C., & Batáry, P. (2021). Beyond organic farming—harnessing biodiversity-friendly landscapes. *Trends in Ecology & Evolution, 36*, 919–930. https://doi.org/10.1016/j.tree.2021.06.010.
43. Turner Skoff, J. B., Johnson, L. R., Stefferud, E., Stratman, P., & Santos, K. (2024). Reimagining relationships with resources as a public garden: Case studies of Longwood Gardens' sustainability and stewardship practices. https://www.semanticscholar.org/paper/bf2ac5b6ff8ef7a562225227a7d8315de6691da0 (accessed on 5 December 2024).
44. von der Lippe, M., Buchholz, S., Hiller, A., Seitz, B., & Kowarik, I. (2020). CityScapeLab Berlin: A research platform for untangling urbanization effects on biodiversity. *Sustainability, 12*(6), 2565. https://doi.org/10.3390/su12062565.
45. Wang, J. (2024). The impacts of Arctic climate change on polar bear biodiversity and associated analysis. *Semantics Scholar*. https://www.semanticscholar.org/paper/b3d9aa56fba305bd287d594be05f043430dc22c8 (accessed on 5 December 2024).
46. Wiegleb, G., Dahms, H. U., Byeon, W., & Choi, G. W. (2017). To what extent can constructed wetlands enhance biodiversity? *International Journal of Environmental Science and Development, 8*(8), 561–569. https://doi.org/10.18178/ijesd.2017.8.8.1016.
47. Willett, W., Rockström, J., Loken, B., Springmann, M., Lang, T., Vermeulen, S., Garnett, T., Tilman, D., DeClerck, F., Wood, A., Jonell, M., Clark, M., Gordon, L. J., Fanzo, J., Hawkes, C., Zurayk, R., Rivera,

J. A., De Vries, W., Majele Sibanda, L., Afshin, A., Chaudhary, A., Herrero, M., Agustina, R., Branca, F., Lartey, A., Fan, S., Crona, B., Fox, E., Bignet, V., Troell, M., Lindahl, T., Singh, S., Cornell, S. E., Srinath Reddy, K., Narain, S., Nishtar, S., & Murray, C. J. L. (2019). Food in the Anthropocene: The EAT-Lancet Commission on healthy diets from sustainable food systems. *The Lancet, 393*(10170), 447–492. https://doi.org/10.1016/S0140-6736(18)31788-4.

48. Xu, L., Rolf, E., Beery, S., Bennett, J., Berger-Wolf, T., Birch, T., Bondi-Kelly, E., Brashares, J., Chapman, M. S., Corso, A. W., Davies, A., Garg, N., Gaylard, A., Heilmayr, R., Kerner, H., Klemmer, K., Kumar, V., Mackey, L. W., Monteleoni, C., Moorcroft, P., Palmer, J., Perrault, A., Thau, D., & Tambe, M. (2023). Reflections from the Workshop on AI-assisted decision making for conservation. https://www.semanticscholar.org/paper/c231220c2274015b2d979cdb1743818fefcaf5fb (accessed on 5 December 2024).

49. Yang, Z. (2024). Impact of hydropower on river biodiversity and strategies for recovery. https://www.semanticscholar.org/paper/ba876e9d3017da6ecbd1cc4c122f15221d1616b7 (accessed on 5 December 2024).

50. Zeng, Y., Koh, L. P., & Wilcove, D. (2022). Gains in biodiversity conservation and ecosystem services from the expansion of the planet's protected areas. *Semantics Scholar*. https://www.semanticscholar.org/paper/08314fa5b9431da92090c55128d8bd73352fc1d4 (accessed on 5 December 2024).

CHAPTER 6

ENVIRONMENTAL HAZARDS AND HUMAN HEALTH

Contents

INTRODUCTION

The intricate relationship between our environment and human health has garnered increasing attention in recent years, as scientists and policymakers alike recognize the profound impacts that environmental hazards can impose on well-being. Climate change, pollution, and biodiversity loss are not only pressing environmental issues but are also pivotal factors affecting public health. For instance, air quality, significantly deteriorated by industrial emissions, becomes a critical concern when examining respiratory ailments and cardiovascular diseases. As communities grapple with these hazards, it becomes essential to acknowledge the broader implications of environmental degradation, necessitating an interdisciplinary approach to health education and public policy initiatives. An understanding of environmental hazards requires exploration of both natural and anthropogenic factors that contribute to health risks. Natural disasters, such as hurricanes and wildfires, can result in immediate health threats, including injuries and disruptions to medical care. In contrast, human-induced factors like chemical spills and urban industrialization lead to chronic health issues, including exposure to toxic substances. Furthermore, socio-economic disparities often exacerbate these risks, as marginalized communities disproportionately face environmental toxins, leading to higher rates of illness. Through this lens, one can appreciate how interconnected our surroundings and health truly are, prompting urgent calls for comprehensive strategies to mitigate environmental threats. Addressing environmental hazards necessitates not only scientific understanding but also collaborative efforts across sectors. Mobilizing public awareness and political action can foster grassroots movements aimed at enforcing stricter regulations to protect vulnerable populations. For example, initiatives promoting renewable energy sources serve not only to combat climate change but also to create healthier living conditions by reducing air pollution. Additionally, community engagement can spur local solutions to environmental problems, encouraging sustainable practices and resource management. Ultimately, the intersection of health and the environment highlights the need for a collective responsibility, driving forward an agenda that prioritizes the health of both people and the planet for the future.

6.1. DEFINITION OF ENVIRONMENTAL HAZARDS

Understanding environmental hazards requires a comprehensive examination of their various features and implications for human health. These hazards encompass a wide range of natural and anthropogenic influences that can adversely affect the health of individuals and communities. Natural hazards, such as earthquakes, floods, and wildfires, are inherently unpredictable and can cause immediate harm through physical destruction and the disruption of services. In addition, there are chronic environmental hazards like air and water pollution, which arise from human activities. They often lead to long-term health consequences, including respiratory diseases and cardiovascular issues. Such complexities illustrate that environmental hazards are not merely isolated events but part of a broader system that interacts with social, economic, and health factors in meaningful ways. The consequences of these hazards are pronounced in the context of public health, where exposure to environmental toxins or ecological disruptions can result in a variety of health issues. For instance, as highlighted in the work of Burris et al. (2011), social determinants such as nutrition, sanitation, and safe housing are influenced by environmental conditions, thereby affecting overall population health. As environmental hazardous substances permeate ecosystems, they can lead to food and water scarcity, fundamentally undermining the quality of life.

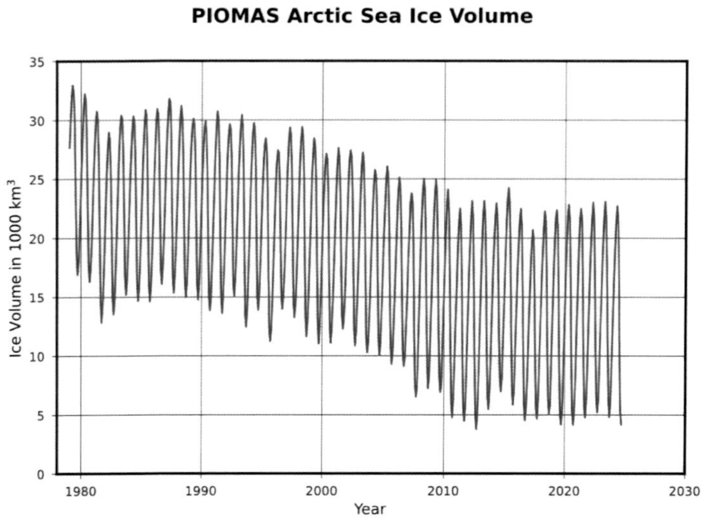

Figure 6.1. *Environmental hazards.*

Source: By Dominiklenne – {{Volume time series and uncertainties: Schweiger, A, R. Lindsay, J. Zhang, M. Steele, H. Stern, Uncertainty in modeled arctic sea ice volume, J. Geophys. Res, doi:10.1029/2011JC007084, 2011, Plot: own work, done with Python, Numpy, Matplotlib}}, based on published data of Polar Science Center, University of Washington, http://psc.apl.washington.edu/research/projects/arctic-sea-ice-volume-anomaly/data/, Public Domain, https://commons.wikimedia.org/w/index.php?curid=11321499.

Moreover, the prominence of non-communicable diseases has grown in part due to sedentary lifestyles exacerbated by unsustainable urban environments and the marketing of unhealthy food systems, illustrating the intricate link between environmental hazard exposure and health disparities. Addressing environmental hazards necessitates a coordinated response that includes effective policy, education, and community engagement to mitigate their impact. Public health frameworks must be adaptable, incorporating evidence-based laws that govern industries contributing to pollution and habitat destruction. According to the Lancet Commission, law has a crucial role in advancing global health initiatives and ensuring justice in health equity (Gostin et al., 2019). From mitigating climate change to promoting sustainable agricultural practices, understanding environmental hazards can guide policymakers in crafting legislation that both protects ecosystems and public health. By fostering collaboration between environmental scientists, public health officials, and legal experts, we can forge pathways to healthier communities while addressing the pressing challenges presented by global environmental changes (Figure 6.1).

6.1.1. Importance of Studying the Relationship Between Environmental Hazards and Human Health

Understanding the connection between environmental hazards and human health is vital for creating effective public policy and safeguarding community well-being. As industrial activity and urbanization continue to rise, the potential for exposure to harmful substances increases, raising significant public health concerns. In particular, the widespread use of herbicides such as glyphosate has raised alarms regarding their long-term effects on human health. Studies have illuminated the toxicity and potential carcinogenicity of glyphosate and its formulations, emphasizing the need for detailed risk assessments that encompass both active ingredients and related adjuvants. The insights gained are essential for designing regulations that protect vulnerable populations, consequently aligning agricultural practices with sustainable development goals while minimizing adverse health outcomes (Gandhi et al., 2021). Moreover, environmental contaminants are not limited to agriculture; they can also emerge from industrial processes, as evident in the examination of the Huainan coal mining area. Research focused on the accumulation of harmful elements like arsenic, nickel, and chromium in surface water reveals the pervasive risk these substances pose to both ecosystems and human health.

The elevated risk of carcinogenic exposure linked to these elements underscores the necessity of thorough environmental monitoring and health risk assessments (Chen et al., 2019). By understanding these relationships, health professionals and policymakers can implement more effective public health strategies that prioritize the reduction of exposure to hazardous materials and develop contingency plans for communities at risk. In addition, the study of environmental hazards fosters awareness and education regarding the implications of environmental practices on human health. When communities are informed of the potential risks posed by pollutants or toxic chemicals, they are better equipped to advocate for safer practices and demand accountability from industries. This empowered public engagement can lead to policy changes that promote sustainability and health. The studies showcasing the health risks from various pollutants emphasize the importance of transparency in environmental health research. As such, integrating these findings into public discourse not only encourages preventive measures but also strengthens community resilience against health threats posed by environmental hazards, ultimately contributing to a healthier society.

6.2. TYPES OF ENVIRONMENTAL HAZARDS

The categorization of environmental hazards is essential for understanding their impact on human health and ecosystems. Among these hazards, biological threats pose significant risks, particularly through exposure to pathogens such as bacteria, viruses, and parasites. These agents can lead to diseases that not only affect individual health but also strain public health systems. The effects of biological hazards are especially pronounced in vulnerable populations, including children and the elderly. Early exposure during critical developmental windows can produce lasting negative health outcomes, emphasizing the need for vigilant monitoring and interventions (Amit Kumar et al., 2020). Combating these hazards requires robust public health strategies, ensuring clean water and sanitation, as well as effective vaccination programs. Chemical hazards represent another serious category of environmental threats. These hazards can arise from industrial activities, agricultural practices, or household products, resulting in exposure to heavy metals, pesticides, and other toxic substances. For instance, cadmium (Cd) is a notable contaminant that poses long-term health risks due to its non-biodegradable nature and potential for bioaccumulation in food chains (Amit Kumar et al., 2020). This persistent heavy metal can lead to severe health conditions such as kidney damage and respiratory illnesses. Additionally, chemical hazards can disrupt ecosystems, leading to biodiversity loss. Mitigating the impact of chemicals in the environment requires strict regulatory frameworks to minimize exposure and encourage sustainable practices among industries. Physical hazards, encompassing a wide range of natural and anthropogenic forces, present another layer of risks to human health. Natural disasters such as earthquakes, floods, and hurricanes can result in both immediate harm and long-term public health crises due to displacement, infrastructure damage, and loss of essential services. These events often exacerbate existing social inequalities, disproportionately affecting marginalized communities. Moreover, the psychological effects of experiencing such traumas can lead to chronic conditions, including anxiety and post-traumatic stress disorder (Nelson et al., 2020). Addressing physical hazards necessitates comprehensive disaster preparedness and an emphasis on resilient community planning. By enhancing resilience and response strategies, societies can better shield themselves from the multifaceted impacts of physical environmental hazards.

6.2.1. Chemical Hazards

Chemical hazards represent a significant concern in the realm of environmental health, influencing both ecological systems and human well-being. The production processes of various materials, particularly plastics, are fraught with risks as they often involve the release of harmful substances into the environment. For instance, the Minderoo-Monaco Commission highlights that more than 98% of plastics are derived from fossil fuels, with a vast array of toxic additives incorporated to enhance their performance. These chemicals can leach into ecosystems, leading to contamination of air and water, whilst posing direct health threats to nearby communities. As the use of plastics has skyrocketed, resulting in staggering volumes of waste, these chemical hazards are perpetuated, contributing to a cycle of environmental degradation and human health risks (Landrigan et al., 2023). In the context of food security, chemical hazards also extend to the consumption of edible insects, which are posited as a sustainable food source for the future. However, the presence of pathogenic microorganisms in these insects poses potential health risks, necessitating stringent safety measures throughout the food chain. Studies have documented varied microbial ecosystems in commonly consumed species like mealworm larvae and grasshoppers, revealing a concerning presence of harmful bacteria, including those resistant to antibiotics. The integration of edible insects into the food supply adds complexity to the discussion of chemical hazards since the need for appropriate safety and hygiene law is highlighted.

Increasing awareness and regulatory frameworks are vital for ensuring that while exploring sustainable food alternatives, human health is not compromised by microbial hazards and chemical contaminants (Garofalo et al., 2019). Addressing chemical hazards requires a multifaceted approach that combines regulation, education, and innovation. The establishment of comprehensive safety guidelines for the production, handling, and disposal of chemical substances is crucial to mitigate their impact. For instance, the Minderoo-Monaco Commission advocates for a Global Plastics Treaty that not only limits plastic production but also mandates stringent safety measures for chemical additives. Incorporating public awareness campaigns and training programs can empower communities to recognize and manage chemical risks effectively. Moreover, ongoing research into alternative materials and sustainable practices is essential for reducing reliance on hazardous chemicals. Given the interconnection between environmental pollutants and health outcomes, a proactive stance on chemical hazards can promote safer communities and healthier ecosystems, ensuring better quality of life for future generations.

6.2.2. Biological Hazards

In recent decades, the proliferation of biological hazards has posed significant challenges to human health and the environment. These hazards, predominantly stemming from pathogens such as bacteria, viruses, fungi, and parasites, can lead to various diseases that threaten public health. The transmission pathways of these biological agents are diverse, often linked to environmental factors such as climate change and globalization. For instance, warmer temperatures can facilitate the spread of diseases like malaria and dengue fever, as mosquitoes expand their habitats into previously uninhabitable regions. Such changes underscore the interconnectedness of environmental factors and human health, necessitating vigilant public health responses to mitigate these threats. The impact of biological hazards extends beyond mere infection rates; they also strain healthcare systems and disrupt economic stability. Outbreaks of infectious diseases can lead to widespread panic and a significant economic burden, as seen during the COVID-19 pandemic. The intricate relationship between environmental deterioration and the emergence of new pathogens, as noted by the Minderoo-Monaco Commission, highlights the urgent need for comprehensive strategies that address both health and environmental concerns (Landrigan et al., 2023). This interconnected approach fosters resilience in societies vulnerable to biological hazards and encourages the integration of health considerations into environmental policies. By recognizing the socioeconomic consequences of these hazards, governments can prioritize funding for research and infrastructure that enhance public health readiness. Moreover, vulnerable populations often bear the brunt of biological hazards, further exacerbating existing social inequities. Communities with limited access to healthcare services, clean water, and sanitation are more susceptible to the impacts of infectious diseases. This aspect highlights the critical need for environmental justice, as marginalized groups disproportionately affected by hazardous conditions must be prioritized in public health strategies. The IPCC's findings on climate change not only illuminate how environmental degradation contributes to health risks but also emphasize the global responsibility to protect those most at risk (Intergovernmental Panel on Climate Change, 2023). To effectively combat biological hazards, comprehensive solutions are required that address both environmental sustainability and social equity, fostering environments where all communities can thrive and remain resilient against pathogens (Figure 6.2).

Figure 6.2. Environmental biological hazards.

Source: https://img.freepik.com/premium-vector/signs-warning-ecological-biological-hazards-effect-radiation-environment-nature-trees-litter-sign_223337-18935.jpg.

6.2.3. Physical Hazards

Environmental threats often manifest in various forms, one of which includes physical hazards that contribute significantly to human health issues. These hazards can arise from natural disasters, such as earthquakes, hurricanes, and floods, all of which can cause immediate physical harm and long-term health repercussions. For instance, the aftermath of such events often sees a rise in infectious diseases, as stagnant water serves as a breeding ground for pathogens. Additionally, the structural collapse of buildings due to extreme weather can lead to injuries or fatalities, underscoring the necessity of proper urban planning and infrastructure resilience. The complex interplay between environmental hazards and public health emphasizes the need for integrated strategies aimed at mitigating the risks associated with natural phenomena, thereby safeguarding communities and enhancing overall resilience against future threats. Microplastic pollution represents a more insidious physical hazard, with potential impacts lingering longer in the environment and affecting human health indirectly. The prevalence of microplastics—tiny plastic particles less than 5 mm—has escalated, raising concern about their interaction with various biological systems. Studies indicate that these microplastics can physically embed themselves in tissues, potentially causing cellular damage and inflammation, which may result in chronic health conditions over time (Xi Yang et al., 2022). This is particularly concerning given the growing body of evidence suggesting that these pollutants are pervasive in food and water supplies, resulting in widespread human exposure. Furthermore, the accumulation of microplastics in the food chain raises questions about their long-term health effects, calling for immediate research and public policy efforts to address this emerging physical hazard. In addition to microplastics, the physical hazards stemming from the release of toxic additives contained in plastic products pose a critical risk to both environmental and human health. While much attention has been directed toward the broader impacts of microplastics, the unique properties of each specific polymer and their associated chemical additives have been largely overlooked (Srinidhi Sridharan et al., 2022). These substances can leach into ecosystems, affecting wildlife and potentially entering the human food supply. The environmental behavior and toxicity of these additives are not well understood, yet they are essential components of the risk associated with microplastics. Understanding the intricacies of these physical hazards requires a multi-faceted approach, involving scientific research to evaluate the leachability and ecological impact of plastic additives, as well as an urgent need for comprehensive regulations governing plastic production and disposal. By prioritizing such initiatives, societies can better manage the complex web of hazards posed by plastic pollution and protect human health.

6.3. AIR POLLUTION AND ITS HEALTH EFFECTS

The effects of air pollution on human health are profound and multifaceted, affecting various systems within the body. Prolonged exposure to polluted air, particularly in urban environments, is linked to increased rates of respiratory and cardiovascular diseases. Pollutants such as particulate matter, nitrogen dioxide, and sulfur dioxide can penetrate deep into lung tissue, leading to chronic conditions such as asthma and bronchitis.

Figure 6.3. *Air pollution.*

Source: https://en.wikipedia.org/wiki/File:Alfred_Palmer_Smokestacks.jpg.

The consequences extend beyond immediate respiratory issues, as air pollution is associated with systemic inflammation and cardiovascular events, including heart attacks and strokes. Furthermore, the impact of these pollutants is exacerbated by socioeconomic factors, with disadvantaged populations experiencing higher morbidity rates due to limited access to healthcare and clean air. Thus, understanding the intricacies of such health impacts is crucial for promoting public health and informing policy changes aimed at cleaner air.

In addition to respiratory and cardiovascular effects, air pollution poses serious risks to mental health and cognitive function. Evidence suggests that exposure to toxic air pollutants can lead to neurodegenerative diseases, including Alzheimer's and Parkinson's, as well as increased rates of anxiety and depression. Research indicates that micro-particles from polluted air can cross the blood-brain barrier, leading to neuroinflammation and neuronal damage, which significantly affects mental health outcomes. A comprehensive study reported that regions with high levels of air pollution correlate with higher incidences of mental health disorders, suggesting a direct link that warrants further investigation. As coal continues to be a primary source of energy in various parts of the world, its combustion releases invisible particles that can be inhaled, contributing to this growing public health concern (Juciano Gasparotto et al., 2020). These findings highlight the urgent need to address air quality to improve not only physical but also mental health outcomes. Mitigating air pollution requires robust policy interventions and collective action at both local and global levels. Initiatives such as transitioning to renewable energy sources and implementing stricter emissions regulations can significantly reduce airborne pollutants, promoting healthier environments. Investment in public transportation systems and urban green spaces can also serve to diminish reliance on fossil fuels, subsequently decreasing emissions. Given that air pollution is a significant environmental hazard impacting human health, governments must prioritize air quality improvement strategies in their public health agendas. The health effects of air pollution, coupled with the urgent climate crisis, necessitate a proactive approach to foster cleaner air for vulnerable populations, as air quality improvements can prevent millions of premature deaths and foster healthier, more resilient communities (Romanello et al., 2021). Addressing this critical issue is not just an environmental concern; it is a matter of public health equity and the human right to breathe clean air.

6.3.1. Sources of Air Pollution

Air pollution emerges from many sources, significantly impacting human health and the environment. Industrial activities, particularly fossil fuel combustion, contribute immensely to airborne contaminants. Factories release particulate matter, sulfur dioxide, and volatile organic compounds that degrade air quality and pose respiratory health risks to nearby communities. In addition, emissions from transportation—cars, trucks, and airplanes—further complicate the situation by introducing nitrogen oxides and carbon monoxide into the atmosphere. Trends indicate that as manufacturing and urban activities expand, the volume of pollution also rises, making regulatory measures essential to mitigate these emissions.

Agricultural practices are another major source of air pollution, as they release ammonia and methane into the atmosphere. Livestock farming, particularly, generates significant methane emissions, while the overuse of fertilizers leads to the release of ammonia. These pollutants not only affect air quality but also contribute to the formation of secondary pollutants like particulate matter and ground-level ozone. Such agricultural emissions could exacerbate health issues, particularly in vulnerable populations lacking access to effective healthcare. Furthermore, increased agricultural productivity is often linked to deforestation and land-use changes, disrupting natural ecosystems and contributing further to air quality degradation (Landrigan et al., 2023). Natural sources of air pollution, such as wildfires and dust storms, also play a crucial role in influencing air quality. As climate change intensifies, the frequency and severity of these events are likely to increase, exacerbating air pollution levels. For instance, wildfires release large quantities of carbon dioxide, nitrogen oxides, and particulate matter, aggravating respiratory conditions and cardiovascular diseases in affected populations (Romanello et al., 2021). Moreover, dust storms can carry harmful substances, including pesticides and heavy metals, over long distances, impacting regions far from the pollution source. Addressing air pollution thus requires a comprehensive understanding of both anthropogenic and natural sources, ensuring effective strategies that protect public health while considering the diverse influences on air quality (Figure 6.4).

Figure 6.4. *Sources of air pollution.*

Source: By Bobak – Own work, CC BY-SA 2.5, https://commons.wikimedia.org/w/index.php?curid=2190112.

6.3.2. Short-Term Health Effects

The exposure to environmental pollutants poses immediate health risks, particularly affecting vulnerable populations such as children and the elderly. Short-term health effects can manifest in various forms, notably respiratory illnesses, which exacerbate conditions like asthma. Evidence shows that just a modest increase in concentrations of pollutants like ozone (O_3) and nitrogen dioxide (NO_2) correlates with an uptick in emergency room visits and hospital admissions for asthma-related complications. Specifically, studies indicate that a 10 $\mu g/m^3$ rise in average 24-hour NO_2 exposure leads to a 1.014 pooled relative risk for asthma exacerbations, illustrating the acute sensitivity of respiratory health to air quality (Zheng et al., 2021). Thus, it becomes apparent that controlling environmental pollutants is vital in safeguarding immediate human health and preventing the escalation of chronic conditions. The immediate health consequences of environmental hazards extend beyond respiratory issues, encompassing cardiovascular problems. Short-term exposure to pollutants like particulate matter (PM) can trigger acute cardiac events, fundamentally altering blood pressure and heart rate. In urban areas, where pollution levels frequently exceed safe thresholds, the general population experiences higher incidences of heart attacks and strokes in response to environmental toxins. This correlation underlines how short-term exposure to air pollution translates into severe health outcomes, warranting the attention of public health officials. Furthermore, the legal frameworks surrounding environmental safety, as discussed in the context of addressing international and national health challenges, highlight the necessity for robust policies to mitigate these health risks effectively (Gostin et al., 2019). Implementing stricter regulations could significantly contribute to reducing these short-term health effects for vulnerable groups. In addition to respiratory and cardiovascular effects, short-term exposure to pollutants can impair overall health by impacting cognitive functions. Recent research indicates that environmental toxins can negatively influence neurological health, particularly in children whose brains are still developing. For instance, pollutants can lead to immediate cognitive impairment, affecting attention and learning capabilities. This cognitive decline often creates a cascade of long-term educational and social repercussions. As society grapples with these impacts, advancing public health initiatives that promote clean air and diminish pollution exposure becomes critical. Moreover, promoting legal and institutional frameworks—suggested by global health commissions—would ensure that the legal determinants of health actively guard against short-term environmental hazards, fostering a healthier, more equitable society (Gostin et al., 2019). Therefore, recognizing, and addressing the immediate health effects of environmental pollution is essential for safeguarding future generations.

6.3.3. Long-Term Health Consequences

Environmental pollutants, including volatile organic compounds (VOCs) and excessive nitrogen emissions, pose significant long-term health risks to human populations. Exposure to VOCs, which are emitted by both industrial processes and agricultural practices, has been linked to various respiratory ailments, neurological disorders, and even cancer. The persistent presence of these chemicals in the atmosphere raises concerns about their chronic effects on health, as they can infiltrate indoor environments and accumulate over time. Research indicates that the proliferation of such pollutants demands immediate attention, as they threaten not only individual health but also the overall well-being of communities. Addressing these issues requires a multi-faceted approach that includes stricter regulations and increased public awareness regarding the sources and consequences of VOC exposure.

The agricultural sector is a major contributor to nitrogen emissions, primarily through the use of fertilizers. An alarming aspect of this issue is the doubling of atmospheric ammonia (NH_3) levels compared to previous centuries, causing detrimental effects on both human health and the environment (Dammers et al., 2019). Long-term exposure to elevated nitrogen levels can lead to respiratory conditions, aggravating pre-existing illnesses such as asthma. Additionally, the role of ammonia in forming secondary particulate matter in the atmosphere exacerbates cardiovascular problems, thus presenting public health challenges that need addressing. As these emissions continue to grow unchecked, their interaction with ecosystems can further lead to soil degradation and water quality issues, which can cyclically impact human health via contaminated food and water sources.

Considering the complex interplay between environmental hazards and human health, it becomes clear that the ramifications of pollution extend far beyond immediate physical symptoms. Chronic exposure to pollutants, including those derived from agricultural practices, can have lasting effects on mental and physical well-being. The integration of biological control approaches, such as the use of microorganisms to combat plant pathogens instead of traditional chemical fungicides, could mitigate some of these long-term health issues (Tilocca et al., 2020). By reducing reliance on harmful pesticides and fertilizers, we can potentially lower the risks associated with chemical exposure. A shift toward sustainable agricultural practices can foster a healthier environment and, in turn, a healthier population, emphasizing the urgency for policy reform and community engagement in addressing environmental health concerns.

6.4. WATER CONTAMINATION AND HUMAN HEALTH

Contamination of water resources has emerged as a critical issue affecting human health on a global scale. The presence of hazardous chemicals, such as pesticides and industrial pollutants, in drinking water can lead to a wide range of health problems. For instance, agricultural runoff contributes significantly to the degradation of water quality, introducing toxins that can disrupt endocrine functions and compromise reproductive health (de Jager et al., 2024). Persistent exposure to contaminated water may result in chronic health conditions, including developmental disorders, hormonal imbalances, and increased susceptibility to various diseases. With the rising dependence on polluted water sources, communities, particularly in agricultural regions, face the dire consequences of inadequate water quality, reflecting a broader need for systemic solutions to safeguard public health. In addition to agricultural pollutants, industrial waste poses substantial risks to human health through water contamination.

Many industrial processes release a variety of chemicals into nearby water bodies, including heavy metals and endocrine-disrupting chemicals (EDCs), which accumulate over time and enter the human food chain (Verma, 2024). Exposure to these contaminants can impair immune responses, increase the risk of cancer, and lead to neurological damage. The synergistic effects of multiple contaminants in water further exacerbate these risks, complicating the health outcomes for exposed populations. Efforts to monitor and regulate these pollutants effectively are essential for preventing adverse health impacts and ensuring safe drinking water for all, particularly in regions with limited access to resources and healthcare.

Moreover, the burden of contaminated water extends beyond immediate health effects, influencing social and economic dimensions as well. Communities reliant on polluted water sources often face increased healthcare costs, decreased productivity, and reduced quality of life. Vulnerable populations, such as children and the elderly, are particularly at risk, as they may experience amplified health effects from chemical exposure. Sustainable strategies for water management and pollution mitigation are paramount to addressing these challenges. To facilitate long-term improvements in human health, governments, and organizations must prioritize clean water initiatives, enforce regulations on pollutants, and promote public awareness about the significance of water quality (de Jager et al., 2024). Only through collective action can society combat the multifaceted impacts of water contamination and foster a healthier future for all.

6.4.1. Common Pollutants in Water Sources

Water sources around the world face a barrage of pollutants that pose significant risks to both human health and ecosystems. Among the most prevalent are heavy metals such as lead, mercury, and cadmium, often introduced through industrial discharge and agricultural runoff. These toxins accumulate in aquatic environments, leading to bioaccumulation in the food chain, which can ultimately affect human consumers. The toxic effects of heavy metals are well-documented, including neurological damage and developmental issues in children. The increasing frequency of industrial activities, especially in developing regions, raises alarming concerns about the quality of drinking water and the safety of recreational water bodies, underscoring the urgent need for stringent regulations and better waste management practices. In addition to heavy metals, microplastics are rapidly becoming one of the most insidious pollutants in freshwater and marine environments.

These tiny plastic particles, which originate from a variety of sources such as degraded larger plastics and synthetic textiles, have been detected in numerous water bodies. They not only hinder the health of aquatic organisms but also introduce harmful chemicals into the food chain, with potential adverse effects on human health as well. The persistence of microplastics in the ecosystem, combined with their ability to absorb other toxic substances, exacerbates their threat. Studies show that microplastics can become vectors for toxins, impacting both freshwater biodiversity and human health through contaminated drinking water sources. Understanding and addressing the sources and impacts of microplastics are crucial for effective environmental management and public health policies (Horton et al., 2017).

Moreover, agricultural practices contribute substantially to water pollution through nutrient runoffs, primarily nitrogen and phosphorus. These chemicals, often found in fertilizers, lead to eutrophication in bodies of water, resulting in harmful algal blooms that create dead zones devoid of oxygen. Such ecological imbalances can lead to declines in fish populations and the loss of aquatic biodiversity. Furthermore, the toxins produced by some algal species can contaminate drinking water supplies, causing serious health risks for humans and animals alike. The interplay of agricultural activities and water pollution illustrates the need for sustainable farming practices that minimize nutrient runoff while ensuring food security. As agricultural productivity must be balanced with the protection of water resources, adopting integrated nutrient management strategies becomes essential for fostering environmental health and human welfare (Intergovernmental Panel on Climate Change, 2023) (Figure 6.5).

Figure 6.5. *Source of water pollution.*

Source: By CNRC – Calexico New River Committee (CNRC), Public Domain, https://commons.wikimedia.org/w/index.php?curid=796398.

6.4.2. Health Risks Associated with Contaminated Water

Contaminated water poses significant health risks, affecting millions worldwide. Various pollutants infiltrate water supplies, including heavy metals, pathogens, and agricultural runoff. Exposure to substances such as lead and mercury from industrial and agricultural waste can lead to severe health issues, including neurological disorders and kidney damage (Manisalidis et al., 2020). Additionally, waterborne pathogens like bacteria and viruses can spread diseases such as cholera and dysentery, particularly in underdeveloped regions lacking adequate sanitation. The infiltration of these contaminants not only affects drinking water quality but also impacts agricultural irrigation, thereby magnifying the potential for disease transmission through food sources. Consequently, the cycle of contamination sustains a public health crisis, as communities that rely on tainted water supplies face higher rates of illness and mortality. Furthermore, the repercussions of contaminated water extend beyond immediate health effects, contributing to long-term public health challenges. Communities exposed to polluted water often experience increased healthcare costs, lost productivity, and diminished quality of life. For example, children exposed to lead-contaminated water may suffer from cognitive impairments and developmental delays, which can hinder their educational and economic prospects.

Additionally, chronic exposure to heavy metals, such as those outlined in the environmental hazards of contaminated ecosystems, can lead to various cancers and chronic diseases—notably respiratory illnesses (Manisalidis et al., 2020). As populations continue to grow and the demand for clean water intensifies, the risks associated with water contamination exacerbate socio-economic disparities, particularly in marginalized communities disproportionately affected by environmental hazards. To address the health risks posed by contaminated water, it is imperative to implement comprehensive strategies aimed at pollution prevention and public health education. This entails enforcing stricter regulations on industries that discharge pollutants and investing in infrastructure to improve water quality. Public awareness campaigns can also empower communities to advocate for their rights to safe drinking water and sanitation. Moreover, interdisciplinary approaches involving environmental scientists, public health officials, and policymakers are crucial for developing innovative solutions that mitigate the impact of contamination. Establishing monitoring systems to regularly assess water quality and the presence of hazardous elements is vital, as it directly relates to the overall health and resilience of populations (Hazrat Ali et al., 2019). By prioritizing clean water access, society can take significant strides toward reducing health risks and ensuring a healthier future for all.

6.4.3. Vulnerable Populations Affected by Water Contamination

Access to clean and safe drinking water is a fundamental human right, yet vulnerable populations often bear the brunt of water contamination issues. Low-income communities, racial, and ethnic minorities, and marginalized groups frequently reside in areas with decaying infrastructure and proximity to industrial sites. These conditions can lead to exposure to harmful contaminants, such as lead, per-, and polyfluoroalkyl substances (PFAS), and other hazardous chemicals, which jeopardize their health and well-being. For example, evidence shows that plastics and related chemical pollutants disproportionately affect low-income families and communities of color, contributing to poor health outcomes and exacerbating existing health disparities (Landrigan et al., 2023). Consequently, these populations face an increased risk of acute and chronic health conditions, including developmental disorders, cancers, and respiratory illnesses, which are often unaddressed due to a lack of resources and accessible healthcare. Children and pregnant women represent two of the most vulnerable groups affected by water contamination. The developing bodies of children are particularly susceptible to the adverse effects of toxins present in contaminated water. Exposure to heavy metals like lead or chemical pollutants such as PFAS during critical growth periods can lead to long-lasting developmental issues, including cognitive deficits and behavioral problems. Studies indicate that even low levels of lead in drinking water can affect a child's intelligence and academic performance, with significant implications for their future opportunities (Bălan et al., 2021). Additionally, pregnant women exposed to contaminated water may face serious risks, including low birth weight, premature birth, and developmental delays in their offspring. These health ramifications not only affect individuals but can also impose significant societal costs, including increased healthcare expenditures and lost productivity. The compounded effects of water contamination on vulnerable populations highlight the urgent need for policy interventions and community advocacy. Comprehensive measures must be implemented to address the environmental injustices faced by these at-risk communities. Policies should focus on improving water infrastructure, enforcing stricter regulations on industrial waste, and ensuring equitable access to safe drinking water for all. Enhancing public awareness about the sources and dangers of water contamination is equally crucial to empower communities to advocate for their health and well-being. By prioritizing the needs of vulnerable populations, we can begin to dismantle the systemic inequities that perpetuate water contamination and its deleterious effects on human health, ultimately fostering a healthier and more just society for all.

6.5. SOIL POLLUTION AND ITS IMPACT

Contaminants in the soil, particularly heavy metals and petroleum products, pose significant challenges to both environmental integrity and human health. The introduction of heavy metals such as cadmium (Cd), which can originate from industrial activities, poses a threat to soil quality and agricultural productivity. Further, Cd toxicity disrupts essential plant physiological functions, including water interaction and mineral uptake, resulting in reduced crop yields and potential food insecurity (Zulfiqar et al., 2022). Furthermore, the ability of plants to bioaccumulate cadmium can lead to increased levels of this toxic element in edible crops, thereby posing direct health risks to consumers. The contamination not only affects plant growth but also has cascading effects on entire ecosystems, disrupting food chains and diminishing biodiversity. Oil spills exemplify another critical aspect of soil pollution, with extensive implications for both the environment and public health. The degradation caused by petroleum products results in the loss of soil fertility and the introduction of toxic compounds that can bioaccumulate organisms, further endangering human health. According to (Ukhurebor et al., 2021), the impact of these spills can be devastating, not just on land but also on aquatic ecosystems, significantly affecting local agriculture and fishing industries. As communities increasingly rely on these sectors for their livelihood, the occurrence of petroleum pollution represents a multifaceted challenge that endangers economic stability, food security, and overall community health. The lack of effective and permanent solutions to the cleanup of oil spills underscores the urgency for the implementation of sustainable remediation strategies. Addressing soil pollution requires a multifaceted approach that integrates scientific research, community awareness, and environmental policy reforms. The emphasis should be on innovative bioremediation technologies that harness natural processes to mitigate the effects of hazardous pollutants. Legislative frameworks must be strengthened to protect against environmental degradation caused by industrial activities, as established in various studies noted in (Ukhurebor et al., 2021). Efforts should also concentrate on public health assessments that gauge the extent of soil pollution and its impact on human health. By promoting sustainable agricultural practices and employing effective remediation measures, society can not only combat soil pollution but also safeguard future generations from its deleterious effects. Fulfilling this imperative is essential for fostering resilient ecosystems and promoting public health in the face of ongoing environmental challenges (Figure 6.6).

Figure 6.6. *Soil pollution.*

Source: By CAWST (Centre for Affordable Water and Sanitation Technology) – https://www.flickr.com/photos/gtzecosan/19529117189, CC BY 2.0, https://commons.wikimedia.org/w/index.php?curid=41604768.

6.5.1. Sources of Soil Contamination

Pollution from industrial activities is a primary contributor to soil contamination, as factories often release hazardous byproducts and waste materials into the environment. These pollutants, which can include heavy metals, solvents, and chemicals, seep into the soil and can persist for years, posing significant risks to both human health and ecosystems. Such contamination can lead to reduced agricultural productivity, compromising the safety of food supplies. The growth and reproduction of soil organisms are drastically affected by these pollutants, as they disrupt the natural balance within soil ecosystems. Consequently, the contaminated soil not only degrades the immediate environment but also poses health risks through the bioaccumulation of toxins in plants and animals, ultimately impacting human health. Addressing industrial pollution through regulatory measures is crucial for protecting soil quality and ensuring safe agricultural practices. Agricultural practices also significantly contribute to soil contamination, primarily through the extensive use of pesticides and fertilizers. While these chemicals are intended to increase crop yields, their over-application can lead to the leaching of harmful substances into the soil, resulting in long-term contamination. Pesticides, specifically, are known to contain toxic compounds that disrupt soil microbiomes and can accumulate, affecting food safety and human health. Furthermore, runoff from agricultural fields can transport these chemicals into nearby water sources, exacerbating the problem of environmental contamination overall. As articulated in research, the relationship between agricultural practices and environmental health cannot be overlooked (Shokat et al., 2023). Sustainable farming methods need to be prioritized to mitigate these adverse impacts, as exploring alternatives like organic farming can significantly reduce soil contamination and promote healthier ecosystems. Furthermore, the improper disposal of waste, including industrial and municipal waste, poses another significant source of soil contamination. Landfills and illegal dumpsites often leak contaminants into the surrounding soil, leading to the introduction of various hazardous substances, including plastics, heavy metals, and organic pollutants. As waste decomposes, it releases toxins that can infiltrate the soil ecosystem, affecting soil health and the organisms that inhabit it. The consequences of this contamination on human health can be severe, as these substances can enter the food chain and ultimately pose risks through ingestion or skin contact. Research suggests that technologies like Verm biochar could be utilized to reduce heavy metal concentrations in contaminated soils, showcasing innovative approaches to counteract these destructive practices (Rehman et al., 2023). Addressing waste management issues is vital for maintaining soil integrity and protecting public health from the harmful effects of soil contamination (Figure 6.7).

Figure 6.7. *Soil contamination.*

Source: Public Domain, https://commons.wikimedia.org/w/index.php?curid=16396.

6.5.2. Effects on Food Safety and Nutrition

Environmental changes and hazards have profound implications for food safety and nutrition, as fluctuations in climate and ecosystems directly affect the availability and quality of food sources. For instance, rising temperatures and extreme weather events can lead to the proliferation of bacteria and pathogens in food supplies, increasing the risks of foodborne illnesses. Changes in precipitation patterns may also hinder agricultural productivity, thereby compromising the nutritional content of crops. As food becomes less safe and less nutritious, vulnerable populations—particularly in developing nations—are often disproportionately impacted, leading to malnutrition and food insecurity. This multidimensional challenge underscores the urgent need for effective adaptation strategies that can enhance resilience in food systems, ensuring that public health is prioritized amidst environmental upheaval. The intricate link between climate change, biodiversity, and food safety cannot be overstated. The decline in biodiversity, driven by environmental hazards, negatively influences pollination and soil health, which are critical for robust agricultural output. As plant and animal species diminish, agricultural diversity is also reduced, leading to monocultures that are less resilient to pests and diseases. This situation not only threatens food safety but may also lead to a reliance on chemical pesticides and fertilizers, potentially increasing health risks for consumers. The working group's findings emphasize that integrating ecological health and agricultural practices fosters food systems that are both sustainable and secure, ultimately safeguarding nutrition.

The interaction among climate, ecosystems, and agriculture further necessitates a holistic approach to food safety that considers ecological impacts, reinforcing the importance of proactive measures in the face of changing environmental conditions (Intergovernmental Panel on Climate Change, 2023). Nutrition quality often suffers as food systems face the repercussions of environmental hazards. Nutrient depletion in crops due to poor soil health and the overuse of synthetic inputs can lead to a diminished food supply that lacks essential vitamins and minerals. Additionally, climate change disrupts seasonal growing patterns, leading to inconsistencies in food availability and variety, which are crucial for a balanced diet. Food safety measures must, therefore, expand beyond simply preventing contamination; they should also prioritize the nutritional value of food produced. Importantly, awareness, and education on the relationship between environmental health and food safety play pivotal roles in mitigating these issues. The Food and Agriculture Organization (FAO) emphasizes that the coordination of sustainable agricultural practices with public health initiatives is vital for addressing these intertwined challenges effectively (Pier Paolo Gatta, 2022). As such, a concerted effort is necessary to adapt and innovate within our food systems to ensure safety and nutritional adequacy for future generations.

6.5.3. Health Implications for Communities Near Contaminated Sites

The presence of contaminated sites has profound health implications for nearby communities, often exacerbating existing vulnerabilities. Industrial contamination can lead to an array of environmental hazards, ranging from soil and water pollution to airborne toxins, which collectively elevate health risks for local populations. Residents in these areas frequently experience higher rates of chronic diseases, such as respiratory disorders and cancers, attributed to both direct exposure and the cumulative effects of pollutants over time. Moreover, socioeconomic factors often intertwine with environmental risks, resulting in marginalized communities enduring enhanced health threats and reduced access to healthcare resources. Addressing these inequalities is essential to mitigate the adverse health outcomes associated with living near contaminated sites. The concept of environmental injustice is especially pertinent in understanding the health implications for communities adjacent to contaminated areas. Many of these populations are subject to procedural injustices where they are often excluded from decision-making processes regarding site management and cleanup efforts. This lack of inclusion can stagnate initiatives aimed at remediating environmental hazards and improving public health. For instance, the SENTIERI project in Italy highlights how procedural injustices lead to the marginalization of communities affected by pollution, creating cycles of deprivation and health disparities. In communities facing environmental degradation, the interplay of health risks from contaminants and socioeconomic disadvantages results in significant barriers to health equity, thus necessitating urgent interventions to empower affected populations. Effective communication strategies are crucial for fostering a better understanding of health risks among communities living near contaminated sites. Engaging local residents and stakeholders in health communication initiatives can enhance environmental health literacy, empowering them to advocate for change. By establishing participatory processes that involve community input, like those implemented in the SENTIERI initiative, public health officials can build trust and encourage informed participation in health-related discussions. This approach informs residents about the potential health risks they face and promotes procedural justice by ensuring their voices are heard in environmental decision-making. Ultimately, enhancing community capacity through targeted communication can lead to strengthened responses to environmental hazards and improved health outcomes for affected populations, paving the way for actionable change in urban and rural settings alike.

6.6. CLIMATE CHANGE AND HEALTH RISKS

The implications of climate change for human health are both immediate and profound, as evidenced by increased exposure to extreme weather conditions. Rising global temperatures are leading to more frequent and intense heatwaves, which have been linked to higher mortality rates, particularly among vulnerable populations such as the elderly and children. In 2020 alone, the Lancet Countdown reported a staggering increase of 3.1 billion heatwave exposure days among individuals aged 65 and older, with populations in developing countries experiencing the most severe vulnerability due to inadequate infrastructure and limited access to cooling mechanisms (Romanello et al., 2021). Such direct health risks emphasize the urgent need for comprehensive adaptation strategies that prioritize community resilience, particularly in regions lacking the necessary resources to combat heat-related illnesses, thereby underscoring the intersection between climate change and public health. The intricate relationship between environmental shifts and infectious diseases further complicates the public health landscape. For instance, the changing climate has expanded the transmission potential of vector-borne diseases such as malaria and dengue fever. Over the past few decades, suitable conditions for malaria transmission in densely populated highland areas of low human development index (HDI) countries significantly increased, threatening populations that were previously at lower risk (Romanello et al., 2021). Similarly, the uptick in extreme weather events facilitates the spread of pathogens and strains healthcare systems, particularly in developing regions where resources are limited.

This scenario underscores the critical need for targeted research and intervention strategies to mitigate the health impacts associated with climate change-induced disease dynamics, as vulnerable populations may face severe and compounding health challenges. Addressing the health risks posed by climate change also requires an examination of food and water security as vital components of public health.

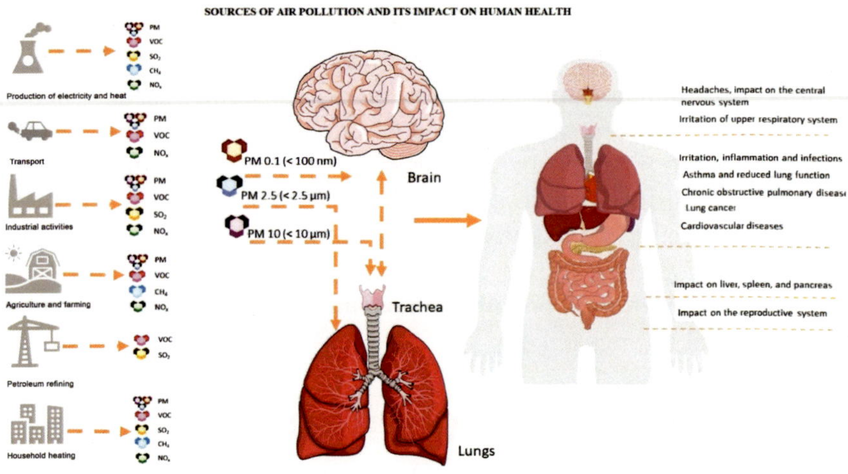

Figure 6.8. *Impact of climate change on human health.*

Source: Zorena K, Jaskulak M, Michalska M, Mrugacz M, Vandenbulcke F. Air Pollution, Oxidative Stress, and the Risk of Development of Type 1 Diabetes. Antioxidants. 2022; 11(10):1908. https://doi.org/10.3390/antiox11101908.

The increased frequency of extreme weather events, such as droughts and flooding, has resulted in significant crop failures and disruptions to food supply chains. For example, up to 19% of the global land surface encountered extreme drought conditions in 2020, an alarming increase compared to previous decades.

Such environmental stressors jeopardize food security, particularly in low and medium HDI countries, and exacerbate nutritional deficiencies that threaten overall health. As food systems become increasingly susceptible to climatic shifts, it is essential to develop sustainable agricultural practices and effective disaster response mechanisms that can protect vulnerable populations and ensure access to nutritious food. Recognizing and addressing these interconnected health risks is essential for fostering resilience against the impacts of climate change on human health (Figure 6.8).

6.6.1. Impact of Climate Change on Environmental Hazards

The increasing severity and frequency of environmental hazards are predominantly driven by the repercussions of climate change. Extreme weather events, such as hurricanes, floods, and droughts, have become more common due to rising global temperatures and shifting weather patterns. For instance, the Lancet Countdown report highlights how record-breaking temperatures have led to significant increases in heat exposure, particularly among vulnerable populations like the elderly and infants. This warming affects human health directly, resulting in heat-related illnesses and fatalities, and exacerbates other environmental hazards, including air pollution and water scarcity. The compounded effects of these hazards create a perilous cycle, where communities become increasingly vulnerable to health risks resulting from an unpredictable climate, thereby highlighting the critical link between climate-driven environmental hazards and public health outcomes. Moreover, climate change places immense strain on agricultural systems, showcasing the interconnectedness between environmental hazards and food security. According to the same report, alterations in temperature and precipitation patterns threaten crop yields, which could reverse years of progress in mitigating food insecurity. As staple crops suffer from reduced yield potential due to rising temperatures, the risk of malnutrition and foodborne illnesses increases, particularly among populations already facing socioeconomic challenges. Additionally, changing climate conditions foster an environment that is more conducive to the spread of pests and diseases, further jeopardizing agricultural stability. By recognizing the adverse impacts of climate change on agriculture, it becomes evident that strategies to combat environmental hazards must also encompass sustainable farming practices and policies aimed at protecting food resources. Furthermore, the interplay between climate change, environmental hazards, and public health underscores the need for comprehensive adaptation strategies. Despite overwhelming evidence regarding the health impacts of a changing climate, many countries struggle to implement effective national health emergency frameworks due to inadequate resources and varying levels of human development. As reported by the Lancet Countdown, over 63% of countries lack high-level implementation of these essential frameworks, leaving them ill-prepared for climate-related health emergencies. Addressing these disparities is crucial; equitable access to resources, improved infrastructure, and public health interventions can mitigate the effects of environmental hazards exacerbated by climate change. As nations contemplate recovery from recent global crises, investments in climate-resilient strategies can pave the way for healthier and more sustainable futures, consequently enhancing resilience against multifaceted environmental hazards (Boko et al., 2018; Romanello et al., 2021).

6.6.2. Increased Prevalence of Vector-Borne Diseases

The accelerated frequency and intensity of climate change have created ideal conditions for the proliferation of vector-borne diseases, significantly impacting public health across the globe. Warmer temperatures expand the habitats of disease-carrying vectors, such as mosquitoes and ticks, facilitating their spread into new geographic areas. As a result, previously unaffected populations now face increased risks of diseases like malaria, dengue fever, and Lyme disease, challenging health infrastructures that are often ill-equipped to respond. The Lancet Countdown indicates that the number of months with suitable conditions for malaria transmission has risen dramatically in highland areas of low HDI countries, which further threatens vulnerable populations who have historically been less exposed to such risks (Romanello et al., 2021). This alarming trend underscores the urgent need for public health responses and proactive measures to mitigate the effects of climate change on vector dynamics. Addressing the rise in vector-borne diseases requires a multifaceted approach that encompasses both environmental and health strategies. Vector control programs, community education, and robust healthcare responses play crucial roles in combating outbreaks linked to climate change. As the evidence mounts, public health entities must adapt their strategies to the shifting landscape of infectious diseases. The increasing suitability for the transmission of pathogens like *Plasmodium falciparum*, which causes malaria, signifies a public health emergency and highlights systemic inequalities in health responses. Health systems in low HDI regions may struggle under the burden of additional diseases, exacerbating existing health disparities (Watts et al., 2020). Collaborative efforts between national and international organizations are essential to implement effective, adaptive control measures tailored to local conditions, thereby reducing morbidity and mortality rates. Moreover, combating the surge of vector-borne diseases will require significant investments in research and innovation to develop new tools for surveillance and intervention. Advances in technology can help predict outbreaks and assess the risk of disease transmission, allowing for timely responses. Health systems must be strengthened to ensure that they are prepared for the challenges posed by climate change. As the evidence increasingly points to the significant health risks associated with rising global temperatures, it becomes imperative for policymakers to prioritize adaptation strategies that include health resilience in the face of vector-borne diseases. By aligning climate action with public health initiatives, there is an opportunity to foster greater cooperation across sectors, ultimately safeguarding communities from the compounding effects of environmental changes on human health and well-being (Romanello et al., 2021; Watts et al., 2020).

6.6.3. Mental Health Effects Related to Climate Anxiety

The profound psychological toll of climate anxiety has emerged as a critical mental health issue in contemporary society. As individuals increasingly comprehend the scale of environmental deterioration, feelings of helplessness and despair proliferate, culminating in heightened stress and anxiety. This phenomenon, often referred to as climate anxiety, manifests through worries about future ecological catastrophes, such as extreme weather events and biodiversity loss. The emotional distress is further exacerbated by media coverage of climate events and the profound implications for global ecosystems and human livelihoods. With research indicating that disruptions caused by climate change lead to increased risks of mental health disorders, many individuals experience chronic fears and negative emotional states that hinder their ability to cope with daily life. The connection between climate change and mental health extends beyond mere anxiety, as it encompasses an array of psychological impacts stemming from environmental disasters. Events such as floods, droughts, and heatwaves can lead to psychological trauma, driving up the prevalence of disorders like post-traumatic stress, depression, and anxiety disorders (Kumar et al., 2023).

Furthermore, exposure to problematic environmental factors, such as air pollution or inadequate access to green spaces, significantly contributes to the deterioration of mental health. Societal factors, including displacement from homes due to climate-related disasters, amplify the psychological burden, further emphasizing the interdependence between environmental hazards and mental well-being. Consequently, addressing mental health concerns linked to climate change must be prioritized to ensure holistic responses to ecological crises. Recognizing the potential economic ramifications of climate-related mental health disorders is crucial for effective policy formulation and intervention strategies. Projections indicate that the global cost of mental health issues exacerbated by climate change could reach approximately $47 billion annually by 2030, escalating to an alarming $537 billion by 2050, if current trends continue without intervention (Kumar et al., 2023). These staggering figures underscore the necessity for policymakers to integrate mental health considerations into climate resilience frameworks. By fostering greater awareness of the mental health impacts linked to environmental degradation, communities can develop targeted interventions that minimize anxiety and promote psychological well-being. Strengthening the evidence base for climate anxiety is essential, paralleling efforts to mitigate physical health risks associated with climate change, thus promoting a comprehensive approach to enhance human health in the face of environmental hazards (Figure 6.9).

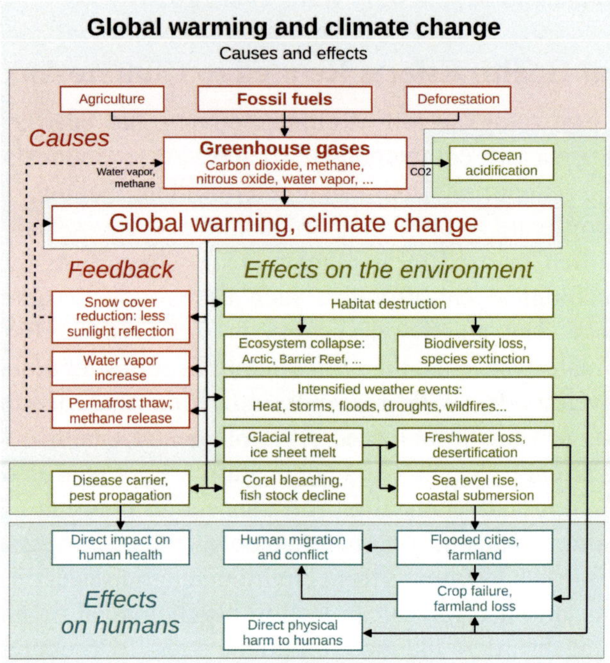

Figure 6.9. *Climate impacts mental health.*

Source: By RCraig09 – Own work Also, successive versions were developed through collaboration with users Efbrazil, Femkemilene, and J. Johnson on en.wikipedia, CC BY-SA 4.0, https://commons.wikimedia.org/w/index.php?curid=113471903.

6.7. MITIGATION STRATEGIES FOR ENVIRONMENTAL HAZARDS

In addressing the escalating prevalence of environmental hazards, it is crucial to implement comprehensive mitigation strategies that target the immediate effects and the underlying causes of pollution. Integrating innovative remediation techniques can significantly enhance the management of hazardous materials, particularly in regions affected by oil spills and heavy metal contamination. Techniques such as bioremediation have shown promise due to their eco-friendly nature and cost-effectiveness, providing a viable alternative to traditional methods that may inadvertently exacerbate environmental degradation. For instance, bacteria with metal resistance can be utilized to remove heavy metals from polluted sites, showcasing how biological management can restore the ecological balance without imposing further risks on human health and the environment (Mathivanan et al., 2021). Moreover, regulatory frameworks and public awareness play a vital role in reinforcing these mitigation strategies. Constructing a robust legal apparatus to govern industrial practices is essential for preventing environmental degradation. In regions like the Niger Delta, where petroleum spillages are frequent, the lack of stringent regulations has led to devastating consequences on both human health and local ecosystems. Awareness campaigns can mobilize community involvement, compelling industries to adopt safer practices, and promoting the development and application of greener technologies. By highlighting the significance of sustainable practices, stakeholders can work collaboratively to create a healthier environment that prioritizes restoration and protection from ongoing pollution (Ukhurebor et al., 2021). Finally, a multi-faceted approach that combines technological innovation, regulatory reforms, and community engagement is indispensable for addressing environmental hazards effectively. Continuous research into adaptive and resilient strategies, including the exploration of advanced bioremediation technologies, can provide crucial insights into managing pollutants. As future challenges loom, these strategies must evolve to mitigate the impacts of environmental hazards on human health. This holistic perspective, emphasizing the interconnectedness of environmental health and human well-being, serves as a foundation for developing sustainable solutions. Ultimately, fostering a culture of responsibility among industries, governments, and individuals alike is essential to ensure a healthier planet for future generations.

6.7.1. Policy Measures and Regulations

Policy measures and regulations play a crucial role in mitigating environmental hazards that adversely impact human health. Effective legislation can establish standards for air and water quality, promote sustainable land use practices, and regulate waste management systems. For instance, establishing legal frameworks that mandate environmental impact assessments can ensure that proposed projects account for their potential effects on public health. Such regulations help safeguard vulnerable populations disproportionately affected by environmental degradation. Furthermore, laws that govern the production and management of chemicals can significantly reduce exposure to toxic substances, thereby fostering healthier communities. As noted by the Lancet

Commission, "law can help build and implement Universal Health Coverage (UHC), which is a crucial element of sustainable development," emphasizing the interconnectedness between legal standards and public health outcomes (Gostin et al., 2019). The interaction between public health and environmental policy is fundamental to fostering a healthier society. Regulatory measures can address immediate public health concerns and long-term environmental sustainability. For example, policies aimed at reducing carbon emissions contribute to improved air quality, which is directly linked to respiratory and cardiovascular health. Additionally, food systems present both challenges and opportunities.

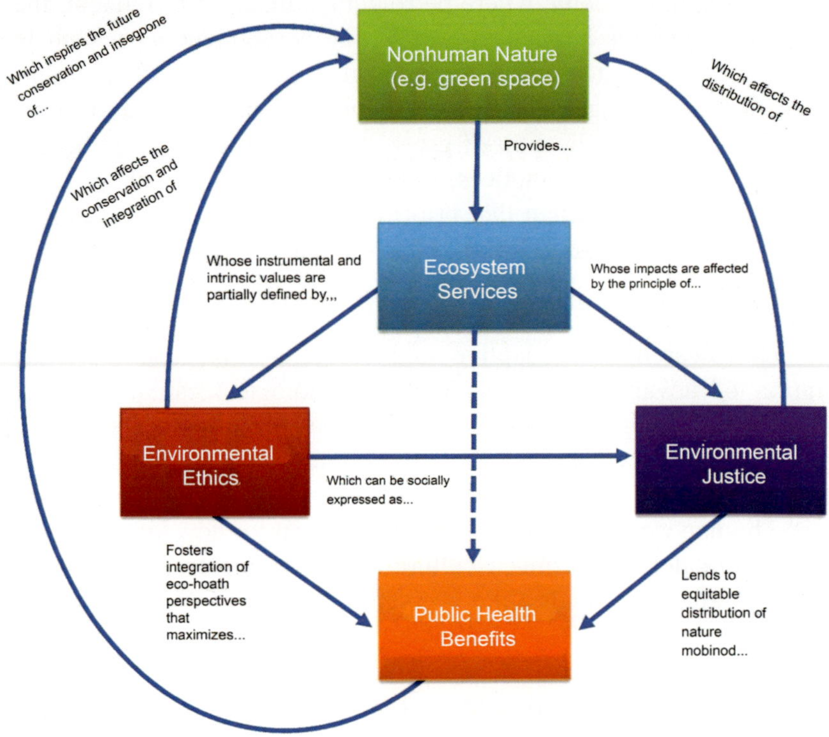

Figure 6.10. *Environmental health promotion and policy.*

Source: By Viniece Jennings, Jessica Yun, Lincoln Larson – Viniece Jennings et al. "Finding Common Ground: Environmental Ethics, Social Justice, and a Sustainable Path for Nature-Based Health Promotion," Healthcare doi:10.3390/healthcare4030061, CC BY 4.0, https://commons.wikimedia.org/w/index.php?curid=75480372.

A well-regulated food industry can ensure access to safe, nutritious food while minimizing the environmental impacts of agricultural practices. As highlighted in the Lancet-ONeill Institute report, "law can translate a vision into action on sustainable development, thus illustrating the critical role of policy in guiding sustainable practices within the food sector (Willett et al., 2019). Comprehensive regulations that support sustainable food systems can be instrumental in combating diet-related diseases, underscoring the importance of policies that integrate health and environmental objectives. Moreover, the enforcement of existing regulations is as vital as their creation. Without proper enforcement mechanisms, policies may fail to deliver intended benefits, leading to persistent environmental hazards that threaten human health. Governments must prioritize compliance through monitoring and evaluation systems that ensure adherence to environmental standards. Furthermore, engaging local communities in the policy-making process can enhance accountability and promote more effective regulation. Public participation helps identify specific environmental health concerns and fosters a sense of ownership regarding community health outcomes. As the Lancet Commission indicates, the aim is to enhance the global health community's understanding of law, regulation, and the rule of law as effective tools to advance population health and equity (Gostin et al., 2019). This perspective underscores that robust policy measures coupled with strong enforcement and community involvement are instrumental in addressing the environmental determinants of health (Figure 6.10).

6.7.2. Community Awareness and Education

A significant aspect of enhancing community resilience against environmental hazards is fostering a culture of awareness and education. Communities that actively engage in learning about local environmental risks are better equipped to respond to disasters and mitigate their impacts. The link between environmental health and human well-being is often underappreciated, leading to preventable health crises. According to recent reports, disasters induced by climate change have disproportionately affected vulnerable populations, including women and children. Education initiatives that focus on these connections can empower residents to advocate for safer practices and policies. Moreover, incorporating sustainable healthcare education within community programs aids in understanding the intricate relationship between ecosystems and health, ultimately fostering a more proactive community dedicated to enhancing public health and environmental safety. Effective community education programs must address the diverse needs and knowledge levels of their populations. Utilizing interactive and scenario-based learning can significantly enhance engagement and retention of information about environmental hazards. For instance, as highlighted by the Sustainable Healthcare Education network, case studies from various institutions demonstrate the success of integrating environmental health education into curricula. These programs can inspire collective action, encouraging community members to participate in discussions about local environmental issues and solutions.

Involving citizens in creating localized environmental policies and plans promotes responsibility and emphasizes the importance of community input in addressing public health challenges related to environmental hazards. This grassroots approach can empower individuals to understand their roles in fostering resilience and advocates for positive change. In today's interconnected world, the responsibility for environmental education is not solely on community leaders but shared among various stakeholders, including governments, non-profits, and local businesses. Collaboration among these entities can lead to more comprehensive education strategies that reach a wider audience and address specific local challenges. For instance, partnerships can be formed to provide workshops on disaster preparedness or initiatives focused on reducing exposure to environmental toxins. As disasters increasingly affect communities globally, often with devastating economic consequences, proactive education becomes crucial in risk reduction. The concerted effort to inform and engage citizens creates a well-rounded community capable of managing environmental issues effectively, ultimately leading to improved public health outcomes and increased resilience in the face of future hazards (Walpole et al., 2019).

6.7.3. Technological Innovations for Hazard Reduction

Addressing environmental hazards and their impact on human health necessitates the adoption of innovative technologies designed for hazard reduction. These innovations range from early warning systems for natural disasters to advanced air quality monitoring devices. For instance, smart sensors can be strategically placed in urban areas to detect and provide real-time data on air pollutants, such as particulate matter and nitrogen oxide. This technology enables swift action to mitigate exposure, thereby protecting public health. As more cities integrate these monitoring systems, they can respond effectively to pollution spikes and implement public health advisories, fostering a healthier environment. Such proactive measures are crucial, especially given the links between pollution and respiratory conditions, underscored by the detrimental effects of air quality on health as noted in (Ioannis Manisalidis et al., 2020). Moreover, advancements in agriculture and food production technologies play a pivotal role in hazard reduction. Sustainable farming practices, supported by precision agriculture, utilize technology like drones and satellite imagery to efficiently monitor crop health and resource usage. This reduces the risk of over-reliance on harmful pesticides and enhances food security and nutritional quality. As emphasized in research, while global food production is rising, more than 820 million people still face food insufficiency. By employing innovations in food systems, it becomes possible to create more sustainable practices that enhance both environmental health and human nutrition (Willett et al., 2019). Thus, focusing on technological advancements in agriculture is essential for reducing environmental hazards and, consequently, the related health risks that stem from inadequate diets. Finally, public awareness and education must keep pace with technological advancements to maximize their effectiveness in hazard reduction. The implementation of new technologies is significantly enhanced when communities understand their benefits and operations. Education initiatives that inform citizens about air quality monitoring apps or sustainable food practices empower individuals to make better choices for themselves and their families. Consequently, a multidisciplinary approach, involving scientific experts and community leaders, is essential to facilitate this integration. Raising public awareness ensures that technological innovations are not only adopted but utilized effectively in everyday life. The convergence of technology and education is vital to combat environmental hazards and mitigate their impact on human health, ultimately leading to a more informed and healthier population.

6.8. CONCLUSION

The intricate relationship between environmental hazards and human health underscores the urgency of addressing the ongoing challenges posed by pollution and toxic substances. As highlighted in various studies, including those concerning oil spillages in sensitive regions like the Niger Delta, Nigeria environmental degradation has far-reaching effects on human health, agriculture, and ecosystems. These interactions reveal a stark reality: inadequate management of environmental risks threatens biodiversity and poses significant health hazards to populations exposed to pollutants. Consequently, addressing these issues demands a holistic understanding of the implications of our actions on both the environment and public well-being. Sustained efforts in identifying and implementing sustainable remediation techniques are crucial for mitigating these risks and fostering an environment that promotes health rather than deterioration. The proliferation of chemicals such as glyphosate further exemplifies the dire consequences of neglecting environmental health. This ubiquitous herbicide, widely utilized in agricultural practices, has been scrutinized for its potential toxicity and carcinogenicity, raising alarms about its long-term effects on human health and ecosystems. Many international agencies have conducted assessments, leading to a complex classification based on accumulating evidence regarding its hazards. As research indicates, the incorporation of adjuvants in glyphosate formulations often goes unnoticed, complicating risk assessments and regulatory measures (Kingsley Eghonghon Ukhurebor et al., 2021). To protect public health, policymakers must consider these additives and promote transparent labeling, ensuring that stakeholders are informed about the risks associated with pesticides. This also reinforces the need for more comprehensive toxicological studies that address both primary and ancillary ingredients in pesticide formulations. In conclusion, fostering a sustainable relationship between human activity and environmental health requires rigorous research, transparent policies, and public engagement. The alarming findings associated with pollutants such as glyphosate and the repercussions of oil spillages reveal that current remediation efforts are insufficient to resolve the pervasive threats we face (Gandhi et al., 2021; Kingsley Eghonghon Ukhurebor et al., 2021). It is vital to adopt a proactive approach in addressing these challenges, promoting innovative strategies such as bioremediation and robust public health assessments. Such measures will help mitigate existing hazards and prevent future occurrences, ultimately enhancing the quality of life within affected communities. As we continue to navigate the complexities of environmental hazards and their implications, it becomes increasingly clear that sustainable development must be pursued as an integral part of public health initiatives.

6.8.1. Summary of Key Points Discussed

Environmental hazards pose significant risks to human health, a theme that resurfaces throughout discussions on climate change and biodiversity. The Intergovernmental Panel on Climate Change (IPCC) emphasizes the interconnectedness of climate, ecosystems, and human societies, particularly in its Sixth Assessment Report. This report highlights how climate-resilient development can mitigate the impacts of climate change while fostering adaptation strategies that promote sustainability. Essentially, this convergence seeks to improve human living conditions while conserving the natural world, addressing the pressing need for action to combat environmental hazards. A notable takeaway is that coordinated adaptation efforts alongside greenhouse gas emission reductions are crucial for maintaining the delicate balance between human needs and ecological integrity, making organizations like the IPCC invaluable as global guides for policy and action in the face of climate-related challenges (Intergovernmental Panel on Climate Change, 2023). The deterioration of ecosystems, as documented in recent assessments, can have dire consequences for human well-being. Ecosystem changes, driven by population growth and economic demands, have resulted in significant declines in species abundance, habitat integrity, and the variety of local ecosystems. Such transformations undermine the myriad benefits that nature provides, including clean air, freshwater, and food security. Evidence suggests that the costs associated with these losses are not distributed equally, affecting marginalized communities disproportionately. The pressing need to reverse these trends requires collective action targeting economic, social, and technological frameworks to address the root causes of environmental degradation. The urgency of this issue is exacerbated by the interlinked crises of health and environmental sustainability, emphasizing that immediate and transformative measures are necessary to ensure the quality of life for future generations (Díaz et al., 2019). In summary, the interplay between environmental hazards and human health underscores a critical call to action. As illustrated through the findings of organizations like the IPCC and recent ecological studies, the implications of neglecting the health of our planet extend well beyond ecological concerns; they threaten the foundational aspects of human prosperity. The concept of climate resilient development encapsulates a necessary synergy between adaptation strategies and mitigation efforts, aiming to foster a sustainable relationship with our natural environment. Recognizing the urgency of this situation is imperative for creating effective policies and initiating transformative actions that prioritize both environmental restoration and public health. By addressing the systemic issues that underlie environmental hazards, we can work towards a healthier, more equitable future for all) (Díaz et al., 2019; Intergovernmental Panel on Climate Change, 2023).

6.8.2. The Importance of Proactive Measures

In an era marked by rapid environmental change and increasing health risks, adopting proactive measures is crucial for safeguarding both ecosystems and public health. The alarming decline in the integrity of ecosystems and wild species highlights the urgent need for preventive actions. As articulated in Díaz et al (2019), the benefits derived from nature are diminishing, threatening the quality of life for future generations. By implementing proactive strategies, such as sustainable resource management and conservation efforts, communities can mitigate the detrimental effects of environmental hazards. These measures not only preserve biodiversity but also ensure the continuous availability of natural resources essential for human survival. The shift from reactive to proactive policies creates a resilient framework that empowers societies to address challenges before they escalate, ultimately fostering a healthier environment. Moreover, proactive measures can enhance global supply chain sustainability, which is increasingly critical in the face of globalization. As mentioned in research, blockchain technology represents a promising tool for improving transparency and traceability within supply chains, facilitating efforts to minimize environmental impacts (Saberi et al., 2018). By adopting such technological advancements, companies can anticipate potential disruptions and environmental challenges, allowing for more effective resource allocation and waste reduction. This proactive stance encourages businesses to engage in sustainable practices that benefit both the environment and their bottom line.

Furthermore, as consumers become more environmentally conscious, businesses that prioritize sustainability through proactive commitments will likely gain a competitive advantage. Therefore, investing in innovative solutions like blockchain can be a vital step towards fostering sustainable practices and upholding corporate responsibility. In conclusion, the importance of proactive measures cannot be overstated when addressing environmental hazards and their implications for human health. Preventative strategies, whether through policy changes, technological innovations, or community engagement, are essential for curbing the negative effects of human activity on ecosystems. The ongoing deterioration of the natural world, as highlighted in research, underscores the urgency for immediate action to reverse these trends (Díaz et al., 2019). Proactivity empowers societies to combat existing challenges and equips them with the tools necessary to anticipate and address future threats. By collectively prioritizing proactive approaches, we can create a healthier, more sustainable world that preserves both human well-being and ecological integrity for generations to come.

6.8.3. Call to Action for Individuals and Policymakers

Addressing environmental hazards requires a concerted effort from both individuals and policymakers, as the intertwined nature of human activities and ecological well-being calls for comprehensive action. Individuals can begin by making conscious lifestyle choices that reduce their carbon footprints, such as opting for public transport, reducing waste, and supporting sustainable products. Every small action contributes to a larger movement toward improved air quality, reduced water contamination, and overall environmental health. Moreover, grassroots initiatives can be invaluable; facilitating community engagement and raising awareness about local environmental issues can empower citizens to demand change from lawmakers. Therefore, an informed and committed populace can drive the momentum needed for substantial reforms while cultivating a culture of environmental responsibility. Policymakers play a pivotal role in shaping the regulatory landscape that governs environmental protection. They must prioritize legislation that addresses the root causes of environmental hazards, such as industrial pollution and unsustainable land use practices. Committing to research and investing in clean technologies is crucial, as these advancements can mitigate risks associated with climate change and resource depletion. Additionally, integrating community feedback when formulating policies can ensure that those most affected by environmental hazards have their voices heard in the decision-making process. By fostering collaboration among different stakeholders, from governmental agencies to private businesses and non-profit organizations, policymakers can create a framework that addresses current environmental challenges and anticipates future ones, ensuring a sustainable and healthy ecosystem for generations to come. A dual approach that combines individual responsibility with decisive policy action is essential in combating environmental crises effectively. As individuals become more aware of their environmental impact and actively participate in advocacy, the collective voice can influence policymakers to enact meaningful change. Conversely, when government initiatives prioritize ecological sustainability, they set a standard that encourages citizens to adopt more sustainable practices. Education plays a critical role here, as a well-informed public is more likely to engage in civic activities and support policies that reflect their values. Ultimately, without a united front comprising both proactive individuals and dedicated policymakers, addressing the myriad of environmental hazards effectively remains an uphill battle. Thus, each person and every government entity must take decisive action, ensuring the preservation of our planet and the health of future generations.

REFERENCES

1. Ali, H., Khan, E., & Ilahi, I. (2019). Environmental chemistry and ecotoxicology of hazardous heavy metals: Environmental persistence, toxicity, and bioaccumulation. *Environmental Toxicology, 2019*, Article 6730305. https://doi.org/10.1155/2019/6730305.
2. Bălan, S., Mathrani, V. C., Guo, D. F., & Algazi, A. M. (2021). Regulating PFAS as a chemical class under the California Safer Consumer Products Program. *Environmental Health Perspectives, 129*. https://doi.org/10.1289/ehp7431.
3. Boko, M., Niang, I., Nyong, A., Vogel, A. I., Githeko, A. K., Medany, M. A., Osman-Elasha, B., Tabo, R., & Yanda, P. Z. (2018). Africa Climate Change 2007: Impacts, adaptation and vulnerability: Contribution of Working Group II to the Fourth Assessment Report of the Intergovernmental Panel on Climate Change. *Cambridge University Press*.
4. Chen, G., Wang, X., Wang, R., & Liu, G. (2019). Health risk assessment of potentially harmful elements in subsidence water bodies using a Monte Carlo approach: An example from the Huainan coal mining area, China. *Ecotoxicology and Environmental Safety, 171*, 737–745. https://doi.org/10.1016/j.ecoenv.2018.12.101.
5. Dammers, E., McLinden, C. A., Griffin, D., Shephard, M. W., van der Graaf, S., Lutsch, E., Schaap, M., Gainairu-Matz, Y., Fioletov, V., Van Damme, M., Whitburn, S., Clarisse, L., Cady-Pereira, K., Clerbaux, C., Coheur, P., & Erisman, J. W. (2019). NH emissions from large point sources derived from CrIS and IASI satellite observations. *Atmospheric Chemistry and Physics, 19*(19), 12261–12293. https://doi.org/10.5194/acp-19-12261-2019.
6. Díaz, S., Settele, J., Brondízio, E. S., Ngo, H. T., Agard, J., Arneth, A., Balvanera, P., Brauman, K. A., Butchart, S. H. M., Chan, K. M. A., Garibaldi, L. A., Ichii, K., Liu, J., Subramanian, S. M., Midgley, G. F., Miloslavich, P., Molnár, Z., Obura, D., Pfaff, A., Polasky, S., Purvis, A., Razzaque, J., Reyers, B., Roy Chowdhury, R., Shin, Y. J., Visseren-Hamakers, I. J., Willis, K. J., & Zayas, C. N. (2019). Pervasive human-driven decline of life on Earth points to the need for transformative change. *Science, 366*(6471), 1035–1052. https://doi.org/10.1126/science.aax3100.
7. Gandhi, K., Khan, S., Patrikar, M., Markad, A., Jothi Kumar, N., Choudhari, A., Sagar, P., & Indurkar, S. (2021). Exposure risk and environmental impacts of glyphosate: Highlights on the toxicity of herbicide co-formulants. *Environmental Chemistry, 4*, 100149. https://doi.org/10.1016/j.envc.2021.100149.

8. Garofalo, C., Milanović, V., Cardinali, F., Aquilanti, L., Clementi, F., & Osimani, A. (2019). Current knowledge on the microbiota of edible insects intended for human consumption: A state-of-the-art review. *Food Research International, 125*, 108527. https://doi.org/10.1016/j.foodres.2019.108527.
9. Gasparotto, J., & Martinello, K. D. B. (2020). Coal as an energy source and its impacts on human health. *Environmental Geosciences, 2*, 113–120. https://doi.org/10.1016/j.engeos.2020.07.003.
10. Gatta, P. P. (2022). *The State of World Fisheries and Aquaculture 2022*. https://doi.org/10.4060/cc0461en.
11. Gostin, L. O., Monahan, J., Kaldor, J., DeBartolo, M. C., Friedman, E., Gottschalk, K., Kim, S. C., Alwan, A., Binagwaho, A., Burci, G. L., Cabal, L., DeLand, K., Evans, T., Goosby, E., Hossain, S., Koh, H. K., Ooms, G., Roses Periago, M., Uprimny, R., & Yamin, A. E. (2019). The legal determinants of health: Harnessing the power of law for global health and sustainable development. *The Lancet, 393*(10170), 1857–1910. https://doi.org/10.1016/s0140-6736(19)30233-8.
12. Horton, A. A., Walton, A., Spurgeon, D. J., Lahive, E., & Svendsen, C. (2017). Microplastics in freshwater and terrestrial environments: Evaluating the current understanding to identify the knowledge gaps and future research priorities. *Science of the Total Environment, 586*, 127–141. https://doi.org/10.1016/j.scitotenv.2017.01.190.
13. Intergovernmental Panel on Climate Change. (2023). *Climate Change 2022–Impacts, Adaptation and Vulnerability*. https://doi.org/10.1017/9781009325844.
14. Jager, C., Patrick, S., Aneck-Hahn, N. H., & Bornman, R. (2024). SPL18 endocrine disrupting chemicals: Impact on the environment and human health. *Occupational Medicine, 74*. https://doi.org/10.1093/occmed/kqae023.0029.
15. Kumar, A., Subrahmanyam, G., Mondal, R., Cabral-Pinto, M. M. S., Shabnam, A. A., Jigyasu, D. K., Malyan, S. K., Fagodiya, R. K., Khan, S. A., Kumar, A., & Yu, Z. G. (2021). Bio-remediation approaches for alleviation of cadmium contamination in natural resources. *Chemosphere, 268*, 128855. https://doi.org/10.1016/j.chemosphere.2020.128855.
16. Kumar, P., Brander, L., Kumar, M., & Cuijpers, P. (2023). Planetary health and mental health nexus: Benefit of environmental management. *Annals of Global Health, 89*, Article 49. https://doi.org/10.5334/aogh.4079.
17. Landrigan, P. J., Raps, H., Cropper, M., Bald, C., Brunner, M., Canonizado, E. M., et al., (2023). The Minderoo-Monaco commission on plastics and human health. *Annals of Global Health, 89*(1), 23. https://doi.org/10.5334/aogh.4056.

18. Manisalidis, I., Stavropoulou, E., Stavropoulos, A., & Bezirtzoglou, E. (2020). Environmental and health impacts of air pollution: A review. *Frontiers in Public Health, 8*. https://doi.org/10.3389/fpubh.2020.00014.

19. Marsili, D., Pasetto, R., & Iavarone, I. (2023). Environmental public health communication to engage stakeholders and foster social capacity in poorly involved communities living in industrial contaminated sites: The case study of Porto Torres (Italy). *Frontiers in Communication, 8*, 1217427. https://doi.org/10.3389/fcomm.2023.1217427.

20. Minx, J. C., Lamb, W. F., Callaghan, M., Fuss, S., Hilaire, J., Creutzig, F., Amann, T., Beringer, T., de Oliveira Garcia, W., Hartmann, J., Khanna, T., Lenzi, D., Luderer, G., Nemet, G. F., Rogelj, J., Smith, P., Vicente Vicente, J. L., Wilcox, J., & Zamora Dominguez, M. M. (2018). Negative emissions—Part 1: Research landscape and synthesis. *Environmental Research Letters, 13*(6), 063001. https://doi.org/10.1088/1748-9326/aabf9b.

21. Nelson, C. A., 3rd, & Gabard-Durnam, L. J. (2020). Early adversity and critical periods: Neurodevelopmental consequences of violating the expectable environment. *Trends in Neurosciences, 43*(3), 133–143. https://doi.org/10.1016/j.tins.2020.01.002.

22. Pasetto, R., & Marsili, D. (2023). Environmental justice promotion in the Italian contaminated sites through the national epidemiological surveillance system SENTIERI. *Environmental Health, 47*(1–2 Suppl 1), 375–384. https://www.semanticscholar.org/paper/b454524e83d32773b71a706638c5ff397c5d6998 (accessed on 5 December 2024).

23. Patrício, J., Elliott, M., Mazik, K., Papadopoulou, N., & Smith, C. (2016). DPSIR—Two decades of trying to develop a unifying framework for marine environmental management? *Frontiers in Marine Science, 3*, 177. https://doi.org/10.3389/fmars.2016.00177.

24. Romanello, M., McGushin, A., Di Napoli, C., Drummond, P., Hughes, N., Jamart, L., et al., (2021). The 2021 report of the Lancet Countdown on health and climate change: Code red for a healthy future. *The Lancet, 398*(10311), 1619–1662. https://doi.org/10.1016/S0140-6736(21)01787-6.

25. Saberi, S., Kouhizadeh, M., Sarkis, J., & Shen, L. (2018). Blockchain technology and its relationships to sustainable supply chain management. *International Journal of Production Research, 57*(1), 1–19. https://doi.org/10.1080/00207543.2018.1533261.

26. Shokat, S., Iqbal, R., & Yaqub, A. (2023). Molecular effect of pollutants on human diversity in Pakistan. https://www.semanticscholar.org/

paper/c4929f792516e0e4d5b5789d8c274a890858783e (accessed on 5 December 2024).

27. Sridharan, S., Kumar, M., Saha, M., Kirkham, M. B., Singh, L., & Bolan, N. (2022). The polymers and their additives in particulate plastics: What makes them hazardous to the fauna? *Science of The Total Environment, 824*, 153828. https://doi.org/10.1016/j.scitotenv.2022.153828.
28. Su, R., De Castro, F., Marini, P., Aprile, A., Benedetti, M., & Fanizzi, F. P. (2023). Vermibiochar: A novel approach for reducing the environmental impact of heavy metals contamination in agricultural land. *Sustainability, 15*(12), 9380. https://doi.org/10.3390/su15129380.
29. Szabo, S., Brondízio, E. S., Renaud, F. G., Hetrick, S., Nicholls, R. J., Matthews, Z., Tessler, Z. D., Tejedor, A., Sebesvári, Z., Foufoula Georgiou, E., Fonseca da Costa, S. M., & Dearing, J. A. (2016). Population dynamics, delta vulnerability and environmental change: Comparison of the Mekong, Ganges–Brahmaputra, and Amazon delta regions. *Sustainability Science, 11*, 539–554. https://doi.org/10.1007/s11625-016-0372-6.
30. Tilocca, B., Cao, A., & Migheli, Q. (2020). Scent of a killer: Microbial volatilome and its role in the biological control of plant pathogens. *Frontiers in Microbiology, 11*, 41. https://doi.org/10.3389/fmicb.2020.00041.
31. Ukhurebor, K. E., Athar, H., Adetunji, C. O., Aigbe, U. O., Onyancha, R. B., & Abifarin, O. (2021). Environmental implications of petroleum spillages in the Niger Delta region of Nigeria: A review. *Journal of Environmental Management, 293*, 112872. https://doi.org/10.1016/j.jenvman.2021.112872.
32. United Nations Office for Disaster Risk Reduction. (2018). *Sendai Framework for Disaster Risk Reduction 2015–2030*. https://doi.org/10.1163/2210-7975_hrd-9813-2015016.
33. Verma, N. (2024). Sustainable approaches used in pesticides contaminated agricultural waste water remediation. https://www.semanticscholar.org/paper/95fde4e6b2aa09d2b1c113b9ee4e20eb6cedfc4f (accessed on 5 December 2024).
34. Walpole, S., Barna, S., Richardson, J., & Rother, H. A. (2019). Sustainable healthcare education: Integrating planetary health into clinical education. *The Lancet Planetary Health, 3*(1), e6–e7. https://doi.org/10.1016/s2542-5196(18)30246-8.
35. Willett, W. C., Rockström, J., Loken, B., Springmann, M., Lang, T., Vermeulen, S. J., et al., (2019). Food in the Anthropocene: The EAT–Lancet Commission on healthy diets from sustainable food systems. *The Lancet, 393*(10170), 447–492. https://doi.org/10.1016/s0140-6736(18)31788-4.

36. Yang, X., Man, Y. B., Wong, M. H., Owen, R. B., & Chow, K. L. (2022). Environmental health impacts of microplastics exposure on structural organization levels in the human body. *Science of The Total Environment, 825*, 154025. https://doi.org/10.1016/j.scitotenv.2022.154025.

37. Zheng, X., Orellano, P., Lin, H., Jiang, M., & Guan, W. (2021). Short-term exposure to ozone, nitrogen dioxide, and sulphur dioxide and emergency department visits and hospital admissions due to asthma: A systematic review and meta-analysis. *Environment International, 150*, 106435. https://doi.org/10.1016/j.envint.2021.106435.

38. Zulfiqar, U., Jiang, W., Wang, X., Hussain, S., Ahmad, M., Maqsood, M. F., Ali, N., Ishfaq, M., Kaleem, M., Haider, F. U., Farooq, N., Naveed, M., Kučerík, J., Brtnický, M., & Mustafa, A. (2022). Cadmium phytotoxicity, tolerance, and advanced remediation approaches in agricultural soils: A comprehensive review. *Frontiers in Plant Science, 13*. https://doi.org/10.3389/fpls.2022.773815.

CHAPTER 7

WATER AVAILABILITY AND USE

Contents

INTRODUCTION

The importance of water as a fundamental resource cannot be overstated; it is the lifeblood of all living organisms and an essential component of human civilization. From sustaining agricultural practices to providing the energy needed for industrial processes, water is intricately linked to economic development and public health. With the population of the world expected to grow significantly in the coming decades, the demand for freshwater resources will only intensify. As such, understanding the dynamics of water availability and use is crucial for policymakers, scientists, and citizens alike, as it informs strategies for sustainable management and conservation efforts that are indispensable in addressing future challenges. Many regions across the globe face significant disparities in water availability, which can lead to conflict and social strife. Areas that experience chronic water scarcity grapple with myriad issues, such as reduced agricultural productivity, impaired health outcomes, and economic stagnation. Conversely, regions with abundant water resources often struggle with pollution and overexploitation, leading to environmental degradation. This dichotomy highlights the need for comprehensive and equitable water management practices that consider both availability and use, recognizing that effective governance can mitigate the risks associated with water scarcity and foster sustainable development. Addressing these complex issues requires an interdisciplinary approach that integrates environmental science, socio-economic factors, and cultural considerations. As the challenges related to water availability emerge more prominently in the discourse surrounding climate change and population growth, it becomes increasingly imperative to adopt innovative methods for conservation and use. Technological advancements, such as drip irrigation, water recycling, and desalination offer valuable solutions to enhance water efficiency. Additionally, policy frameworks that promote community engagement and stewardship of water resources play a pivotal role in building resilience against water-related challenges. Ultimately, informed dialogue about water availability and use is a prerequisite for creating sustainable communities that can thrive in harmony with nature, ensuring that future generations have access to this vital resource.

7.1. DEFINITION OF WATER AVAILABILITY

Water availability is a multifaceted concept that encompasses both the physical presence of water resources and the accessibility of those resources for human use. It is not merely about the volume of water present in a given region, but also about the dynamics that govern its distribution and utilization. Factors such as climate, topography, and human infrastructure significantly influence how water is available for various purposes. Regions experiencing an abundance of rainfall may seemingly have ample resources; however, poor infrastructure may hinder effective collection and distribution, rendering water availability inadequate for agricultural or domestic use. Consequently, understanding this definition includes acknowledging the complexities surrounding how water is managed and distributed, ultimately shaping the use and sustainability of this vital resource.

In addition to physical presence, water availability is closely tied to socio-economic factors that determine access for different communities. The agricultural sector, encompassing around 70% of global freshwater use, significantly relies on water availability for crop production and livestock management. In contrast, urban areas need substantial freshwater resources for drinking, sanitation, and industrial processes. The competition for water resources between these sectors can lead to significant challenges, particularly in arid regions where water scarcity is prevalent. If institutions and policymakers do not adequately manage competing demands, disparities in access may arise, resulting in severe socio-economic implications.

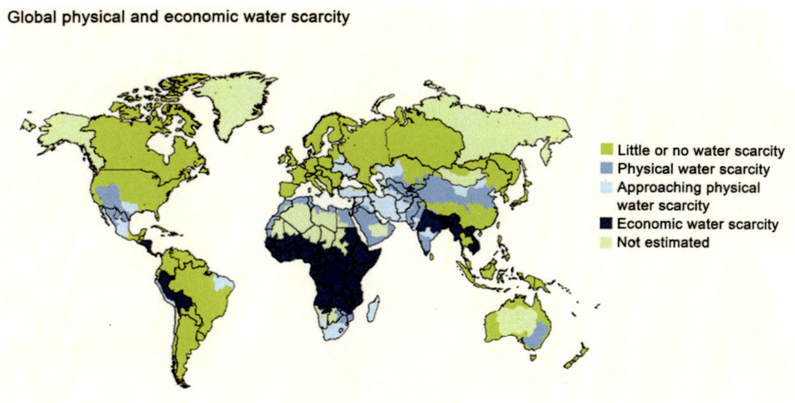

Figure 7.1. *Water availability on earth and its importance.*

Source: By World Water Assessment Programme (WWAP) – http://www.un.org/waterforlifedecade/scarcity.shtml, Public Domain, https://commons.wikimedia.org/w/index.php?curid=74305886.

A holistic understanding of water availability thus must include not only the environmental dimensions but also the equity and governance issues surrounding water distribution. Ultimately, water availability is both a quantitative measure and a qualitative experience that reflects the social, economic, and governance dynamics at play within any given region. This concept emerges as critical in discussions about sustainable water management, especially in light of increasing pressures from population growth, climate change, and industrialization. In various sectors, the onset of technological advancements offers promising solutions for improving water use efficiency, but these will only be successful if they align with a comprehensive understanding of local water availability conditions.

As highlighted by experts, the interdependence of ecological health and water resource management is imperative; unchecked rates of usage could lead to resource depletion and ecological degradation. Thus, the definition of water availability must encompass both current realities and future sustainability considerations, ensuring that effective management practices are established to safeguard this precious resource for generations to come (Figure 7.1).

7.1.1. Importance of Water in Daily Life

Water serves as a cornerstone for human health and well-being, influencing both physical and mental functions. Adequate hydration is crucial for maintaining bodily processes, as water plays a vital role in digestion, nutrient transport, and temperature regulation. Insufficient water intake can lead to dehydration, which negatively impacts cognitive function, leading to issues such as fatigue, confusion, and irritability. Addressing the well-being of individuals with dementia, for instance, timely hydration can help maximize cognition, decrease distressing symptoms, and improve overall quality of life. As highlighted in research, effective interventions can transform living conditions, underscoring the essential role of water not only in sustaining life but also in enhancing its quality (Livingston et al., 2017).

Moreover, water is indispensable in various sectors, including agriculture, industry, and sanitation. In agricultural contexts, clean water is essential for irrigation processes, significantly influencing crop yield and food security. As a result, managing water resources effectively can directly impact the availability of food staples for communities. In industrial settings, water is utilized in manufacturing processes, cooling systems, and waste management. Furthermore, adequate sanitation relies heavily on water availability, significantly affecting public health. Failure to manage these resources properly can lead to detrimental societal outcomes, perpetuating cycles of poverty and poor public health. Finally, access to safe drinking water is a critical component of social equity and environmental sustainability.

Marginalized communities often face disparities in access to clean water, which exacerbates health issues and limits opportunities for economic growth. Strategies for improving water availability must prioritize equitable distribution to ensure that all populations benefit. The evolution of understanding languages through comparative studies exemplifies the need for collaborative efforts in addressing global water challenges. By approaching water availability as a shared responsibility, societies can better advocate for policies that enhance access and safeguard this essential resource for future generations (Brugmann et al., 2017).

7.1.2. Overview of Global Water Distribution

The vastness of the Earth's water resources is primarily concentrated in specific regions, with approximately 97.5% of all water on the planet being saline, found in oceans and seas. This significant portion is not directly available for human consumption or agricultural use. The remaining 2.5% of freshwater, essential for drinking, irrigation, and maintaining ecosystems, is unevenly distributed. Most freshwater is locked in glaciers, ice caps, and deep underground aquifers, with only a small fraction accessible in rivers, lakes, and reservoirs. This imbalance creates challenges in managing water resources, particularly for regions that rely on limited freshwater sources for their needs. Geographic and climatic conditions contribute significantly to the disparities in water availability across the globe. For example, arid, and semi-arid regions often face severe water scarcity, impacting both human populations and agricultural productivity. Simultaneously, areas with abundant rainfall, such as tropical rainforests, experience excess water that can lead to flooding and soil erosion.

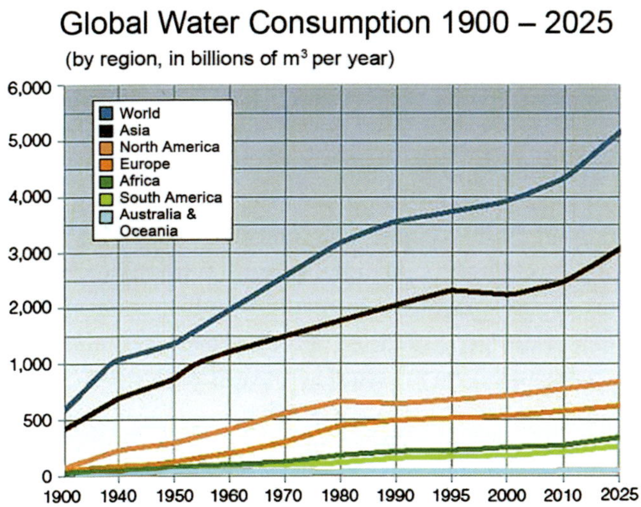

Figure 7.2. *Water distribution on earth.*

Source: By Sampa – Own work, CC BY-SA 4.0, https://commons.wikimedia.org/w/index.php?curid=65212874.

This uneven distribution necessitates effective water management strategies to ensure sustainable use. Ultimately, climate change exacerbates these challenges, as shifting weather patterns can lead to modified rainfall distribution and intensified droughts or floods, further complicating the already intricate dynamics of global water distribution. Despite the challenges posed by uneven global water distribution, there are ongoing efforts to enhance the efficient use of available resources.

Technological advancements in water conservation, such as improved irrigation techniques and rainwater harvesting, have emerged as potential solutions to address scarcity. Moreover, international collaborations aim to promote equitable water sharing among nations, particularly in transboundary river basins. As highlighted in studies on land use and water management, Bray et al (2018) emphasize the importance of data for making informed decisions, while Immitzer et al (2016) demonstrate the relevance of integrating remote sensing technology to map resources effectively. Such initiatives can lead to better policies and practices that optimize water availability and use, thereby fostering sustainable development in both urban and rural communities (Figure 7.2).

7.2. SOURCES OF WATER

Water is essential for both ecological balance and human survival, with its sources being diverse and crucial for various applications. One primary source of water is precipitation, which includes rain, snow, and sleet, contributing significantly to surface water bodies such as rivers, lakes, and reservoirs. These freshwater sources play a vital role in replenishing groundwater aquifers and supporting ecosystems. Moreover, surface water is often utilized for agricultural, industrial, and domestic purposes, making its management critical for sustainable development. The over-extraction and pollution of these sources, however, pose significant challenges. Without careful stewardship, these vital resources may diminish, impacting both human needs and natural habitats. In addition to precipitation, groundwater remains a fundamental source of water, particularly in arid and semi-arid regions. Aquifers, which are natural underground reservoirs, store significant quantities of freshwater that can be accessed through wells. The reliance on groundwater for irrigation in agriculture is pervasive, as it ensures a steady supply of water during droughts when surface water levels are low. Nevertheless, the excessive withdrawal of groundwater can lead to aquifer depletion, land subsidence, and deterioration of water quality, illustrating the need for responsible management. Protective measures and regulations are paramount to ensure that this critical resource remains sustainable in the face of increasing demand. Another emerging source worth noting is the use of treated wastewater and desalination. As traditional freshwater sources come under pressure, innovations such as wastewater treatment and seawater desalination have gained prominence. Treated wastewater can be recycled for agricultural irrigation and industrial processes, promoting a circular water economy. On the other hand, desalination processes transform seawater into potable water, providing solutions for regions facing severe water scarcity. However, the environmental impacts associated with both methods, including energy consumption and the potential release of pollutants, must be carefully considered. Striking a balance between harnessing these alternative sources and protecting environmental systems will be crucial for future water availability and use.

7.2.1. Freshwater Sources

The availability of freshwater sources is crucial for sustaining life, agriculture, and industry. Among these sources, rivers, lakes, aquifers, and glaciers play essential roles in maintaining the ecosystem and providing water to communities. Rivers and lakes, often replenished by precipitation and melting snow, serve as primary sources for drinking water, irrigation, and industrial processes. Groundwater, stored in aquifers, supplies about one-third of the global drinking water, illustrating the importance of subsurface resources. Conversely, glaciers, which contain a significant portion of the world's freshwater, are facing rapid melting due to climate change, leading to potential shortages for millions who rely on glacial runoff for their water supply. Understanding and managing these freshwater sources is imperative to address the increasing water scarcity experienced by various regions globally. As water demands grow due to population expansion and economic development, the competition for these freshwater sources intensifies. Agriculture, which accounts for about 70% of global freshwater withdrawals, often leads to over-extraction of lakes and rivers, resulting in habitat degradation and diminished water quality. Moreover, industrial activities further stress these resources through pollution and excessive consumption. The situation is exacerbated by the effects of climate change, as altering precipitation patterns and rising temperatures impact the availability and distribution of freshwater. Efficient water management strategies, including the development of alternative technologies such as desalination, could alleviate some pressure on traditional sources. This approach may prove vital in areas heavily dependent on freshwater, particularly as global populations continue to expand and urbanize (Curto et al., 2021). The global cycling of mercury (Hg) has implications for freshwater ecosystems as pollutants can affect water quality and health (Obrist et al., 2018). Contaminants from industrial processes, especially in densely populated regions, can accumulate in freshwater sources, posing risks to aquatic life and human health. Efforts to safeguard freshwater resources must integrate pollution control measures alongside sustainable water use practices. By investing in infrastructure to reduce contaminants and restore degraded freshwater ecosystems, societies can enhance water quality while ensuring that these vital resources remain available for future generations. Thus, the management of freshwater sources encompasses the sustainable withdrawal of water and the protection of water quality from anthropogenic influences, highlighting the need for a holistic approach to addressing the challenges of water availability and use (Figure 7.3).

Figure 7.3. *Fresh water sources.*

Source: By Dotun55 – Own work, CC BY-SA 4.0, https://commons.wikimedia.org/w/index.php?curid=77300103.

7.2.2. Saltwater Sources

The challenge of managing freshwater resources in many regions has prompted increased interest in utilizing saltwater sources, especially given their abundance. Coastal areas, where saltwater is predominant, face the dual issues of rising sea levels and declining freshwater availability. Traditional methods of desalination have emerged as potential solutions, yet they often come with environmental and economic drawbacks, including high energy consumption and the production of brine waste. Innovative approaches are necessary to ensure sustainable practices. For instance, integrating locally sourced materials, such as crushed Quahog shells, into water treatment systems shows promise. The low-cost, environmentally beneficial properties of these naturally occurring substrates could be leveraged to address challenges in saltwater remediation without exacerbating carbon footprints as is common with activated charcoal production. The methods used to treat saltwater sources can significantly impact the environment and the efficiency of water availability. Researchers are exploring various porous materials, including those found naturally, to enhance salt removal from seawater. For instance, using amorphous calcium carbonate derived from crushed Quahog shells has demonstrated effective removal of chlorides, albeit less efficiently than activated charcoal. However, this method can release beneficial calcium into the environment, promoting aquatic life, which is a critical ancillary benefit in saltwater management (Istiyanto et al., 2023).

Additionally, innovative approaches like coastal reservoirs highlight the interaction between salt and freshwater systems. Projects such as those utilizing the Cisadane River's flow prioritize diminishing saltwater intrusion while providing substantial freshwater capacity, showcasing an integrative strategy for managing water resources in a salinity-affected landscape (Istiyanto et al., 2023). Balancing the utilization of saltwater sources with the need for environmental sustainability remains a complex task. As demand for freshwater grows amidst climate change and population growth, exploring alternative sources becomes increasingly vital. The dual focus on efficiency and ecological impact could reshape the future of water availability. As seen in recent studies, innovative materials derived from natural sources not only lend themselves to effective desalination but also contribute positively to surrounding ecosystems (Istiyanto et al., 2023). Emphasizing sustainable practices while harnessing the potential of saltwater systems offers a promising avenue for alleviating water scarcity. The knowledge gained from these explorations can build resilient strategies that leverage both saltwater and freshwater resources to support biodiversity and human needs alike, thus ensuring a more sustainable future for water use globally.

7.2.3. Groundwater Reserves

The sustainability of groundwater reserves is increasingly crucial as global freshwater sources dwindle. Given that groundwater is a primary water supply for agricultural, industrial, and domestic use, understanding its depletion is essential to water management strategies. In regions like Iran, excessive extraction for agricultural needs has led to alarming rates of groundwater depletion, estimated at around 74 km^3 from 2002 to 2015. This depletion not only poses risks to water availability but also exacerbates environmental issues such as soil salinity and land subsidence. Furthermore, it highlights the direct impact of human activities on groundwater reserves, emphasizing the urgent need for sustainable management practices that balance human requirements and ecological preservation. In addition to human influence, climatic factors play a significant role in the health of groundwater reserves. Variability in precipitation patterns, coupled with prolonged droughts, can trigger increased dependence on groundwater and accelerate its depletion. Research indicates that in Iran, groundwater depletion is largely driven by anthropogenic factors, although hydrological droughts intensify this crisis.

Figure 7.4. *Groundwater reserves.*

Source: By NASA – http://spaceflight.nasa.gov/gallery/images/shuttle/sts-83/html/sts083-747-033.html (full resolution), Public Domain, https://commons.wikimedia.org/w/index.php?curid=20762791.

For instance, the potential impacts of such conditions are evident in the Ifni basin in Morocco, where a multi-criteria analysis assessing groundwater potential identified significant areas at risk due to changing hydroclimatic conditions (Ikirri et al., 2023). Thus, understanding the interplay between climatic variations and human activities is vital for developing resilient water management strategies. Effective groundwater management policies must incorporate scientific assessments and community involvement to ensure long-term viability. The establishment of groundwater potential maps, as demonstrated in the Ifni basin study, is one approach that can inform decision-makers on where to focus conservation efforts. Additionally, public awareness and education about groundwater sustainability can foster community engagement in protecting these vital resources. The case of Iran serves as a cautionary tale, illustrating how unchecked water extraction can lead to detrimental environmental impacts that compromise future water security (Ashraf et al., 2021). By recognizing the interconnectedness of groundwater reserves with both human actions and climatic influences, we can devise comprehensive strategies aimed at safeguarding this indispensable resource (Figure 7.4).

7.3. GLOBAL WATER DISTRIBUTION

Access to water is a critical issue that extends beyond simple availability; it encompasses the complex dynamics of usage, management, and equitable distribution. Current statistics reveal that while approximately 71% of the Earth's surface is covered in water, only a fraction—about 2.5%—is freshwater. This limited supply is further compromised by geographical and climatic disparities, which shape how effectively regions can access clean water. Imagine a country like Brazil, endowed with one of the largest freshwater reserves in the world, juxtaposed against numerous regions in Africa and the Middle East where water scarcity is rampant. This inequality often leads to exacerbated social tensions and health crises, highlighting the urgent need for better management practices and awareness of sustainable consumption patterns to ensure that all populations receive fair and adequate water resources. The relationship between water availability and agricultural practices is particularly critical, considering that agriculture accounts for approximately 70% of global freshwater use. Inefficient irrigation methods and unsustainable farming practices have led to significant water depletion and degradation of water quality.

As the demand for food increases, especially in developing regions, the balance between operational efficiency and resource conservation becomes increasingly delicate. For instance, data indicates that regions with severe water scarcity often experience lower agricultural yields, leading to food insecurity and malnourishment. Efforts to implement water-saving technologies and promote sustainable farming practices could bridge this gap, fostering healthier diets while simultaneously protecting water resources. Therefore, understanding this nexus is essential for creating policies that not only enhance food security but also prioritize sustainable water management. Finally, global water distribution does not merely influence human health and food security; it also affects environmental sustainability. The interconnectivity between ecosystems and water supply is vital for preserving biodiversity and supporting ecosystem services. For instance, wetland regions, which act as natural water filters and carbon sinks, are under constant threat from urbanization and industrial pollution, leading to a significant loss in ecological health. Protecting these areas is essential for maintaining clean water sources and for combating climate change. Furthermore, as urban populations grow, the strain on water resources intensifies, often resulting in contamination and resource monopolization by industrial players. Thus, the understanding of water distribution must be coupled with strategies that encourage sustainable practices, thereby fostering environments where both human populations and natural ecosystems can thrive.

7.3.1. Regional Disparities in Water Availability

Water scarcity is not uniformly experienced across regions; disparities in water availability can significantly affect local populations and their agricultural practices. In the Volta River basin of West Africa, projections indicate that climate change will severely impact water resources, particularly under high greenhouse gas emissions scenarios such as RCP8.5. The variability in precipitation and resultant hydrological dynamics underscores the challenge for communities that depend heavily on rainfall for agriculture, further complicating efforts to secure water for both consumption and irrigation. The study notes that increased flood risks and recurrent drought conditions could exacerbate the existing vulnerabilities, disproportionately affecting regions with lower socioeconomic status and limited adaptive capacity to climate impacts (Dembélé et al., 2022). Socioeconomic factors help explain the regional disparities in access to sufficient water resources. Different areas exhibit varying levels of infrastructure development, investment in water management systems, and technological advancements that enhance water use efficiency. For example, metropolitan regions like Addis Ababa benefit from better access to clean water sources and modern sanitation infrastructures, resulting in lower rates of waterborne diseases and higher life expectancy (Misganaw et al., 2022). In contrast, rural regions often lack basic amenities, leading to reliance on unsafe water sources. This inequitable access has profound implications on health, education, and economic productivity, revealing that water scarcity extends beyond mere physical availability to encompass systemic inequalities that affect marginalized communities disproportionately. Efforts to address regional disparities in water availability must be multifaceted and locally tailored. Policymakers need to implement adaptive water management strategies that consider climatic forecasts, population pressures, and the specific needs of various communities. For instance, integrating indigenous knowledge and community involvement in decision-making processes can enhance resilience against water-related challenges. Furthermore, investments in sustainable agricultural practices that improve water retention and efficiency are essential for cultivating food security in water-scarce regions. Ultimately, a holistic approach that combines both technical and policy interventions is crucial in managing water resources effectively, ensuring that all regions can thrive despite the challenges imposed by climate change and uneven access to water (Dembélé et al., 2022; Misganaw et al., 2022).

7.3.2. Impact of Climate on Water Distribution

Climate changes significantly disrupt traditional patterns of water distribution, presenting serious challenges for both natural ecosystems and human communities. As temperatures rise, the rate of evaporation increases, which can exacerbate drought conditions. This situation is particularly alarming for agricultural regions reliant on consistent water availability. For instance, drought stress affects physiological processes within plants, leading to lower productivity and altered species distribution (Kapoor et al., 2020). Consequently, regions that previously supported diverse crop production may experience reduced yields, threatening food security and economic stability. Understanding these dynamics is crucial for developing effective strategies to mitigate the impacts of climate change on water resources. Furthermore, the alteration of precipitation patterns due to climate change contributes to uneven water distribution across different regions. For example, some areas may experience increased rainfall during certain seasons, while others face prolonged dry spells.

This variability can lead to shifts in hydrological cycles, with rivers and aquifers becoming stressed or depleted. In the Karnali-Mohana river basin in Western Nepal, projections indicate changes in precipitation and temperature that will impact water availability, particularly in mountainous areas where vulnerability is heightened (Pandey et al., 2020). Without intervention, these shifts threaten not only local ecosystems but also the livelihoods of communities that depend on stable water supplies.

Addressing the impact of climate on water distribution necessitates comprehensive planning and cooperation at local, regional, and global levels. Adaptation strategies must take into account diverse climatic conditions and their effects on water resources. For instance, developing resilient agricultural practices that can withstand drought conditions is critical to maintaining food security (Kapoor et al., 2020). Additionally, investments in infrastructure that capture and manage variable water supplies can help mitigate the adverse effects of climate change (Pandey et al., 2020). Ultimately, embracing sustainable practices and innovative management strategies will be essential to ensure the availability of water resources in the face of ongoing climate challenges (Figure 7.5).

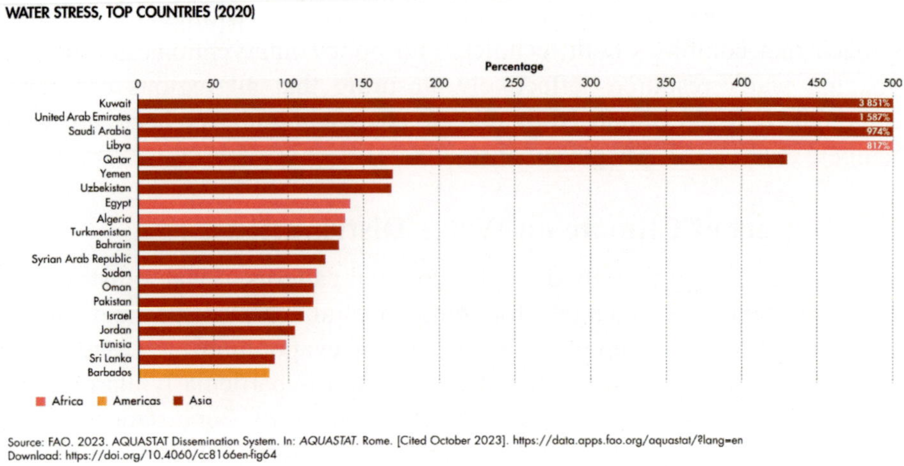

Figure 7.5. *Climate change impacts water resources.*

Source: By FAO – https://www.fao.org/documents/card/en/c/cc8166en, File:World Food and Agriculture – Statistical Yearbook 2023.pdf, CC BY-SA 3.0 igo, https://commons.wikimedia.org/w/index.php?curid=142154584.

7.3.3. Water Scarcity in Arid Regions

Addressing the complexities of water scarcity in arid regions is crucial for sustainable management and future planning. The increasing demands on limited water resources are exacerbated by climate change, leading to significant challenges, particularly for densely populated areas. For instance, the Amu Darya Basin in Central Asia highlights the potential impact of climate variability, with projections indicating shifts in terrestrial water storage (TWS). An increase in TWS by approximately 4 cm is expected in some remote areas, while more densely populated regions face a decline of around –4 cm due to changing precipitation patterns and temperature increases (Salehie et al., 2024). This disparity illustrates how vulnerable populated zones risk facing critical water shortages, underscoring the urgent need for effective resource management and adaptation strategies. Drought further amplifies the water scarcity issue, particularly in regions characterized by arid and semi-arid climates. In Balochistan, Pakistan, drought has devastating socio-economic implications, leading to migration, food insecurity, and heightened competition for dwindling resources (Abdullah et al., 2023). The lack of reliable meteorological and hydrological data complicates efforts to manage these drought conditions effectively. Innovative methodologies that utilize basic meteorological data alongside robust hydrological models, such as MIKE-SHE, can aid in understanding and mitigating drought impacts. By quantifying drought severity through various indices, such as the Palmer Drought Severity Index, watershed management strategies can be developed to better address the unique challenges faced by these vulnerable regions. Addressing water scarcity in arid regions necessitates both immediate and long-term interventions that prioritize sustainable practices. Effective water management policies must consider changing climatic conditions and population dynamics to mitigate risks. Integrating advanced modeling techniques can improve our understanding of water availability and inform decision-making processes. Moreover, enhancing collaboration among governmental and non-governmental organizations is vital for implementing adaptive strategies. Such actions will safeguard essential water resources and promote resilience in communities heavily reliant on these limited supplies. To achieve sustainable water use, it is imperative to combine scientific understanding with culturally appropriate strategies that empower local populations in the face of water scarcity (Figure 7.6).

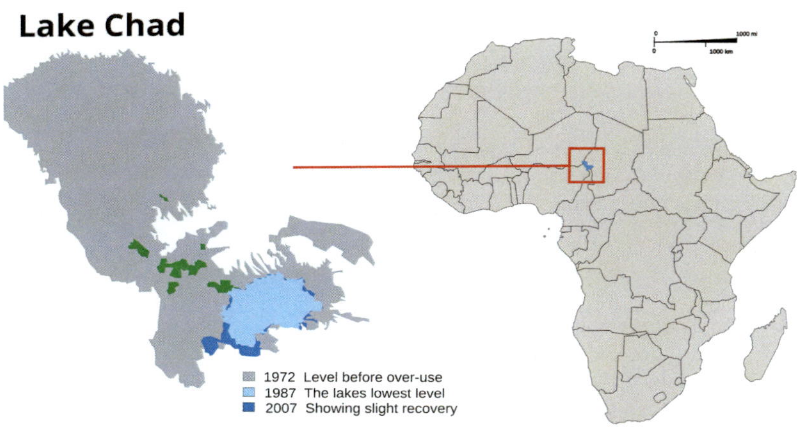

Figure 7.6. *Confronting water scarcity in Africa.*

Source: By Blank_Map-Africa.svg: Andreas 06derivative work: Chaosdruid (talk) – Blank_Map-Africa.svg, Public Domain, https://commons.wikimedia.org/w/index.php?curid=9587823.

7.4. WATER USAGE IN AGRICULTURE

Agricultural practices are significantly influenced by water usage, which is vital for crop production and livestock management. As global demand for food rises, efficient water management becomes increasingly crucial. A 2013-published study highlights the projected need for water resources to meet anthropogenic demands, predicting that, by 2030, up to 160% of the total available water may be necessary for such activities (Jonathan Morris et al., 2021). This demand creates a pressing need for sustainable practices that ensure adequate water supply for agriculture and account for environmental protection. Encouraging farmers to adopt techniques like drip irrigation and rainwater harvesting can mitigate water wastage while optimizing resource use, enabling them to work within the limits of their available water supply. However, the practices surrounding agricultural water use face significant challenges, including a general lack of awareness and proper management techniques among farmers. Research indicates that many farmers rely on misleading information from pesticide dealers regarding pest control methods, further complicating their knowledge of sustainable agricultural practices (Matowo et al., 2020). This gap in knowledge results in practices that contribute to water pollution and the degradation of local water sources.

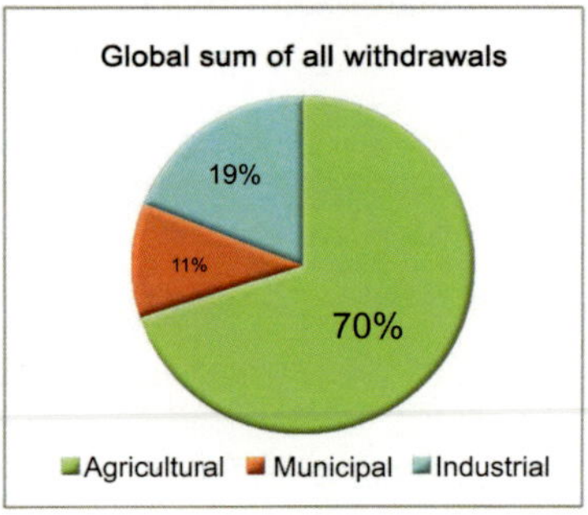

Figure 7.7. *Uses of water.*

Source: By Sampa – Own work, CC BY-SA 4.0, https://commons.wikimedia.org/w/index.php?curid=65334887.

Additionally, the improper disposal of empty pesticide containers and leftover chemicals exacerbates this issue, posing risks to both ecosystems and public health. Coordinated education programs for farmers and retailers can be instrumental in improving understanding and management of agricultural practices, particularly emphasizing the importance of maintaining clean water sources. The integration of waste-water reuse in agriculture emerges as a crucial strategy in response to the growing challenges posed by water scarcity. Despite local initiatives aimed at incorporating treated wastewater into agricultural practices, significant barriers persist, particularly in terms of legal frameworks and societal acceptance. An investigation into these barriers found that political support and public perception are key elements in enhancing the adoption of wastewater usage (Morris et al., 2021). Engaging communities and policymakers in discussions concerning the benefits of wastewater reuse can pave the way for more sustainable agricultural practices. By fostering an understanding of the economic and environmental advantages associated with wastewater resource utilization, agricultural water management can progress toward a circular economy, thereby preserving precious water resources while supporting agricultural productivity (Figure 7.7).

7.4.1. Irrigation Practices and Techniques

Effective irrigation practices are essential for optimizing agricultural productivity, particularly in the face of increasing climate variability and population growth. Various methods are employed across the globe, including surface irrigation, drip irrigation, and sprinkler systems, each with its own benefits and challenges. Surface irrigation is one of the oldest techniques, where water flows over the ground and allows for natural absorption. However, this method may lead to inefficiencies if not managed properly, resulting in water wastage. On the other hand, drip irrigation is more efficient, delivering water directly to the plant roots, which minimizes evaporation and percolation losses. This precision ensures that crops receive the necessary moisture without over-application, which is increasingly critical in water-scarce regions. New advancements in technology are further enhancing irrigation practices, enabling more efficient water management strategies. The integration of remote sensing techniques, such as those discussed in (Dari et al., 2021), allows farmers to monitor soil moisture levels and irrigation extents more accurately. This data-driven approach helps in distinguishing between irrigated and rainfed agricultural areas, offering insights into resource allocation.

As farmers face pressures from climate change, implementing smart agriculture techniques becomes paramount. These innovations, particularly when combined with artificial intelligence, can revolutionize how irrigation is managed. By analyzing past and present conditions, farmers can optimize water usage while improving yields and sustaining the environment. The future of irrigation practices hinges on the adoption of these advanced technologies, which can provide comprehensive data about water usage and crop needs. Research indicates that using remotely sensed soil moisture products substantially enhances irrigation monitoring, allowing for better decision-making regarding water allocation (Dari et al., 2021). As smart agricultural systems become more mainstream, their ability to integrate such technologies will be essential for ensuring food security. Moreover, the potential for minimizing over-irrigation through precise data collection aligns with sustainable water use—a critical concern amidst declining water resources. Ultimately, the evolution of irrigation practices, bolstered by innovation, is about increasing agricultural efficiency and about fostering resilience in water management as global challenges intensify.

7.4.2. Impact of Agriculture on Water Resources

Agriculture is a substantial consumer of freshwater resources, accounting for nearly 70% of global water usage. This high demand can lead to significant depletion of local water supplies, particularly in regions where water scarcity is already a pressing issue. Intensive practices, such as traditional irrigation methods, often exacerbate this situation by facilitating runoff and evaporation, resulting in inefficient water use. As agricultural practices expand to meet the needs of a growing population, the pressure on water resources intensifies, leading to the potential for conflicts between agricultural demands and the necessity for sustainable water management. Innovative techniques, such as drip irrigation and rainwater harvesting, are essential for improving water efficiency in agricultural practices.

Without adopting more sustainable approaches, the agricultural sector may contribute to long-term ecological imbalances that jeopardize future water availability. In addition to direct water use, agriculture significantly impacts water quality within ecosystems. The application of fertilizers and pesticides in farming practices often results in runoff that can carry harmful chemicals into nearby bodies of water. This poses a risk to aquatic ecosystems and to human health, as contaminated water sources can lead to serious health issues for communities relying on these waters for drinking and irrigation. Furthermore, excess nutrients from agricultural runoff can lead to eutrophication, resulting in oxygen-depleted zones that harm aquatic life.

Sustainable agricultural practices, such as organic farming and integrated pest management, can mitigate these effects by reducing chemical inputs and improving soil health. Engaging farmers in education about the impacts of conventional agricultural practices on water quality is crucial for fostering responsible management of water resources. Technological advancements in agriculture are emerging as critical solutions to help conserve water resources without sacrificing productivity. For instance, the integration of biochar into farming systems enhances soil properties and aids in water retention and reduces the need for excessive irrigation (Efficiency of biochar with mineral fertilizers on soil properties. This, in turn, can lead to increased crop resilience amid changing climatic conditions. Additionally, innovative irrigation technologies like pivot sprinkler systems have shown promise in regions facing water scarcity, as they maximize efficiency in water use and increase productivity (Mohammed and Jbara, 2024). By adopting such methods, the agricultural sector can reduce its overall water footprint and contribute to sustainable water management. Ultimately, balancing agricultural productivity with responsible water use is essential for the health of ecosystems and the availability of water resources for future generations.

7.4.3. Sustainable Agricultural Practices

Adopting sustainable agricultural practices is pivotal for enhancing water availability and ensuring long-term food security. Methods such as conservation agriculture (CA), which emphasizes reduced tillage and crop residue retention, have shown promise in improving soil health while simultaneously conserving water. By increasing soil organic matter (SOM) and promoting water infiltration, CA not only addresses issues of soil erosion and nutrient depletion but also significantly enhances the efficiency of water use in agriculture. As noted in the study of rice-wheat cropping systems in the Indo-Gangetic Plains, the adoption of CA has become a practical response to the challenges posed by conventional farming methods, effectively mitigating the risks of environmental degradation and aligning with water conservation objectives (Jayaraman et al., 2020). Thus, integrating CA practices presents a viable pathway for sustaining both agricultural productivity and water resources. Furthermore, the relationship between land-use practices and water quantity underscores the importance of sustainable methods in modern agriculture. A comprehensive analysis of land-use conversion reveals that increased forest cover can substantially improve hydrological parameters within regions like the Tietê Basin, Brazil, providing critical buffers against the adverse impacts of urbanization and climate change. Increased forestation contributes to better water quality and enhances overall water availability, making it a crucial component of sustainable agricultural strategies (Paula Ferreira et al., 2019).

Moreover, the adoption of agroecological practices that promote biodiversity can also help maintain and restore natural water cycles, further emphasizing the necessity of sustainable approaches in mitigating water scarcity issues. Therefore, recognizing the interconnectedness of land management and water resources is essential for crafting effective agricultural policies. In conclusion, the imperative for sustainable agricultural practices has never been clearer, particularly in the context of global food security and water resource management. As highlighted in the challenges facing South Asia, where traditional farming methods often exacerbate environmental degradation, transitioning towards sustainable approaches holds the key to reversing this trend. Implementing practices like conservation agriculture and reforestation addresses the immediate concerns of soil health and crop yields and aligns with broader ecological goals, such as improving water availability and quality. Sustainable agriculture can serve as a model for harmonizing the needs of food production with the imperative of conserving vital water resources, and creating a resilient agricultural system that supports both people and the planet. Through such integrated strategies, the dual crises of food insecurity and water scarcity can be effectively mitigated for future generations (Figure 7.8).

Figure 7.8. *Sustainable farming practices.*

Source: By PermaKulturgut.de – Own work, CC BY-SA 3.0, https://commons.wikimedia.org/w/index.php?curid=18839412.

7.5. WATER USAGE IN INDUSTRY

The industrial sector is a significant consumer of water, using substantial amounts for various processes, including cooling, cleaning, and manufacturing. Industries such as textiles, food processing, and energy generation depend heavily on water, which is essential for maintaining operational efficiency and product quality. However, this high demand places immense pressure on local water resources, often leading to over-extraction and depletion of freshwater supplies. As environmental concerns mount, companies are increasingly exploring sustainable practices to reduce water usage. Implementing water-saving technologies can mitigate the environmental impact and improve the operational sustainability of industrial processes. Such strategies help industries maintain their production capacities and enhance their reputations as responsible water users, aligning with broader corporate social responsibility goals.

In recent years, innovative solutions have emerged to address water scarcity in various industries. One effective approach is the recovery and reuse of wastewater, which enables companies to recycle water within their processes, significantly lowering their overall consumption. For example, in the data center sector, strategies for waste heat recovery illustrate the potential for reducing water usage while maintaining operational efficiency. By cultivating an understanding of the interdependencies between water and energy, industries can devise comprehensive plans to minimize waste and maximize resource efficiency; about 68% of the waste heat in data centers can be captured and repurposed, thus optimizing both energy and water resources (Luo et al., 2019). This reciprocal relationship highlights the importance of integrating innovative practices into industry standards to enhance resource management. Moreover, the economic benefits of reducing water usage in industrial settings cannot be understated. By adopting sustainable water management strategies, businesses can significantly cut costs associated with water consumption and wastewater treatment. Additionally, reduced water usage often leads to lower energy requirements, promoting overall efficiency and sustainability. The emphasis on sustainability has prompted industries to rethink their operations and invest in technologies that enhance water management, such as advanced filtration systems and water-efficient manufacturing processes. As highlighted by recent studies focusing on natural fibers and their applications in composite materials, industries are becoming increasingly aware of the economic advantages tied to resource-efficient practices. The long-term viability of industrial operations will depend on the ability to adapt to water scarcity while maintaining competitive advantages (Jagadeesh et al., 2022).

7.5.1. Industrial Water Consumption Statistics

Water consumption statistics in industrial sectors reveal critical insights into global resource management. In many developed nations, industrial activities are among the largest consumers of freshwater, accounting for approximately 20% of total freshwater withdrawals. Sectors such as manufacturing, energy production, and agriculture significantly contribute to this consumption. The manufacturing of goods, particularly in water-intensive industries like textiles and paper, requires substantial volumes of water for production processes and for cooking and cleaning. These demands raise concerns about sustainability and the need for efficient water management practices that mitigate potential shortages. The statistics surrounding industrial water usage underscore the importance of adopting more sustainable practices. Many industries have begun to implement water-saving technologies that recycle and reuse water within their processes. According to recent analyses, the global adoption of such technologies has shown a reduction in freshwater withdrawal, leading to a notable trend towards circular water usage systems. Policymakers must encourage these innovations and policies that promote best practices in water conservation, as this could alleviate some of the pressures on freshwater supplies while ensuring that industrial operations remain economically viable. The relationship between industrial water consumption and environmental sustainability cannot be overstated. Higher water usage often correlates with increased pollution and ecological degradation. As industries extract water from local sources, they can disrupt ecosystems and reduce water quality, affecting both human and environmental health. The global carbon budget monitoring reported in recent years indicates that better management of industrial water use can contribute to improved ecological outcomes and to the overall health of the carbon cycle (Friedlingstein et al., 2020). Thus, a comprehensive approach to water consumption in industry is imperative, as it encompasses not only resource availability but also broader environmental impacts that extend beyond immediate industrial needs.

7.5.2. Water Recycling and Management in Industries

Effective water recycling and management practices hold significant potential for industries seeking to enhance sustainability while mitigating the impacts of water scarcity. These practices not only help in reducing operational water risks but also contribute to minimizing the environmental burden associated with excessive water consumption and waste generation. For instance, the Oil & Gas sector, known for its substantial water use, can benefit from a structured methodological framework to assess water resources and identify improvement opportunities. A Water Balance Assessment can provide critical insights into the qualitative and quantitative aspects of water use, ensuring that industries strategically manage their resources in a manner that aligns with both regulatory requirements and environmental conservation efforts (Bergamini et al., 2021). Moreover, the integration of recycled water into industrial processes can lead to remarkable economic benefits and reduced environmental impact. Industries that adopt water reuse strategies often increase operational efficiency while decreasing costs associated with freshwater procurement and wastewater treatment. For example, implementing advanced treatment solutions for produced water not only prepares wastewater for reuse within the same production cycle but can also create opportunities for third-party applications, such as supplying local communities with treated water for agricultural purposes. This circular approach not only conserves freshwater resources but also fosters connections between industrial operations and surrounding communities, emphasizing the importance of collaborative efforts in achieving sustainable water management (Bergamini et al., 2021).

The urgent need for sustainable practices is underscored by the alarming statistics surrounding plastic waste, emphasizing the necessity of rethinking resource use across industries. Up to 79% of plastic waste is currently discarded, which shows the dire need for innovative solutions like utilizing solid waste within construction materials. Similar paradigms can be applied to water management, where recycling and reuse can significantly contribute to a circular economy. As industries face increasing scrutiny regarding their environmental footprints, embracing comprehensive water management and recycling strategies is not just a regulatory necessity but also a moral imperative. By leveraging waste materials and optimally managing water resources, industries can reduce costs and environmental impacts while enhancing their reputational standing in an increasingly eco-conscious market.

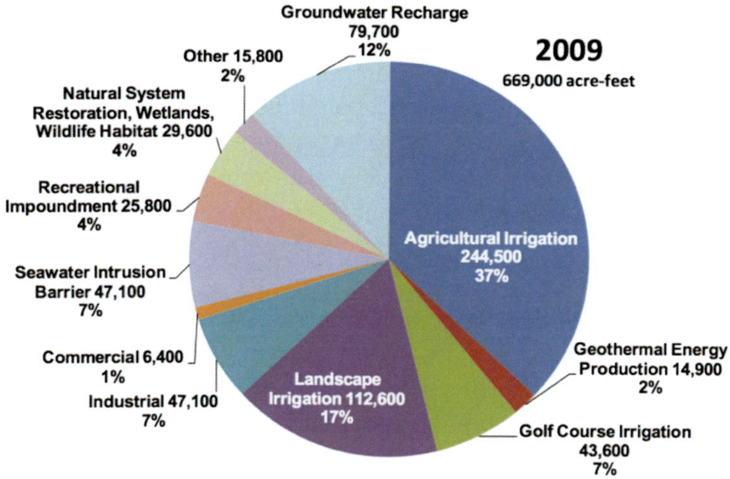

Figure 7.9. *Water recycling system.*

Source: By California State Water Resources Control Board (SWRCB) – U.S. Environmental Protection Agency and U.S. Agency for International Development, Washington, D.C. "Guidelines for Water Reuse." September 2012. EPA/600/R-12/618, CC BY-SA 4.0, https://commons.wikimedia.org/w/index.php?curid=48782827.

7.5.3. Impact of Industrial Activities on Water Quality

Industrial activities significantly undermine water quality through the release of pollutants, which can have dire consequences for ecosystems and human health. The discharge of untreated wastewater directly into water bodies is a predominant issue, particularly in urban areas. For instance, in Pakistan, Karachi generates 472 million gallons of wastewater daily, with only a fraction being treated before being discharged into the sea (Karuppaiah, 2021). This negligence leads to contamination from both domestic and industrial sources, including heavy metals and organic waste, which disrupt aquatic ecosystems and degrade water quality. Inadequate treatment practices exacerbate the severity of the situation, highlighting the urgent need for effective wastewater management systems that prioritize environmental sustainability and safeguard public health. Moreover, industrial activities can introduce a range of toxic substances into water systems, often causing the proliferation of waterborne diseases. Poor water quality from industrial pollution can lead to health crises among populations that depend on these water sources for drinking, sanitation, and agriculture.

As noted in a study regarding urban water quality, the lack of systematic use of Geographic Information System (GIS) technology in monitoring surface water contamination highlights a significant gap in addressing the health implications associated with polluted water (Karuppaiah, 2021). This lack of oversight not only limits the ability to identify contamination sources but also hampers the development of strategic health interventions. Enhanced monitoring and regulation of industrial effluents are crucial to mitigate the health risks posed by waterborne pathogens and pollutants, reinforcing the connection between industrial activities and public health. The cumulative effect of industrial pollution on water quality stresses the importance of regulatory measures and community awareness. In regions like Karachi, unchecked industrial practices contribute to severe water quality degradation, impacting both marine life and local communities reliant on these waters for their livelihoods (Nergis et al., 2021). The absence of robust pollution controls allows industries to discharge harmful pollutants without repercussions, leading to significant environmental harm. Initiatives that promote sustainable industrial practices can alleviate this issue, alongside collaborative efforts between government agencies and industrial sectors to enforce stricter regulations. As communities become more informed about the impacts of industrial activities on local water quality, they can advocate for changes that prioritize both ecological integrity and public health, illustrating the critical role that informed citizenry plays in environmental stewardship.

7.6. DOMESTIC WATER USE

In many households, water usage plays a pivotal role in daily activities, impacting both individual lifestyles and broader environmental systems. Consumption patterns often reflect cultural norms and technological accessibility, influencing how much water is used for various tasks, from drinking and cooking to bathing and cleaning. The increase in synthetic fabric pollution illustrates a pressing issue linked to domestic water use, as the washing of synthetic textiles contributes significantly to microfiber pollution in aquatic ecosystems, accounting for up to 35% of the threat posed to marine environments (Mishra et al., 2020). This highlights how the choices made at the household level have far-reaching implications, for personal health and hygiene and for the overall ecological balance. The management of water resources is increasingly necessary in the face of growing populations and climate variability. Domestic water systems must adapt to ensure that availability keeps pace with rising demand. As urbanization continues, many communities are reevaluating their water use efficiency, seeking strategies to reduce waste and promote conservation. Innovative approaches such as rainwater harvesting, greywater recycling, and the use of efficient fixtures contribute to sustainable domestic water practices, which can alleviate pressure on existing water supplies. In tandem with these strategies, public awareness campaigns about the impacts of water consumption, including those related to synthetic pollutants, are critical to driving behavioral changes at the household level. The interconnection between domestic water use and broader environmental impact cannot be overstated. Increasing awareness of nitrogen management in wastewater treatment further emphasizes this connection, as conventional methods often focus on removing compounds like ammonium without considering potential resource recovery (Cruz et al., 2019). Transitioning to a more resource-oriented approach that emphasizes recovery over destruction presents opportunities for enhanced sustainability. By integrating advanced wastewater treatments and water usage practices, households can actively contribute to a circular economy while minimizing their ecological footprint. An understanding of these dynamics will encourage individuals to rethink their water use habits, fostering a culture of conservation that ultimately benefits both local communities and the planet (Figure 7.10).

Figure 7.10. *Use of rain and well water economically.*

Source: By ReubenGBrewer – Own work, CC BY-SA 3.0, https://commons.wikimedia.org/w/index.php?curid=33635622.

7.6.1. Average Household Water Consumption

Water consumption patterns within households reveal significant insights into broader trends in resource management and sustainability. The average daily water usage per household varies dramatically based on factors such as income levels, demographics, and regional climatic conditions. For instance, a study conducted in Freetown demonstrated that higher-income households with piped connections utilize approximately 133 liters of water per capita per day, whereas lower-income households dependent on alternative water sources average 73 liters per capita per day (Ibrahim et al., 2021). This disparity highlights how accessibility and economic status profoundly impact water consumption, indicating a need for strategic planning in water resource allocation to address inequalities. Such patterns underscore the central challenge of ensuring equitable water access, especially for low-income communities facing severe water scarcity. Examining average household water consumption also reveals how seasonal variations and usage behaviors affect overall demand. The research indicates that per capita water consumption is generally higher during the rainy season due to increased reliance on multiple sources and outdoor activities, marking a 7% increase compared to dry season consumption (Ibrahim et al., 2021). This seasonal fluctuation is not merely a reflection of changing weather but also an indicator of how households navigate water availability.

Understanding these consumption dynamics can inform more effective water management strategies that adapt to seasonal changes and mitigate potential shortages. As climate patterns evolve and water scarcity intensifies, examining the interplay between seasonal demands and household behaviors becomes crucial for sustainable management. Moreover, the implications of water consumption extend beyond individual households, intersecting with global challenges such as water scarcity and pollution. The average household water footprint serves as a microcosm for broader environmental impacts, including the over-extraction of freshwater resources and the implications of consumer behavior. Specifically, research indicates that agriculture accounts for a staggering 92% of the overall water footprint, illustrating the significant role of domestic consumption in the context of agricultural demand. The intimate relationship between household consumption patterns and agricultural practices emphasizes the urgency for sustainable practices to reduce overall water use. As communities grapple with the realities of dwindling water supplies, innovative solutions in household water management will be critical to addressing these global issues effectively.

7.6.2. Water Conservation Practices at Home

Implementing effective water conservation practices at home is essential for addressing the growing challenges of water scarcity. Simple changes in daily routines can have significant impacts on overall water usage. For instance, fixing leaky faucets and toilets can prevent hundreds of gallons of water from being wasted each month. Additionally, adopting water-efficient appliances, such as low-flow showerheads and dual-flush toilets, can greatly reduce indoor water consumption. By becoming more aware of water usage in everyday activities—like limiting shower times or reducing the frequency of lawn watering—households can collectively contribute to a substantial decrease in demand for this precious resource. Such practices align with the need to promote sustainable living, ensuring that both current and future generations have access to clean water. In some regions, the understanding of indoor water usage patterns reveals significant variation, as demonstrated in a study conducted in Gujarat, India, which highlights the importance of targeting specific behaviors to enhance conservation efforts (Kumar et al., 2021). Identifying how much water is used for various household tasks enables families to reassess their consumption habits and pursue more efficient practices. For example, the findings from this study showed discrepancies in average water consumption between two cities, illustrating how geographic and socio-economic factors influence usage. This knowledge can guide tailored interventions to promote water-saving behaviors, incorporating community-specific educational initiatives. By heightening awareness about the particular uses of water at home, individuals can start implementing focused changes that translate into responsible water management and conservation. Moreover, fostering a culture of conservation extends beyond individual households and into the broader community. Encouraging neighbors to adopt water-saving techniques through local workshops or community challenges can create a collective sense of responsibility. As evidenced in the findings regarding socio-demographic factors influencing water consumption, engaging diverse community members and understanding their specific challenges and needs can lead to more effective conservation strategies (Kumar et al., 2021). Additionally, emphasizing the importance of water as a finite resource can inspire collaborative efforts within households and across communities. By sharing successful practices, such as rainwater harvesting or using greywater for irrigation, communities can harness creativity and innovation to address water shortages, ultimately promoting sustainability and resilience in an era of increasing water scarcity and variability.

7.6.3. Role of Technology in Domestic Water Use

Advancements in technology play a crucial role in optimizing domestic water use, impacting both efficiency and accessibility. Innovations such as smart home systems allow households to monitor and control their water consumption in real-time. Smart meters can provide users with immediate feedback on their water use patterns, prompting them to reduce waste. Additionally, technologies such as low-flow fixtures and efficient irrigation systems have dramatically decreased the amount of water consumed in households and gardens. These interventions are especially significant in areas where water scarcity is a prevailing issue, as they help to extend the availability of this vital resource.

Consequently, technology encourages responsible water usage and enhances the ability of individuals and families to contribute to broader conservation efforts. Furthermore, the integration of renewable energy technologies, particularly in solar thermal applications, significantly enhances water heating efficiencies in residential settings. Solar thermal collectors, such as evacuated tube collectors, harness energy from the sun to heat water directly, thus reducing reliance on traditional energy sources. These systems can manage domestic hot water needs efficiently, particularly in regions abundant in sunlight. Research indicates that pairing these solar technologies with heat pipes and phase change materials can optimize performance, making them reliable for residential hot water supply (Chopra et al., 2018). This shift towards using renewable sources reduces energy costs and contributes to a more sustainable domestic water usage model, addressing both efficiency and environmental concerns simultaneously. Moreover, the development and implementation of advanced water treatment technologies contribute to the overall sustainability of domestic water use. Technologies such as reverse osmosis and advanced filtration systems enable households to purify their water, ensuring that it is safe and clean for consumption. Meanwhile, rainwater harvesting systems allow for the collection of rainwater for non-potable uses, reducing the strain on municipal supplies. As noted in the literature, utilizing thermochemical energy storage systems also aids in balancing energy demand with domestic water heating needs, especially during periods of low solar availability (Lin et al., 2020). These technological advancements illustrate a growing trend towards innovative solutions that not only improve the quality and sustainability of domestic water use but also promote environmental stewardship at the individual level.

7.7. CHALLENGES TO WATER AVAILABILITY

One significant challenge to water availability stems from rapid urbanization and industrialization, which escalate water demand while complicating supply management. As cities grow, the infrastructure needed to support increased population density often lags behind the growth rate, resulting in water scarcity for residents. For example, in countries like China, urban expansion has strained water resources, highlighting critical gaps between demand and sustainable supply (Da Xu et al., 2018). Furthermore, industrial activities require substantial water usage for processes such as cooling and cleaning, exacerbating scarcity in regions reliant on finite freshwater sources. Consequently, elevated water demand in urban areas impacts the immediate populace and hinders agricultural practices in surrounding rural regions, as competing priorities for water place immense pressure on the already limited supply. Climate change presents a dire challenge to water availability through both altering precipitation patterns and increasing the frequency of extreme weather events. As global temperatures rise, regions previously abundant in water may experience prolonged drought conditions, while others may face overwhelming floods that challenge existing management systems. The World Health Organization emphasizes that improved water, sanitation, and hygiene (WASH) initiatives are critical for countering the effects of climate change, particularly on neglected tropical diseases (Campbell et al., 2017).

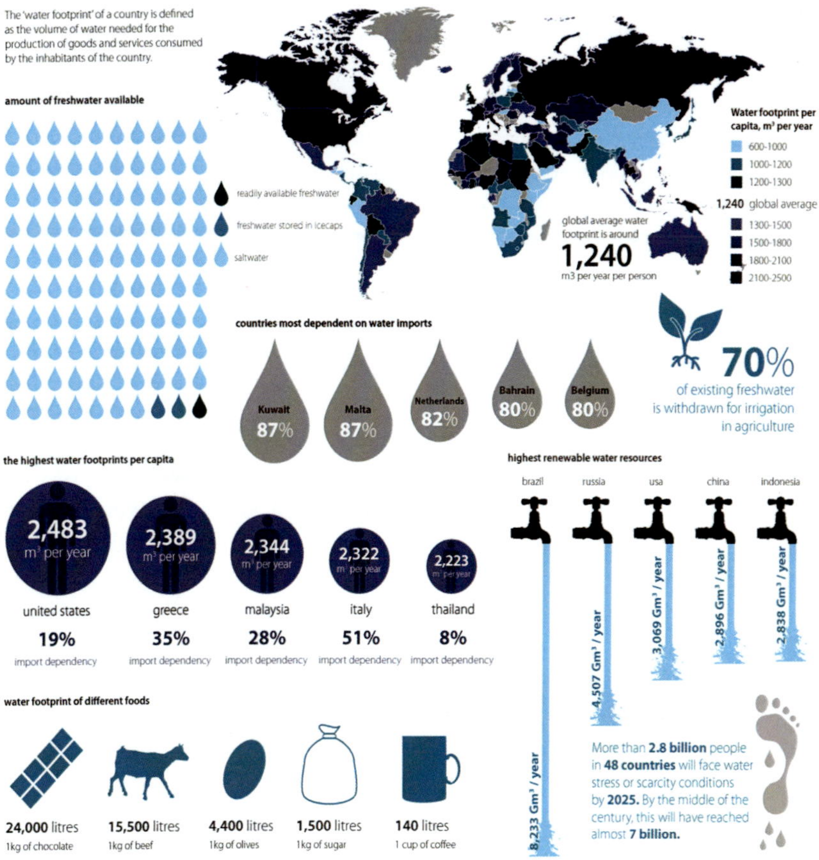

Figure 7.11. *Challenges to water availability.*

Source: By Tiffany Farrant – https://www.flickr.com/photos/gdsdigital/4306320940/, CC BY 2.0, https://commons.wikimedia.org/w/index.php?curid=72143043.

Diminished water availability due to shifting climatic conditions complicates agricultural planning and heightens tensions over water allocation between competing sectors, such as agriculture, industry, and domestic use. Such unpredictable changes demand innovative solutions and policies that can adapt to the evolving landscape of water resources, fostering resilience in both urban and rural settings. Equitable access to water remains a formidable challenge that intertwines with socioeconomic disparities and governance issues. Marginalized communities often bear the brunt of water scarcity due to inadequate infrastructure, lack of political advocacy, and economic constraints. Inequality in water distribution perpetuates cycles of poverty and aggravates public health outcomes, as limited access to clean water facilitates the spread of waterborne diseases (Campbell et al., 2017). Additionally, the absence of effective governance structures can result in poor management of available resources, leading to further depletion and misallocation. A focus on collaborative governance models, community involvement, and equitable resource management can facilitate more sustainable practices that address the needs of all citizens. Ultimately, overcoming challenges to water availability requires a multi-faceted approach that combines efficient resource management with a commitment to social equity (Figure 7.11).

7.7.1. Climate Change and Its Effects

Understanding the intricate relationship between climate change and water availability is essential for developing effective strategies to mitigate its impacts. Rising global temperatures have been linked to increased evaporation rates, resulting in altered precipitation patterns. These changes lead to extreme weather events such as droughts and floods, further straining water resources. Specifically, agricultural ecosystems are profoundly affected, as regions traditionally known for their productivity face declining water supplies. Major food production areas are projected to experience severe reductions in water availability, along with significant impacts on essential crops like wheat, rice, and maize. This threatens food security and exacerbates inequalities in resource distribution, particularly in regions with low human development indices. The looming threat of reduced crop yields due to climate change necessitates innovative agricultural practices to enhance water-use efficiency (Romanello et al., 2021). One promising avenue of research focuses on the manipulation of plant stomata, the pores responsible for gas exchange and transpiration. By understanding how stomatal number and morphology regulate water loss, scientists can develop crop varieties that maintain higher yields with less water input. This effort becomes increasingly critical as drought conditions and water scarcity become more prevalent due to climate change (Bertolino et al., 2019). Moreover, improving drought tolerance through genetic modification may provide agricultural systems with a more resilient response to changing environmental conditions. By prioritizing research into plant adaptations, we can better prepare our food systems for the anticipated challenges posed by climate shifts. In conclusion, the ongoing effects of climate change on water availability underscore the urgent need for adaptive strategies across various sectors. Policymakers must prioritize sustainable water management practices that consider these environmental changes, balancing agricultural needs with conservation efforts. By integrating advancements in crop science with effective water management policies, we can improve the resilience of food systems while addressing the growing concerns surrounding climate adaptation. Ultimately, the intersection of climate change and water availability demands a collaborative and multifaceted approach, ensuring that communities worldwide can secure access to this vital resource amidst an evolving climate landscape.

7.7.2. Pollution and Contamination of Water Sources

The state of our water sources has increasingly become a matter of concern as pollution continues to escalate due to various anthropogenic activities. Common pollutants include heavy metals, agricultural runoff, and untreated wastewater, which significantly degrade the quality of freshwater bodies. Sediment pollution assessments have highlighted the need for detailed studies on various ecosystems, revealing alarming patterns of contamination across different seasons. Research conducted in several wetlands indicates that sediment quality is particularly compromised in the post-monsoon season, suggesting that seasonal variations exacerbate pollution levels and pose ecological risks (Dash et al., 2020). Addressing the problem requires a nuanced understanding of how human actions affect water sources, emphasizing the importance of regular monitoring and the implementation of effective pollution control measures.

Groundwater, often considered a reliable source of drinking water, is also vulnerable to contamination, particularly from elements like arsenic that pose severe health risks. This contamination is primarily geogenic, linked to geological processes that can mobilize harmful substances into aquifers. As noted, more than 230 million people globally face risks related to arsenic in groundwater, with Southeast Asia, including India, being significantly affected (Shaji et al., 2020). The correlation between aquifer type and contamination levels underscores the necessity for targeted interventions. Remedial measures such as filtration systems, artificial recharge, and community awareness programs are critical to mitigate these risks, especially for the rural populations that heavily depend on groundwater but are often unaware of its dangers. Addressing water pollution is not just an ecological issue but a pressing public health concern that necessitates coordinated efforts across multiple sectors. The intersection of water quality and health outcomes underscores the urgency of comprehensive water management approaches. The strategies employed should focus on immediate cleanup efforts and incorporate preventive measures aimed at reducing contaminant inputs. Community education plays a vital role in empowering individuals to understand the implications of water contamination and the importance of safeguarding their local water sources. Collaborative initiatives among governments, non-profits, and local communities can pave the way toward sustainable practices that protect water availability, ensuring that future generations have access to clean and safe water resources.

7.7.3. Population Growth and Increased Demand

The surge in global population continues to exert immense pressure on natural resources, particularly freshwater supplies. As the world population is projected to reach nearly 10 billion by 2050, the water demand will escalate significantly. This increased need comes from various sectors, including agriculture, industry, and domestic consumption, each vying for limited water resources. Studies underscore that current predictions about irrigated areas underestimate the potential extension of irrigation necessary to meet food demands, potentially reaching between 300 and 800 million hectares by 2050 (Puy et al., 2020). While water scarcity has been acknowledged, the challenge lies in balancing this growing demand with sustainable water management practices essential for future generations. Failure to address these challenges risks exacerbating existing inequalities and could lead to severe socio-economic repercussions, particularly in regions already grappling with water stress. Compounding this issue is the increased demand for food resulting from population growth. Though food production has kept pace with caloric needs, a significant proportion of the global population suffers from malnourishment and diet-related health issues. This paradox highlights that simply increasing food supply does not guarantee nutritional equity or health outcomes. Current food systems are not only failing to meet dietary needs but are also linked to environmental degradation (Willett et al., 2019). Sustainable food production must incorporate strategies that optimize water use, as the agricultural sector is a major consumer of freshwater. Thus, the intersection of population growth and food demand necessitates a comprehensive reevaluation of agricultural practices, ensuring that they are both water-efficient and capable of supporting healthier diets for a rapidly growing population. To tackle the dual challenges of population growth and increasing demand, innovative approaches to water management and food systems are essential. Efficient irrigation techniques, such as drip irrigation and rainwater harvesting, must be prioritized to maximize water use efficiency in agricultural practices. Additionally, policies promoting the restoration and conservation of freshwater ecosystems can enhance water availability. Simultaneously, addressing the nutritional quality of food production is critical. As noted, the current global food system produces enough calories; however, it often neglects the essential vitamins and minerals needed for a healthy diet, contributing to widespread dietary deficiencies (Willett et al., 2019). By fostering collaborative efforts among policymakers, farmers, and consumers, it is possible to create a sustainable framework that ensures food security while conserving precious water resources, thereby enhancing resilience in the face of ongoing population growth.

7.8. CONCLUSION

In reflecting on the essential relationship between water availability and its multifaceted uses, it becomes clear that strategic management practices are crucial for sustainability. Access to clean water is not merely a public health issue but a foundational element for agricultural productivity, economic stability, and social equity. For instance, (Pawaskar et al., 2024) outline how inadequate sanitation and water management practices in rural areas of India impede progress toward achieving Sustainable Development Goals (SDGs), particularly Goal 6, which emphasizes clean water and sanitation. The persistence of open defecation and the marginalization of sanitation facilities highlights the urgent need for effective interventions. Thus, addressing these challenges requires an integrated approach that combines education, policy reform, and community engagement to ensure equitable access to water resources and enhance overall societal well-being. Transitioning to agricultural applications, the management of water resources emerges as a critical factor influencing food security and economic growth, particularly in developing regions. The findings from (Kennedy Kalu et al., 2023) illustrate that factors such as socio-economic conditions, climate change, and technological advancements significantly affect water capacity for farming.

By optimizing water management techniques through data-driven analyses and innovative practices, stakeholders can improve agricultural output while conserving precious water resources. This approach not only addresses immediate agricultural challenges but also fosters long-term resilience against climate variability. The evidence supports a shift towards sustainable practices that can mitigate water scarcity and ensure food systems remain robust in the face of growing global demands. Ultimately, prioritizing the sustainable management of water resources serves environmental goals and social and economic objectives. By recognizing the interconnectedness of water availability, agricultural productivity, and community health, leaders, and policymakers can create synergistic strategies that benefit all stakeholders. As evidenced by the stark realities presented the urgency of addressing water issues cannot be overstated. Only through collaborative efforts that unite various sectors—government, agriculture, and community organizations—can we hope to develop a comprehensive solution to the global water crisis. The commitment to innovation, coupled with an understanding of local contexts, will pave the way for achieving sustainable water management and ensuring a stable future for generations to come (Kalu et al., 2023; Pawaskar et al., 2024).

7.8.1. Summary of Key Points

The increasing global demand for water due to population growth and industrialization poses significant challenges for water availability. Currently, over 820 million people worldwide suffer from inadequate access to sufficient food and safe water, which is vital for both human health and productive agricultural systems. This alarming statistic underscores the urgent need for sustainable water management practices, particularly in regions where water scarcity is most pronounced. Additionally, the degradation of freshwater ecosystems due to pollution and over-extraction exacerbates the issue, highlighting the importance of fostering a balanced approach to water utilization that emphasizes conservation and ecosystem health. Understanding the interplay between water availability and agricultural practices is essential for addressing food security and environmental sustainability. The current food systems often lead to excessive water use, posing a threat to both human nutrition and ecological integrity; thus, there is an urgent need to shift towards more sustainable agricultural practices. For example, implementing efficient irrigation techniques and utilizing drought-resistant crops can significantly reduce water consumption while maintaining agricultural productivity. Furthermore, integrative approaches that consider local water resources and limitations will help ensure that food production does not compromise water availability for future generations, making it imperative to prioritize sustainable water management strategies. The importance of water quality cannot be overlooked alongside quantity in discussions about water availability. Clean water is essential not only for drinking but also for maintaining healthy ecosystems that support diverse forms of life. Water pollution from agricultural runoff and industrial waste directly impacts the availability of safe drinking water, leading to adverse health effects on communities. Emphasizing the need for improved regulations and community involvement in water stewardship can aid in addressing this multifaceted issue. By fostering collaborative efforts among governments, NGOs, and local communities, we can ensure equitable access to water and promote practices that safeguard our vital water resources not only today but for future generations as well.

7.8.2. Importance of Sustainable Water Management

Effective water management is crucial for sustaining the delicate balance between human needs and environmental conservation. As global populations grow and urbanization expands, the pressure on water resources intensifies. This growing demand highlights the necessity for strategies that prioritize sustainable management, ensuring equitable access to clean water for present and future generations. Unchecked exploitation leads to catastrophic environmental impacts, such as habitat destruction and diminished biodiversity, threatening the very systems that provide vital ecosystem services. Therefore, implementing sustainable approaches is essential for addressing the immediate concerns of water scarcity and for promoting long-term environmental health. Integrating sustainable practices into agriculture is a significant aspect of water management that has far-reaching effects. Crops such as olives and vineyards, historically associated with high soil degradation, can greatly benefit from sustainable techniques like cover cropping. This practice enhances soil fertility and biodiversity and plays a critical role in runoff reduction and improved soil water retention, as evidenced by research demonstrating that cover crops reduce soil water loss and increase water availability in arid conditions (Novara et al., 2020). Such methods document the importance of sustainable water management in agricultural systems, which directly influences food security and economic stability. Thus, adopting sustainable agriculture practices conserves essential water resources and supports broader environmental and economic goals. In the context of broader societal needs, water security is intricately linked to sustainable water management. It is imperative to ensure reliable access to safe water, as this has become a cornerstone for health, economic productivity, and social stability (Mishra et al., 2021). The constraints imposed by climate change, urbanization, and socio-economic shifts necessitate a comprehensive approach to water resource allocation that harmonizes human, environmental, and economic aspects. Fostering collaboration among stakeholders is key to implementing effective strategies that enhance water availability and quality. By embracing a holistic perspective on water security, societies can navigate the challenges posed by evolving environmental changes and strive for sustainable development that benefits all.

7.8.3. Call to Action for Water Conservation Efforts

The urgent need for collective action emphasizes the importance of water conservation in our daily lives. As global temperatures rise and populations increase, access to clean, potable water is becoming increasingly strained. By adopting sustainable practices, individuals can significantly reduce their water usage. Simple steps, such as fixing leaks, utilizing water-efficient appliances, and taking shorter showers, contribute to a more sustainable future. These efforts not only alleviate the pressure on local water supplies but also foster a culture of responsibility and awareness about the critical role water plays in our ecosystem. Encouraging communities to engage in water-saving initiatives fosters a collective environment where conservation becomes a shared priority. Furthermore, educational programs in schools and communities can enhance awareness of water conservation strategies and their practical applications. Through workshops, informational campaigns, and community challenges, people can learn innovative methods to reduce water consumption. School projects, such as rainwater harvesting or creating drought-resistant gardens, can instill a sense of stewardship in younger generations. When communities come together to understand the importance of water conservation, they create a rallying point that strengthens efforts and builds momentum for broader initiatives. Government policies, when complemented by grassroots education, have the potential to effect meaningful change at local, national, and global levels. Ultimately, the responsibility for water conservation lies with each of us, and we must act now. The consequences of inaction are dire; as water scarcity develops into a pressing global issue, it intensifies competition for this essential resource. By collectively committing to conservation efforts, we can safeguard our water resources and ensure their availability for future generations. Support for policies that incentivize water efficiency, coupled with personal accountability and proactive behavior, will lead to sustainable water management. Failure to act could mean leaving a dwindling water supply for those who follow us, underscoring the necessity for immediate, concerted efforts to preserve what is arguably our most vital resource.

REFERENCES

1. Abdullah, S., Ahmed, S. I., Husain, E., Hasan, F., & Ahmed, S. (2023). Drought assessment of a data-scarce watershed–Quetta Valley, Pakistan. Retrieved from https://www.semanticscholar.org/paper/2e6b5f67091cfd1 60d78a58946acee554fbadb21 (accessed on 5 December 2024).
2. Akinpelu, A. A., Ali, M. E., Johan, M. R., Saidur, R., Qurban, M. A., & Saleh, T. A. (2018). Polycyclic aromatic hydrocarbons extraction and removal from wastewater by carbon nanotubes: A review of the current technologies, challenges, and prospects. *Separation and Purification Technology, 122*, 68–82. https://doi.org/10.1016/j.psep.2018.11.006.
3. Anger, J., Lee, U., Ackerman, A. L., Chou, R., Chughtai, B., Clemens, J. Q., Hickling, D., Kapoor, A., Kenton, K. S., Kaufman, M. R., Rondanina, M. A., Stapleton, A., Stothers, L., & Chai, T. C. (2019). Recurrent uncomplicated urinary tract infections in women: AUA/CUA/SUFU guideline. *Journal of Urology, 202*(2), 282–289. https://doi.org/10.1097/ JU.0000000000000296. Update in: *Journal of Urology, 208*(4), 754–756. https://doi.org/10.1097/JU.0000000000002888.
4. Ashraf, S., Nazemi, A., & AghaKouchak, A. (2021). Anthropogenic drought dominates groundwater depletion in Iran. *Scientific Reports, 11*. https://doi.org/10.1038/s41598-021-88522-y.
5. Bergamini, M. P., Chiavico, M., Bracco, L., Moglia, A., & Buffagni, M. (2021). Water balance assessment for water management in oil & gas operations: A methodological approach. Retrieved from: https://www.semanticscholar. org/paper/a871942f875aa9dc7d33a2d88a666624fffb8c3f (accessed on 5 December 2024).
6. Bertolino, L. T., Caine, R. S., & Gray, J. E. (2019). Impact of stomatal density and morphology on water-use efficiency in a changing world. *Frontiers in Plant Science, 10*. https://doi.org/10.3389/fpls.2019.00225.
7. Bray, F., Ferlay, J., Soerjomataram, I., Siegel, R. L., Torre, L. A., & Jemal, A. (2018). Global cancer statistics 2018: GLOBOCAN estimates of incidence and mortality worldwide for 36 cancers in 185 countries. *CA: A Cancer Journal for Clinicians, 68*(6), 394–424. https://doi.org/10.3322/caac.21492.
8. Brugmann, K., Osthoff, H., Braune, W., Sievers, E., & Paul, H. G. (2017). *Course in General Linguistics*. https://doi.org/10.4324/9781912281732.
9. Campbell, S. J., Biritwum, N. K., Woods, G., Velleman, Y., Fleming, F., & Stothard, J. R. (2018). Tailoring water, sanitation, and hygiene (WASH) targets for soil-transmitted helminthiasis and schistosomiasis control. *Trends in Parasitology, 34*(1), 53–63. https://doi.org/10.1016/j. pt.2017.09.004.

10. Chopra, K., Tyagi, V. V., Pandey, A. K., & Sarı, A. (2018). Global advancement on experimental and thermal analysis of evacuated tube collector with and without heat pipe systems and possible applications. *Applied Energy, 228*, 351–389. https://doi.org/10.1016/j.apenergy.2018.06.067.
11. Cruz, H., Law, Y., Guest, J. S., Rabaey, K., Batstone, D. J., Laycock, B., Verstraete, W., & Pikaar, I. (2019). Mainstream ammonium recovery to advance sustainable urban wastewater management. *Environmental Science & Technology, 53*(21), 11066–11079. https://doi.org/10.1021/acs.est.9b00603.
12. Curto, D., Franzitta, V., & Guercio, A. (2021). A review of the water desalination technologies. *Applied Sciences, 11*(2), 670. https://doi.org/10.3390/app11020670.
13. Dari, J., Quintana-Seguí, P., Escorihuela, M. J., Stefan, V., Brocca, L., & Morbidelli, R. (2021). Detecting and mapping irrigated areas in a Mediterranean environment by using remote sensing soil moisture and a land surface model. *Journal of Hydrology, 596*, 126129. https://doi.org/10.1016/j.jhydrol.2021.126129.
14. Dash, S., Borah, S. S., & Kalamdhad, A. S. (2020). Heavy metal pollution and potential ecological risk assessment for surficial sediments of Deepor Beel, India. *Ecological Indicators, 122*, 107265. https://doi.org/10.1016/j.ecolind.2020.107265.
15. Dembélé, M., Vrac, M., Ceperley, N., Zwart, S. J., Larsen, J., Dadson, S., Mariethoz, G., & Schaefli, B. (2022). Contrasting changes in hydrological processes of the Volta River basin under global warming. *Hydrology and Earth System Sciences, 26*(4), 1481–1506. https://doi.org/10.5194/hess-26-1481-2022.
16. Dwivedi, Y., Hughes, L., Ismagilova, E., Aarts, G., Coombs, C., Crick, T., Duan, Y., Dwivedi, R., Edwards, J., Eirug, A., Galanos, V., Ilavarasan, V., Janssen, M., Jones, P., Kar, A., Kizgin, H., Kronemann, B., Lal, B., Lucini, B., Williams, M., et al. (2019). Artificial intelligence (AI): Multidisciplinary perspectives on emerging challenges, opportunities, and agenda for research, practice, and policy. *International Journal of Information Management, 57*, 101013. https://doi.org/10.1016/j.ijinfomgt.2019.08.002.
17. Efficiency of biochar with mineral fertilizers on soil properties and crop growth. (2024). Retrieved from: https://www.semanticscholar.org/paper/f723fd1d231cc500d2d22bdc18b68018cddbf1b2 (accessed on 5 December 2024).

18. Ferreira, P., van Soesbergen, A., Mulligan, M., de Freitas, M. A. V., & Vale, M. M. (2019). Can forests buffer negative impacts of land-use and climate changes on water ecosystem services? The case of a Brazilian megalopolis. *Science of The Total Environment, 685*, 248–258. https://doi.org/10.1016/j.scitotenv.2019.05.065.

19. Friedlingstein, P., O'Sullivan, M., Jones, M. W., Andrew, R. M., Hauck, J., Olsen, A., et al., (2020). Global carbon budget 2020. *Earth System Science Data, 12*, 3269–3340. https://doi.org/10.5194/essd-12-3269-2020.

20. Gatta, P. P. (2022). *The State of World Fisheries and Aquaculture 2022*. https://doi.org/10.4060/cc0461en.

21. GBD 2019 Ethiopia Subnational-Level Disease Burden Initiative Collaborators. (2022). Progress in health among regions of Ethiopia, 1990–2019: A subnational country analysis for the Global Burden of Disease Study 2019. *The Lancet, 399*(10332), 1322–1335. https://doi.org/10.1016/S0140-6736(21)02868-3.

22. Hogeboom, R. J. (2020). The water footprint concept and water's grand environmental challenges. *One Earth, 2*, 218–222. https://doi.org/10.1016/j.oneear.2020.02.010.

23. Ibrahim, S. A., Memon, F. A., & Butler, D. (2021). Seasonal variation of rainy and dry season per capita water consumption in Freetown City, Sierra Leone. *Water, 13*, 499. https://doi.org/10.3390/w13040499.

24. Ikirri, M., Boutaleb, S., Ibraheem, I. M., Abioui, M., Echogdali, F. Z., Abdelrahman, K., Id-Belqas, M., Abu-Alam, T., El Ayady, H., Essoussi, S., et al. (2023). Delineation of groundwater potential areas using an AHP, remote sensing, and GIS techniques in the Ifni Basin, Western Anti-Atlas, Morocco. *Water, 15*(7), 1436. https://doi.org/10.3390/w15071436.

25. Immitzer, M., Vuolo, F., & Atzberger, C. (2016). First experience with Sentinel-2 data for crop and tree species classifications in Central Europe. *Remote Sensing, 8*(3), 166. https://doi.org/10.3390/rs8030166.

26. Intergovernmental Panel on Climate Change. (2023). *Climate change 2022–Impacts, Adaptation and Vulnerability*. https://doi.org/10.1017/9781009325844.

27. Istiyanto, D. C., Wulandari, I., Aziiz, S. A., Yuniardi, R. C., Suranto, Y. T. D., Harita, A., Hamid, A. B., & Widagdo. (2023). Seepage analysis and the reservoir water pollution potential under vertical dam structure planning. Retrieved from: https://www.semanticscholar.org/paper/ebae01c3f82e8bf58e9368652381113a2d4a0df1 (accessed on 5 December 2024).

28. Jagadeesh, P., Puttegowda, M., Boonyasopon, P., Rangappa, S. M., Khan,

A., & Siengchin, S. (2022). Recent developments and challenges in natural fiber composites: A review. *Polymer Composites, 43*, 2545–2561. https://doi.org/10.1002/pc.26619.

29. Jayaraman, S., Sinha, N. K., Dalal, R. C., Lal, R., Mohanty, M., Naorem, A., Hati, K. M., Chaudhary, R. S., Biswas, A. K., Patra, A. K., & Chaudhari, S. K. (2020). No-till farming and conservation agriculture in South Asia–Issues, challenges, prospects, and benefits. *Journal of Soil and Water Conservation, 39*, 236–279. https://doi.org/10.1080/07352689.2020.1782069.

30. Kalu, C. K., & Sakilu, O. B. (2023). Innovative data-driven analysis of water management for effective agricultural practices. Retrieved from: https://www.semanticscholar.org/paper/81343910c7d6db94477ffb02f045410af2ebac64 (accessed on 5 December 2024).

31. Kapoor, D., Bhardwaj, S., Landi, M., Sharma, A., Ramakrishnan, M., & Sharma, A. (2020). The impact of drought in plant metabolism: How to exploit tolerance mechanisms to increase crop production. *Applied Sciences, 10*(16), 5692. https://doi.org/10.3390/app10165692.

32. Karuppaiah, C. (2021). Identification of vulnerable zones of surface water epidemiology using remote sensing and GIS techniques. Retrieved from: https://www.semanticscholar.org/paper/7d7bea4dcc536dcda3737bb973604192e174482b (accessed on 5 December 2024).

33. Kumar, M., Sharma, A., Tabhani, N., & Otaki, Y. (2021). Indoor water end-use pattern and its prospective determinants in the twin cities of Gujarat, India: Enabling targeted urban water management strategies. *Journal of Environmental Management, 288*, 112403. https://doi.org/10.1016/j.jenvman.2021.112403.

34. Lin, J., Zhao, Q., Huang, H., Mao, H., Liu, Y., & Xiao, Y. (2020). Applications of low-temperature thermochemical energy storage systems for salt hydrates based on material classification: A review. *Solar Energy, 214*, 149–178. https://doi.org/10.1016/j.solener.2020.11.055.

35. Livingston, G., Sommerlad, A., Orgeta, V., Costafreda, S. G., Huntley, J., Ames, D., et al., (2017). Dementia prevention, intervention, and care. *The Lancet, 390*(10113), 2673–2734. https://doi.org/10.1016/S0140-6736(17)31363-6.

36. Luo, Y., Andrésen, J. M., Clarke, H., Rajendra, M., & Maroto-Valer, M. M. (2019). A decision support system for waste heat recovery and energy efficiency improvement in data centres. *Applied Energy, 250*, 1217–1224. https://doi.org/10.1016/j.apenergy.2019.05.029.

37. Matowo, N. S., Tanner, M., Munhenga, G., Mapua, S. A., Finda, M. F., Utzinger, J., Ngowi, V., & Okumu, F. O. (2020). Patterns of pesticide usage in agriculture in rural Tanzania call for integrating agricultural and public

health practices in managing insecticide-resistance in malaria vectors. *Malaria Journal, 19*. https://doi.org/10.1186/s12936-020-03331-4.

38. McCarron, B. (2023). Potential applications of amorphous calcium carbonate (ACC) in water treatment operations (Master's thesis, University of Rhode Island). Retrieved from: https://digitalcommons.uri.edu/theses/2329 (accessed on 5 December 2024).
39. Mishra, B. K., Kumar, P., Saraswat, C., Chakraborty, S., & Gautam, A. (2021). Water security in a changing environment: Concept, challenges and solutions. *Water, 13*, 490. https://doi.org/10.3390/w13040490.
40. Mishra, S., Singh, R., & Rath, C. C. (2020). Synthetic microfibers: Source, transport, and their remediation. *Journal of Water Process Engineering, 10*, 101612. https://doi.org/10.1016/j.wpe.2020.101612.
41. Mohammed, M. S., & Jbara, O. K. (2024). A study of factors affecting adoption of pivot sprinkler irrigation technology for wheat crop producers. *Iraqi Journal of Agricultural Sciences, 55*(4), 1381–1393. https://doi.org/10.36103/0sdxzx58c42a248c703233e4cb7f0598c5a93ffc (accessed on 5 December 2024).
42. Morris, J., Georgiou, I. Y., Guenther, E., & Caucci, S. (2021). Barriers in implementation of wastewater reuse: Identifying the way forward in closing the loop. *Environmental Science and Pollution Research, 1*, 413–433. https://doi.org/10.1007/s43615-021-00018-z.
43. Nergis, Y., Butt, J. A., Sharif, M., & Kalu, C. K. (2021). Assessment of marine coastal water pollution from Karachi Harbour, Pakistan. *International Journal of Economic and Environmental Geology*. Retrieved from: https://www.semanticscholar.org/paper/e4019a96ff4afef5fa8eb9d755ccde76e208d7d3 (accessed on 5 December 2024).
44. Novara, A., Cerdà, A., Barone, E., & Gristina, L. (2021). Cover crop management and water conservation in vineyard and olive orchards. *Soil and Tillage Research, 208*, 104896. https://doi.org/10.1016/j.still.2020.104896.
45. Obrist, D., Kirk, J. L., Zhang, L., Sunderland, E. M., Jiskra, M., & Selin, N. E. (2018). A review of global environmental mercury processes in response to human and natural perturbations: Changes of emissions, climate, and land use. *Ambio, 47*(2), 116–140. https://doi.org/10.1007/s13280-017-1004-9.
46. Pandey, V., Dhaubanjar, S., Bharati, L., & Thapa, B. (2020). Spatio-temporal distribution of water availability in Karnali-Mohana Basin, Western Nepal: Hydrological model development using multi-site calibration approach (Part-A). *Journal of Hydrology: Regional Studies, 29*, 100690. https://doi.org/10.1016/j.ejrh.2020.100690.

47. Pawaskar, R., & Velhal, G. (2024). A cross-sectional study on sanitation and hygiene in rural villages in Southeastern Maharashtra: On our way to achieving the Sustainable Development Goal No. 6. *Bharati Vidyapeeth Medical Journal, 4*, 4–10. https://doi.org/10.56136/BVMJ/2024_02759.
48. Puy, A., Lo Piano, S., & Saltelli, A. (2020). Current models underestimate future irrigated areas. *Geophysical Research Letters, 47*. https://doi.org/10.1029/2020gl087360.
49. Qazi, S., Khawaja, B. A., & Farooq, Q. U. (2022). IoT-equipped and AI-enabled next-generation smart agriculture: A critical review, current challenges, and future trends. *IEEE Access, 10*, 21219–21235. https://doi.org/10.1109/access.2022.3152544.
50. Romanello, M., McGushin, A., Di Napoli, C., Drummond, P., Hughes, N., Jamart, L., et al., (2021). The 2021 report of the Lancet Countdown on health and climate change: Code red for a healthy future. *The Lancet, 398*, 1619–1662. https://doi.org/10.1016/s0140-6736(21)01787-6.
51. Salehie,O.,Jamal,M.H.B.,Ismail,T.,Harun,S.,&Shahid,S.(2024).Projectionof future water availability in the Amu Darya Basin. Retrieved from: https://www.semanticscholar.org/paper/c4db3783758ffa382562a4a3e22203a781560318 (accessed on 5 December 2024).
52. Sborgi, L., Rühl, S., Mulvihill, E., Pipercevic, J., Heilig, R., Stahlberg, H., Farady, C. J., Müller, D. J., Brož, P., & Hiller, S. (2016). GSDMD membrane pore formation constitutes the mechanism of pyroptotic cell death. *The EMBO Journal, 35*, 1766–1778. https://doi.org/10.15252/embj.201694696.
53. Shaji, E., Santosh, M., Sarath, K. V., Prakash, P., Deepchand, V., & Divya, B. V. (2020). Arsenic contamination of groundwater: A global synopsis with focus on the Indian Peninsula. *Groundwater for Sustainable Development, 12*, 101079. https://doi.org/10.1016/j.gsf.2020.08.015.
54. Sharma, S., et al. (2021). J. Phys.: Conf. Ser. *1804*, 012147. https://www.semanticscholar.org/paper/6f750dd323cb296f742f5677d9086ccb9a6ff1ac
55. Willett, W. C., Rockström, J., Loken, B., Springmann, M., Lang, T., Vermeulen, S. J., et al., (2019). Food in the Anthropocene: The EAT–Lancet Commission on healthy diets from sustainable food systems. *The Lancet, 393*(10170), 447–492. https://doi.org/10.1016/s0140-6736(18)31788-4.
56. Xu, L. D., Xu, E., & Li, L. (2018). Industry 4.0: State of the art and future trends. *International Journal of Production Research, 56*, 2941–2962. https://doi.org/10.1080/00207543.2018.1444806.
57. Zhao, H., Shi, J., Qi, X., Wang, W., & Jia, J. (2017). Pyramid scene parsing network. https://doi.org/10.1109/cvpr.2017.660.

CHAPTER 8

FOOD AND HUNGER

Contents

INTRODUCTION

Across the globe, the intricate relationship between food and hunger continues to shape the lives of millions. As societies evolve and face technological advancements alongside climate change, the dynamics of food production and distribution have become more complex. Understanding this relationship requires a multifaceted approach that encompasses various disciplines, including economics, sociology, and environmental science. By examining how food systems operate and their implications for hunger, we can unravel the challenges and solutions that are critical to addressing this pressing issue. The struggle for food security is not just a matter of availability; it intertwines with political decisions, cultural contexts, and economic realities that deserve thorough exploration. In examining the intricate ties between food and hunger, one must acknowledge the stark disparities that exist both within and between nations. While some regions boast abundant resources and culinary diversity, others grapple with insufficient access to basic nutrients, leading to malnutrition and health complications. This unequal distribution of food resources can be attributed to a range of factors, including poverty, inadequate agricultural practices, and global market fluctuations. Furthermore, systemic issues like food waste and environmental degradation exacerbate these disparities. By understanding the root causes of hunger, stakeholders can develop targeted policies and initiatives aimed at fostering equitable access to food, ensuring that no individual suffers from the consequences of hunger. Efforts to address food insecurity necessitate a cohesive strategy that balances local needs with global sustainability. Innovations in farming techniques, supply chain efficiencies, and nutritional education are critical to building resilient food systems. Additionally, community engagement and local involvement can empower individuals to take charge of their food sources, reducing dependency on external systems that may not prioritize their well-being. Collaborative partnerships among governments, non-profits, and private enterprises can further enhance food security initiatives, driving meaningful change. Ultimately, a comprehensive understanding of food and hunger issues fosters a holistic perspective that encourages effective action, ensuring that every person has the opportunity to nourish themselves and thrive.

8.1. DEFINITION OF FOOD AND HUNGER

The concept of food goes beyond mere sustenance; it encompasses the nutritional components necessary for human survival and well-being. Food is defined not just in terms of its caloric content but also in its ability to provide essential vitamins, minerals, and other nutrients vital for maintaining health. Furthermore, cultural, social, and economic factors play a significant role in determining what constitutes food for different communities. Accessibility and availability significantly influence dietary choices, which can lead to disparities in nutrition across various populations. In many instances, the quality of food available equates to health outcomes, emphasizing the necessity for a comprehensive understanding of food that includes all these dimensions. In contrast, hunger manifests as a complex interplay of physical lack and social inequality, often leading to food insecurity. The U.S. Household Food Security Measure serves as an essential tool to assess this phenomenon, illustrating that hunger is about the absence of food and about the uncertainty surrounding food access and the nutritional quality of what is available. According to research conducted by "Partners In Ending Hunger," the alarming statistic reveals that approximately 31 million Americans, including 12 million children, do not have access to sufficient food. This lack of access can lead to severe immediate consequences, such as malnutrition and its associated health issues, as well as long-term effects on individual's physical and cognitive development, ultimately hindering overall societal progress (Rachael Kulick, 2002).

Figure 8.1. *Major differences between hunger and food insecurity.*

Source: By Jens Galschiøt – http://www.aidoh.dk/, GFDL, https://commons.wikimedia.org/w/index.php?curid=18269976.

Addressing hunger requires communal responsibility and intervention. The work of organizations like "Partners" underscores the need for strategic initiatives that define and measure food insecurity accurately. By establishing specific criteria for ending hunger, such as achieving food security for all households over consecutive years, these organizations provide actionable frameworks for communities. Moreover, the integration of food banking efforts with broader community food security initiatives has been shown to create a more inclusive and effective approach to combating hunger. By understanding the definitions and distinctions between food, hunger, and food insecurity, communities can employ targeted strategies to alleviate hunger, ensuring that no individual faces the dire consequences of inadequate access to food (Andy Fisher, 2005) (Figure 8.1).

8.1.1. Importance of Addressing Food Insecurity

The multifaceted impacts of food insecurity extend beyond personal hardship, significantly affecting societal well-being and economic stability. When individuals lack reliable access to sufficient food, not only do their health outcomes suffer, but there is also a detrimental effect on productivity. This cycle of poverty and hunger can perpetuate systemic inequality, particularly in communities of color, where students face disproportionate rates of food insecurity. Addressing food insecurity is crucial for breaking this cycle and fostering a healthier workforce. Implementing effective policies and programs can alleviate hunger and enhance educational outcomes, as well-nourished students are more likely to succeed academically. Hence, strategic intervention in food insecurity is paramount for cultivating a thriving society. Moreover, promoting food security necessitates addressing institutional barriers that disproportionately affect vulnerable populations, such as college students. Many students experience food insecurity while navigating the financial demands of higher education. This scenario is exacerbated by the rising costs of tuition and living expenses, leaving students to rely on under-resourced campus dining facilities or depleting savings on inexpensive meals.

The Health, Hunger-Free Kids Act (HHFKA) introduced valuable provisions, such as the "identified students" mechanism, which highlights the effectiveness of reducing barriers to access to food assistance. Implementing similar adjustments to programs like SNAP for college students can provide essential resources. By reducing these barriers, students can focus on their education rather than worrying about where their next meal will come from, thus contributing to improved academic performance and retention rates in higher education. Lastly, addressing food insecurity is not just a moral imperative; it is an essential component in promoting sustainable economic growth. High levels of food insecurity strain public resources, as government agencies must allocate funds to emergency food assistance programs and health care costs associated with diet-related illnesses. By investing in comprehensive solutions to food insecurity, such as improved access to nutritional education and sustainable food systems, we alleviate immediate hunger and enhance long-term economic resilience. Organizations can form partnerships that strengthen the agricultural sector, improve food distribution networks, and create job opportunities, addressing the root causes of food insecurity. By adopting a holistic approach, society can pave the way toward a future where access to nutritious food is considered a fundamental right rather than a privilege, fostering a healthier, more equitable world for all (Figure 8.2).

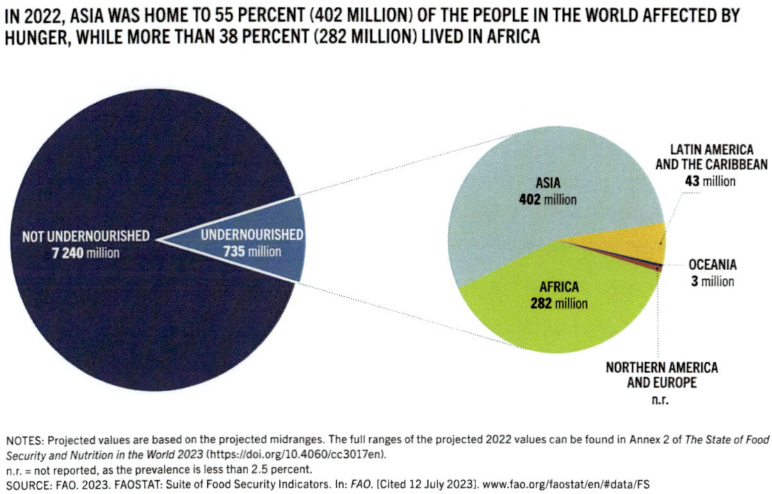

Figure 8.2. *Food insecurity.*

Source: By http://www.fao.org Food and Agriculture Organization – https://doi.org/10.4060/cc3017en The State of Food Security and Nutrition in the World 2023, CC BY-SA 3.0 igo, https://commons.wikimedia.org/w/index.php?curid=144880558.

8.1.2. Overview of the Chapter

The chapter titled Food and Hunger is structured to provide a comprehensive examination of the multifaceted issues surrounding food security and malnutrition. The introductory section sets the stage by defining key concepts and contextualizing the problem within global and local frameworks. Following this introduction, the chapter delves into the various factors contributing to food insecurity, such as economic instability, political discord, and environmental challenges. Each of these elements is explored in detail to demonstrate how they interconnect, highlighting the complexity of addressing hunger. This structured approach prepares readers for the analytical sections that follow, where specific case studies and policy responses are discussed. In the subsequent sections, the chapter transitions into empirical data and theoretical models that underpin the discussion on food security. Cited frameworks, like the International Model for Policy Analysis of Agricultural Commodities and Trade (IMPACT), provide valuable insights into the future scenarios of global food availability and the implications of various policy actions. For instance, Bos et al suggest that progressive policy initiatives can dramatically reduce food insecurity and child malnutrition in developing countries. The analysis encourages readers to critically engage with the data presented, elucidating the potential pathways toward improved global food security. This analytical segment is vital for understanding the effectiveness of different strategies and their real-world applications in combating hunger. Finally, the concluding portion of the chapter synthesizes the insights drawn from earlier sections and proposes actionable recommendations for stakeholders involved in food security initiatives. Emphasizing the importance of coherent policies that integrate agricultural growth, rural development, and poverty reduction, the conclusion underscores the urgent need for collaboration among governments, NGOs, and the international community. Moreover, addressing the challenges faced by transitional and crisis-affected nations is crucial for a holistic approach to food security. This final section summarizes the key findings and inspires ongoing dialogue about practices that can mitigate hunger and improve nutrition globally, leaving the reader with a clear roadmap of future directions in alleviating food insecurity.

8.2. HISTORICAL CONTEXT OF FOOD AND HUNGER

The interplay between food systems and social justice has deep historical roots, particularly within the framework of racial equality movements in the United States. During the long civil rights era, food access became a critical issue that illuminated broader societal inequities. Activists recognized that hunger was not merely a personal or individual challenge but a structural consequence of systemic racism. For instance, the Nation of Islam, under Elijah Muhammad, engaged in dietary reform as a means of promoting spiritual and physical well-being among black Americans. By advocating for healthier food choices and establishing community enterprises, they challenged the oppressive structures of their time while also demonstrating how food could serve as a vehicle for empowerment and autonomy (Potorti et al., 2016). This movement underscored the idea that food access is intrinsically linked to other civil rights, framing health and nutrition as essential components of the liberation struggle. Moreover, grassroots campaigns like the Student Nonviolent Coordinating Committees (SNCC) Food for Freedom initiative revealed the acute connection between hunger and political power. In response to local authorities withholding food aid to suppress voter registration among Mississippi sharecroppers, SNCC mobilized support nationwide, linking the right to food with the right to vote. This awareness of food security as a civil right resonated with broader activist circles, leading to supportive networks like Friends of SNCC that advocated for food access as a means of attaining political voice and agency (Potorti et al., 2016).

These campaigns addressed immediate nutritional needs and challenged the systemic barriers that perpetuated hunger, thus highlighting the need for food as a foundational element of social and racial justice. The legacy of these historical contexts continues to shape contemporary discussions on food and hunger, particularly among marginalized communities. The Khwe San indigenous peoples of Namibia, for example, face food insecurity closely tied to their cultural and historical experiences. Research shows that their food choices are influenced by a complex interaction of individual preferences, cultural taboos, and macroeconomic factors like cost and access. These elements stress the significance of understanding food systems through an ecological lens that includes historical injustices and present-day realities. Addressing these complex layers of food insecurity requires recognizing historical context and fostering sustainable management of traditional knowledge and practices. This approach not only enhances community health but also safeguards nutritional strategies essential for survival amid extreme poverty, echoing the lessons from past struggles against hunger.

8.2.1. Evolution of Food Production Methods

The transition from traditional to modern agricultural practices marks a significant shift in food production methods throughout history. Initially, subsistence farming dominated, where families grew food primarily for themselves with a minimal surplus for trade. This practice relied heavily on manual labor and local ecosystems, sustaining communities in a relatively sustainable manner. However, the advent of mechanization in the late 19th and early 20th centuries revolutionized food production. Farmers began to adopt tractors and other machinery, which increased yield potential and efficiency. Consequently, this mechanization enabled the rise of large-scale farming operations, significantly altering the landscape of agricultural economics and food distribution. The Green Revolution of the mid-20th century further transformed food production by introducing high-yield crop varieties, synthetic fertilizers, and advanced irrigation techniques. These innovations led to unprecedented increases in food production, substantially reducing hunger in many parts of the world. However, the approach brought about significant challenges, including environmental degradation and the socio-economic displacement of smallholder farmers. Critically, the consequences of these methods have spurred discussions about sustainability and the need for alternative systems. As noted in the findings on food security, developing countries exhibit diverse experiences and responses to food crises (Fan et al.(n.d)), highlighting that generalized approaches to food production may not adequately address local needs. Contemporary discussions on food sovereignty and urban agriculture reflect an evolution toward more localized and participatory food production methods. Community-driven initiatives have emerged, focusing on the belief that food production should meet the needs of local populations rather than serving multinational corporations. This paradigm shift promotes food security and an enhanced understanding of the social and environmental factors influencing food systems. By empowering communities to grow their food, these practices encourage individuals to cultivate critical consciousness regarding their food choices (Davila et al., 2015). Thus, the evolution of food production methods reflects an ongoing dialogue between technological advancements and the imperative to develop sustainable, community-centered practices that address hunger and promote food justice (Figure 8.3).

Figure 8.3. *Different types and techniques are involved in food processing.*

Source: By No machine-readable author provided. Matthias Kabel assumed (based on copyright claims). – No machine-readable source provided. Own work assumed (based on copyright claims), CC BY-SA 3.0, https://commons.wikimedia.org/w/index.php?curid=246468.

8.2.2. Major Famines and Their Impacts on Society

The historical record of major famines highlights their profound and often devastating impacts on societies. While famine is primarily characterized by severe food shortages, its effects extend far beyond mere hunger. Increased mortality rates serve as a critical measure in understanding the magnitude of such crises. For instance, the Great Irish Famine of the 1840s caused widespread starvation and caused significant demographic changes, including unprecedented migration and shifts in population dynamics. Critics assert that modern famines possess distinct characteristics when compared to those of the past, which complicates our understanding of their implications on societal structures (Vanhaute et al., 2009). As populations respond to food scarcity, whether through displacement or adaptation, the ramifications are felt in economic, social, and political realms, underscoring the interconnectedness of food insecurity and societal stability. Shifts in societal behavior often accompany major famines, as communities are forced to navigate the survival landscape. Historical examples indicate a clear connection between famine and socio-economic transformations, including the decline of agrarian lifestyles. The process referred to as depeasantisation emerges during such crises, wherein traditional farming communities dissolve, leading to urban migration and disruption of local economies (Vanhaute, 2009). This transition changes individual livelihoods and alters societal fabric, leading to increased vulnerability for those who remain. Furthermore, understanding this dynamic allows for greater insight into how populations adapt and reorganize in response to food scarcity.

The disintegration of established social structures during famines also raises questions about long-term resilience and adaptation strategies for affected communities, emphasizing the critical need for comprehensive food systems that mitigate such crises. Additionally, the impacts of famines are often felt multigenerationally, as the consequences ripple through the years, affecting health, education, and economic stability. The direct mortality associated with famine is only part of the story, as the reductions in birth rates, altered familial structures, and shifts in societal roles can create long-lasting effects. Children born into famine conditions are frequently at a higher risk of poor health and malnutrition, compounding the crisis over generations (Cormac Ó Gráda and Dyson, 2001.). Moreover, the sociopolitical landscape can shift dramatically as a response to widespread hunger, often leading to social unrest and changes in governance. In some instances, power dynamics can shift, with marginalized groups rising to prominence as a result of changing circumstances. Understanding these complex interrelations is vital for developing effective policies and interventions aimed at preventing famine and mitigating its devastating impacts on society.

8.2.3. The Role of Agriculture in Civilization Development

The emergence of agriculture marked a pivotal transformation in human society, facilitating the shift from nomadic lifestyles to settled communities. This transition enabled people to cultivate crops and domesticate animals, leading to surplus food production. As communities began to grow, specialization of labor became possible, which allowed individuals to pursue various trades beyond mere subsistence. The surplus of resources contributed significantly to the development of social structures, trade routes, and eventually, complex civilizations. This shift not only supported population growth but also fostered cultural and technological advancements that would ultimately shape human history. As civilizations flourished, the relationship between agriculture and governance became increasingly pronounced. Governments recognized the importance of food security in maintaining social stability and economic prosperity. For instance, in regions affected by water scarcity, such as Pakistan, the management of water resources directly influences food security and agricultural productivity (Fahim et al. (n.d)). Effective governance involved implementing agricultural policies that ensured efficient water utilization while promoting sustainable farming practices—an essential aspect for adapting to climate change and securing food supplies. Consequently, agriculture became not only the backbone of economies but also a driving force behind political and social organization, thereby enhancing communal resilience. Furthermore, the role of agriculture extends beyond mere sustenance; it is intrinsically linked to ethical considerations surrounding humanitarian issues, such as food inequality and hunger relief. The increasing awareness of suffering risks associated with agricultural practices highlights the need for change in how food systems operate (Gloor et al., 2017). As superintelligent technologies emerge, there is potential to improve agricultural outputs sustainably and equitably, addressing hunger and supporting global food security. Thus, recognizing the multifaceted contribution of agriculture to civilization underscores the necessity for policies that promote sustainable practices while also considering the potential risks projected by advancements in technology and climate-related challenges. By focusing on these areas, society can work toward a future where food accessibility is a fundamental right, and hunger is effectively tackled.

8.3. GLOBAL FOOD SYSTEMS

While addressing food security on a global scale, the critical role of sustainable practices in food systems cannot be overstated. The interconnected challenges of overcapacity in fisheries, as well as the impact of perverse subsidies, directly affect the availability and sustainability of food resources. Reports emphasize the significance of innovative solutions to these issues, especially in fisheries, where poor governance and lack of data lead to ineffective management. By analyzing case studies demonstrating successful interventions in sustainable fisheries, it becomes evident that these practices can significantly enhance food availability while protecting oceanic ecosystems. Ultimately, fostering a sustainable fishing industry is essential not only for meeting immediate food needs but also for ensuring the resilience of future generations against hunger (Harris et al., 2013). Moreover, the impacts of climate change significantly exacerbate global food insecurity, making adaptive strategies crucial for maintaining food systems. Recent studies highlight the necessity for vulnerability assessments that incorporate both climatic risks and socio-economic factors to improve food security outcomes. This multifaceted approach recognizes that agriculture must navigate a complex landscape of risks, demanding effective partnerships among meteorological services, agricultural researchers, and policymakers.

Figure 8.4. *Global food systems, culture, and nutrition.*

Source: By travel way of life – Baking Bread at Communal Oven auf flickr, CC BY-SA 2.0, https://commons.wikimedia.org/w/index.php?curid=56839418.

Providing infrastructure for accurate weather data and seasonal forecasts is vital in equipping farmers with the knowledge needed to make informed decisions that enhance food production. Addressing these challenges through targeted technological interventions and governance reforms is essential to bolster food systems against climatic uncertainties and contribute to a more stable food supply globally (Aggarwal et al., 2010). Finally, a transformative shift in global food systems must prioritize equitable access to resources and technologies. Ensuring food security is not solely about increasing food production; it also requires addressing disparities in food distribution and access. By integrating social equity considerations into food policy, we can mitigate hunger and improve livelihoods, particularly in vulnerable populations. Effective governance structures that promote fair trade practices and eliminate wasteful consumption are key. Collaborative efforts among governments, NGOs, and the private sector are needed to foster resilient food systems that address both environmental sustainability and social justice. Ultimately, the goal of a robust global food system hinges on the intersection of sustainability, climate resilience, and equitable access, paving the way for a future with reduced hunger and improved food security for all (Figure 8.4).

8.3.1. Overview of the Global Food Supply Chain

Understanding the intricacies of the global food supply chain is essential for addressing critical issues such as food security and nutrition. This complex network encompasses various stakeholders, including farmers, food processors, distributors, and consumers, each playing a vital role in how food is produced, transported, and consumed across the globe. Structural changes within the supply chain, influenced by globalization and technological advancements, have transformed traditional practices into more sophisticated systems of food distribution. For instance, as indicated in (Capron et al., 2017), the coordination of agricultural production from rural areas with urban demand highlights the interdependencies that characterize modern food systems. Such a multi-faceted approach is crucial for managing the supply chain effectively, ensuring fresh and accessible food options to urban populations.

Furthermore, the rising interconnections among stakeholders complicate the dynamics of this supply chain, particularly in light of the United Nations Sustainable Development Goals (SDGs). Each aspect of the food supply chain contributes to feeding the global population and impacts broader goals related to sustainability and health. As outlined in Bala et al (2021), various stakeholders—from producers to consumers—must collaboratively work towards achieving SDG 2 (Zero Hunger) and SDG 12 (Responsible Consumption and Production). These goals emphasize the necessity of ensuring equitable food distribution and responsible consumption practices, which are fundamental in combatting issues like hunger and malnutrition. In this context, partnerships among all stakeholders become invaluable for synergizing efforts to address these challenges effectively. The challenges faced by the global food supply chain are further aggravated by factors such as climate change, economic disparities, and population growth. Climate change directly affects agricultural productivity, altering planting seasons and impacting food quality, thus exacerbating food insecurity. Additionally, the economic divide between nations complicates equitable access to food, leading to stark differences in nutrition and health outcomes globally. The epidemic of non-communicable diseases associated with overnutrition illustrates the consequences of systemic failures within the global food supply framework, emphasizing the need for comprehensive policy interventions. To realize an efficient and sustainable food supply system, the integration of innovative solutions and stakeholder collaboration is necessary to foster resilience against these challenges while providing nutritious food that meets the needs of a diverse population (Capron et al., 2017).

8.3.2. Factors Affecting Food Distribution

In assessing the intricacies of food distribution, one must consider both physical and systemic barriers that complicate access to nourishment. Infrastructure plays a crucial role, as regions with inadequate transportation networks often face significant food scarcity, particularly in rural areas. Inefficient logistics may hinder the movement of agricultural products from farms to consumers, exacerbating hunger issues. Moreover, market access is not equal across regions; disparities in pricing and availability can result in food deserts that leave many communities without sufficient access to nutritious food options. Achieving equitable food distribution requires investment in infrastructure and an emphasis on improving supply chain efficiency to ensure that all populations can obtain the resources they need to thrive. Economic factors also heavily influence food distribution patterns, dictating who has access to food and at what cost. Global trade agreements and local market policies often determine food prices, affecting availability to the most vulnerable populations. The relationship between agriculture and prevailing economic conditions can exacerbate food insecurity when markets fail to provide affordable options. Additionally, as highlighted in the Nereus Program report, progress in achieving sustainable development goals can have a significant impact on food distribution by tackling broader issues such as climate change and socio-economic inequality, which can directly affect agricultural productivity and accessibility.

By fostering economic resilience, policymakers can help dismantle barriers to equitable food access. Finally, cultural, and social dynamics significantly shape food distribution systems. Preferences and culinary traditions can affect demand patterns, potentially disadvantaging certain food groups or agricultural practices. This cultural lens must be considered when developing food distribution strategies to ensure they align with community values and needs. The SAARC countries report mentions the importance of adapting agricultural policies to local contexts, emphasizing the need for research and development focused on regional food preferences. Understanding the interplay between culture and food distribution can lead to more effective policies that enhance access to food and promote local agricultural practices, ensuring that distribution systems are both sustainable and nourishing to local populations (Figure 8.5).

Figure 8.5. *Many factors affect food development.*

Source: By User: Rainer Zenz – Own work, CC BY-SA 3.0, https://commons.wikimedia.org/w/index.php?curid=669540.

8.3.3. The Impact of Globalization on Local Food Systems

The phenomenon of globalization has significantly transformed local food systems, leading to both opportunities and challenges for communities across the globe. One of the most notable effects is the increased availability of a diverse array of food products, often sourced from various international markets. This shift allows consumers to enjoy foods that were once limited to specific geographic regions. However, this accessibility often comes at a cost—the prioritization of large-scale agribusinesses, which can marginalize local farmers and producers. The dominance of these corporations over local markets can contribute to the erosion of traditional food practices, diminishing the cultural significance and nutritional value of local diets as individuals gravitate towards cheap, mass-produced options. Moreover, the intertwining of local food systems with global trade networks raises significant concerns regarding sustainability and food security. As globalization drives the demand for uniformity and efficiency in food production and distribution, local ecosystems can suffer. The reliance on industrial agriculture and the depletion of natural resources can lead to a decline in local food sovereignty, as communities become dependent on imported foods. This jeopardizes local economies and amplifies food insecurity, particularly in urban areas experiencing rapid growth, as highlighted in recent discussions surrounding the United Nations Sustainable Development Goals. These shifts necessitate a critical examination of the role of international trade agreements, which often prioritize corporate interests over the needs of local populations striving for equity and sustainability in their food systems. Lastly, the impact of globalization is further exacerbated by the complexity of urban food systems in the Global South. As cities expand and populations surge, the challenges of feeding urban dwellers become increasingly pressing. The supermarket revolution has reshaped food retail, often sidelining smaller, informal vendors who play a crucial role in local economies and community well-being. Simultaneously, the informal food economy is vital in providing affordable options to those facing economic constraints. The tension between formal and informal sectors must be acknowledged, as both can co-exist and even support each other in addressing food insecurity. As neighborhoods transform and food practices evolve, there is an urgent need to prioritize local food solutions that recognize and enhance the interconnectedness of community, culture, and sustainable practices, ultimately contributing to a more resilient food system in the face of global pressures.

8.4. CAUSES OF HUNGER

A multifaceted web of factors contributes to the persistent issue of hunger across the globe. Primarily, economic instability and poverty serve as significant barriers to food access. In low-income regions, families often lack the financial resources necessary to purchase adequate and nutritious food. The intertwined relationship between poverty and hunger creates a vicious cycle; as individuals struggle to afford food, their health deteriorates, limiting their ability to work and improve their economic situation. This economic strain ultimately exacerbates food insecurity, demonstrating that addressing poverty is critical to alleviating hunger. Effective interventions must focus on improving income levels and economic opportunities to break this cycle and enhance food accessibility. Moreover, insufficient agricultural development plays a crucial role in perpetuating hunger, particularly in developing countries. Despite advancements in agricultural practices, many regions still grapple with outdated techniques, lack of resources, and insufficient market access. Agricultural research and technology can greatly enhance productivity, yet barriers to implementation persist, including inadequate infrastructure and climate change impacts. As Pandya-Lorch et al. suggest while significant progress was made in past decades to improve food quality and accessibility, recent developments highlight the urgent need for a renewed focus on sustainable and inclusive agricultural practices. Investing in agricultural development not only addresses immediate hunger concerns but also fosters long-term economic growth, making it an essential component of food security strategies. Finally, political instability and conflict can exacerbate food insecurity, leading to widespread hunger in affected regions. In areas where governance is weak or where civil unrest exists, food distribution systems can break down, causing shortages and escalating prices. In these environments, families are often forced to flee their homes, further disrupting their access to food sources. The interplay between hunger and political unrest underscores the necessity of addressing governance issues to create stable climates for food production and distribution. Strategies that include international cooperation and conflict resolution are vital to ensure that hunger is systematically addressed indicates the need for coordinated individual, community, and national actions to effectively combat the root causes of hunger and poverty (Burke et al., 2008).

8.4.1. Economic Factors Contributing to Food Scarcity

Addressing food scarcity necessitates a thorough exploration of various economic factors that intertwine with agricultural productivity and resource management. One crucial aspect is the rising cost of agricultural inputs, such as fertilizers, seeds, and water, driven by fluctuating global market prices and increasing competition for resources. For instance, as pressures mount on water supplies due to climate change and over-extraction, the cost of accessing and maintaining irrigation systems rises, further straining the budgets of farmers, particularly in developing countries. These high costs can result in reduced crop yields and, consequently, lower food availability, creating a cycle of scarcity that disproportionately impacts low-income populations. Acknowledging these economic barriers is essential for devising effective interventions to enhance food security and address the growing issue of hunger. Additionally, market dynamics and trade policies play a significant role in shaping food availability and access. Economic globalization has led to increased interdependence among nations, wherein fluctuations in one region can have ripple effects elsewhere. Countries that rely heavily on food imports are particularly vulnerable to price volatility and trade restrictions, which can exacerbate food insecurity. For example, in Pakistan, water scarcity directly affects agricultural output, leading to reduced food stocks and higher prices. Such inflation not only compromises food security but also perpetuates poverty among already vulnerable populations (Fahim et al.).Thereby promoting stability and resilience within food systems by critically analyzing trade policies and ensuring equitable market access, governments can mitigate the impacts of economic factors contributing to food scarcity. Finally, the interconnectedness of economic growth, employment opportunities, and food security cannot be overlooked. High unemployment rates and low wages in agricultural sectors can diminish purchasing power, making it difficult for individuals and families to afford essential food supplies. Furthermore, large-scale agricultural practices often prioritize profits over local food production, resulting in food deserts in economically disadvantaged communities. This disconnection between local economies and food systems emphasizes the need for sustainable practices that support both farmers and consumers. Localized food production can alleviate dependence on external markets, enhance food sovereignty, and encourage equitable distribution of resources. Stakeholders can work collaboratively toward combating hunger and fostering long-term food security in vulnerable populations by focusing on creating economic opportunities within food systems (Andrés Cisneros-Montemayor et al., 2017) (Figure 8.6).

Figure 8.6. *The role of income growth in mitigating food insecurity.*

Source: By Photographic Collection – originally posted to Flickr as Women working in a cannery, CC BY 2.0, https://commons.wikimedia.org/w/index.php?curid=10573265.

8.4.2. Political Instability and Its Effects on Food Access

Political instability significantly disrupts food access, affecting supply chains and market functionality. When governments struggle with unrest, the resulting chaos can lead to the breakdown of essential infrastructure, illustrating how stability is crucial for efficient food distribution. Farmers may be unable to transport their goods to markets due to violence or blocked roads, causing food shortages even in areas with abundant agricultural production. Consequently, food prices can skyrocket, disproportionately affecting low-income households that already face difficulties meeting their nutritional needs. In many cases, such instability also leads to panic buying and hoarding, exacerbating scarcity and ensuring that vulnerable populations remain at risk of hunger. Moreover, the interplay between political turmoil and economic conditions creates a cycle that perpetuates food insecurity. Economic growth often stalls in times of political unrest, undermining the ability of governments to invest in agricultural development and infrastructure. As outlined in the analysis of food security dynamics, regions experiencing political instability often see food security policies shift from long-term development strategies to short-term crisis management (Timmer, 2004). Such shifts can stifle investments in sustainable agriculture and fail to address the issues contributing to hunger, perpetuating a cycle of dependency on aid rather than fostering self-sufficiency. The consequences of political instability extend beyond immediate food access issues; they also lead to broader societal impacts. As food scarcity grows, tensions can escalate, triggering conflicts that further destabilize regions. This dynamic was observed during the Arab Spring, where rising food prices sparked widespread protests, revealing the critical link between food security and social unrest. Ultimately, by understanding the intricacies of how political instability influences food access, it becomes clear that addressing governance and stability must be part of any comprehensive strategy to combat hunger. Prioritizing stability not only contributes to immediate food availability, but also enhances long-term economic resilience and societal well-being.

8.4.3. Environmental Challenges, Including Climate Change

Environmental challenges have far-reaching implications for food security and hunger, particularly as climate change intensifies. As temperatures rise and weather patterns become increasingly unpredictable, agricultural productivity faces significant threats. Farmers often struggle with droughts, floods, and other extreme weather events that disrupt crop yields. A decrease in reliable food production not only exacerbates hunger but can also destabilize communities, leading to economic challenges and reduced productivity. This interaction between climate change and agricultural productivity is critical, as decreased agricultural output directly affects food availability. Thus, addressing environmental challenges is essential for protecting food systems and ensuring that communities have consistent access to necessary resources. The interplay between climate change and global conflicts further complicates food security. Environmental degradation linked to ongoing conflicts undermines agricultural systems and reduces food supply (Khanal et al., 2023). In conflict-ridden regions, resources are often scarce, and agricultural infrastructure is compromised, leading to increased food insecurity. The COVID-19 pandemic has magnified these issues, disrupting supply chains and destabilizing economies. This situation creates a vicious cycle where conflict exacerbates environmental challenges, which then jeopardizes food systems and heightens the risk of malnutrition.

The consequences are particularly significant for vulnerable populations, who may lack the means to adapt to these shifting challenges. To combat these intertwined threats, comprehensive strategies are necessary. Addressing climate change demands immediate action within agricultural policy frameworks (Badiane et al., 2022). Policymakers must prioritize sustainable practices that mitigate environmental harm while enhancing agricultural resilience. This could involve investing in innovative technologies, promoting crop diversification, and implementing environmentally friendly farming techniques. Additionally, international cooperation is crucial to fostering stability in regions affected by acute environmental challenges and conflict. By understanding the connections between climate change, conflict, and food security, we can develop more effective policies that secure food systems and enhance resilience for communities worldwide (Figure 8.7).

Figure 8.7. *Food and environmental issues.*

Source: By OpenIDUser2 – http://trashwiki.org/en/File:Luxembourg_dumpster.jpeg, GFDL, https://commons.wikimedia.org/w/index.php?curid=26747239.

8.5. FOOD SECURITY

Ensuring food security is a multifaceted challenge that intertwines agricultural productivity, economic stability, and environmental sustainability. As the global population continues to rise, projected to reach 9.2 billion by 2050, it becomes increasingly imperative to adapt to urbanization, changing diets, and resource constraints to ensure that everyone has access to sufficient and nutritious food. Innovations in agricultural chemistry, such as drought-resistant crops and sustainable farming methods, play a critical role in enhancing food production. These developments not only address the immediate need for increased yield but also help combat the adversities posed by climate change and resource depletion, which threaten to exacerbate hunger and malnutrition globally (Lugonja et al., 2023). Therefore, investment in agricultural research and education remains essential to develop strategies that can effectively counteract these challenges and secure food for future generations. Food loss and waste represent another major hurdle in achieving food security, as an estimated one-third of global food production is wasted annually. With 1.3 billion metric tons of food discarded every year, addressing this issue is crucial to maximize the potential of existing food resources.

A comprehensive understanding of food loss requires analyzing various factors throughout the food supply chain, from farmers unable to harvest their crops to consumers discarding food without consideration, highlighting the complexity of this issue (Ekwall et al., 2014). Implementing policies that promote reduction initiatives and educating consumers on responsible consumption could significantly mitigate food loss. By examining case studies such as the loss of maize in Kenya, we can identify effective strategies that not only improve food availability but also contribute to sustainability and reduce environmental impacts associated with food production. Collaboration among stakeholders is vital in tackling food security effectively. Governments, businesses, non-profits, and individuals each play an essential role in creating a comprehensive framework to address hunger and malnutrition. Developing policies that prioritize food security, along with fostering partnerships within the agricultural sector, can greatly improve food distribution and accessibility. Additionally, continuous dialogue between scientists, policymakers, and the private sector is necessary to share knowledge and implement sustainable practices that benefit all stakeholders involved in the food supply chain. Ultimately, reducing food loss and enhancing agricultural practices can create a holistic approach to combating hunger that not only preserves resources and improves nutrition but also fosters economic development and ensures a more equitable food system for all.

8.5.1. Definition and Components of Food Security

The concept of food security encompasses far more than merely having enough food available; it involves a complex interplay of access, utilization, stability, and sustainability. At its core, food security is defined as a condition wherein individuals have physical, social, and economic access to sufficient, safe, and nutritious food that meets their dietary needs for a healthy life. This holistic definition suggests that food security must be viewed through multiple lenses: the availability of food, the ability of individuals to access it, the quality and nutritional value of food consumed, and the resilience of the food system itself in the face of fluctuations, such as climate change or market disruptions. As such, understanding food security requires an analysis of these interconnected components to tackle the broader implications of hunger in diverse contexts. In examining the individual components of food security, it is essential to recognize how these factors influence overall well-being and societal stability. Availability pertains to the production and distribution of food supplies, which can be affected by agricultural productivity, trade policies, and even climate conditions. Access relates to the economic and physical opportunity to obtain food, often contingent upon economic stability and income levels. Utilization refers to how food is prepared and consumed, highlighting the importance of nutrition and food safety. Stability encapsulates the resilience of these elements over time, suggesting that temporary food surpluses do not equate to long-term food security. Addressing these components is crucial in crafting effective policies, particularly in vulnerable regions, where disparities can lead to profound implications for health and social order (Timmer, 2004). Critically, the relationship between food security and social equity speaks to the urgent need for inclusive policies that address the underlying causes of hunger, particularly in the Global South. Current research indicates that climate change exacerbates food insecurity, especially in urban areas where populations are rapidly growing and poverty is prevalent (Frayne et al., 2011). In this regard, effective interventions must not only secure food availability but also ensure equitable access, emphasizing support for marginalized communities. Sustainable agricultural practices, improved infrastructure, and social safety nets are vital components for fostering resilience against shocks. By understanding and addressing the intricate components of food security, policymakers can more effectively combat hunger and promote health equity, ultimately contributing to economic sustainability and development in the broader context of global food systems.

8.5.2. Measurement of Food Security Levels Globally

Understanding the measurement of food security levels globally requires consideration of various indicators that reflect the complex interplay of economic, social, and environmental factors influencing access to food. Traditional metrics often rely on direct measures like calorie availability, malnutrition rates, and household food expenditure, offering a snapshot of food accessibility. However, these indicators can sometimes mask underlying disparities. For instance, urban populations, particularly in rapidly growing cities of the Global South, experience unique challenges such as inadequate infrastructure and rising food prices, which traditional metrics may not fully capture. As highlighted in discussions about urban food systems, there is a pressing need for more nuanced measurements that address the realities of urban food insecurity, where market dynamics are drastically changing due to the emergence of supermarkets and informal food economies (Crush et al., 2016).

Beyond standard measures, a broader understanding of food security as a public good introduces another layer of complexity. Viewing food security as a global public good helps underscore its collective importance, as it suggests that efforts to enhance food access and availability benefit all of humanity (Timmermann et al., 2018). This perspective emphasizes the need for shared action and cooperation among nations to develop effective policies and interventions. For instance, international partnerships can lead to the sharing of best practices in agricultural productivity, food distribution, and nutrition education. However, the unique sociopolitical context of each nation must be considered to ensure that policies are inclusive and equitable, addressing specific local needs. Thus, the measurement of food security should also reflect the normativity and benevolence aspects of public goods, advocating for the ethical imperative to ensure no one is excluded from access to adequate food. Moreover, assessing food security requires ongoing adaptation to diverse and evolving challenges, such as climate change, economic instability, and geopolitical tensions. These factors can exacerbate disparities in food access, emphasizing the importance of robust data collection and analysis. For example, food insecurity rates are often higher in marginalized communities, which indicates the necessity of targeted interventions. The integration of qualitative data around personal experiences of food insecurity can also enhance quantitative measurements, providing a more comprehensive understanding of the issue. Innovative approaches, such as combining satellite imagery with big data analytics, can help visualize food distribution and access across geographically diverse regions. Comprehensive food security measurements must not only capture these dimensions but also drive global policy efforts aimed at alleviating hunger and promoting sustainable food systems, ultimately ensuring that food security becomes a shared reality for all (Crush et al., 2016).

8.5.2. The Role of Governments and Organizations in Promoting Food Security

The complex landscape of food security underscores the significant role of governments and organizations in shaping sustainable agricultural policies. By prioritizing investments in rural development, these entities can uplift communities reliant on agriculture, thereby addressing extreme poverty and hunger on a global scale. Initiatives focused on improving agricultural productivity and ensuring fair trade practices are essential. For instance, implementing credit systems allows farmers to access necessary inputs and invest in adaptive measures to withstand climate impacts. This approach could be instrumental in meeting the first Millennium Development Goal of eradicating extreme hunger and poverty, as highlighted in (Bäckman et al., 2009). Without these government-led initiatives, the rural poor remain at the mercy of unpredictable markets and environmental challenges, perpetuating cycles of poverty and food insecurity. Moreover, international organizations play a pivotal role in confronting the multifaceted challenges affecting global food security. Inefficiencies in existing frameworks, such as the WTO, hinder timely responses to pressing issues like climate change and its impact on agriculture. Current governance structures struggle to address the connections between agricultural sustainability and broader socio-economic challenges. This inadequacy necessitates the development of new governance systems that can effectively tackle the intricate relationship between food security, agricultural practices, and global public goods. Enhanced collaboration between governments and these organizations can facilitate innovative approaches, ensuring that food systems are resilient and capable of supporting the diverse needs of populations worldwide, especially as climate-related pressures intensify. Lastly, coordinated efforts between national governments and international entities are crucial in establishing policies that promote food security at all levels. This involves not only addressing the immediate needs of vulnerable populations but also creating long-term strategies that prioritize sustainable agricultural practices. By implementing policies that safeguard natural resources and support environmentally friendly approaches, governments can enhance food system resilience. Organizations dedicated to monitoring and evaluating these policies play an essential role, as ongoing assessments are necessary to adapt to evolving challenges. As we face the reality of heightened food scarcity and the ramifications of climate change, united action between governments and organizations becomes imperative for fostering food security. Only through diligent collaboration will we be able to mitigate the risks of hunger and develop more equitable food systems for future generations.

8.6. NUTRITIONAL ASPECTS OF FOOD

The relationship between food and nutrition is intricate and multifaceted, encompassing various biological, economic, and psychological factors. Nutritional aspects significantly influence not only individual health but also community well-being. A holistic understanding of nutrition requires an appreciation of its broad implications, particularly as concerns regarding undernourishment and hunger persist globally. Efforts to address these issues cannot merely be simplified to food availability; they must also consider the nutritional quality of the food consumed. As experts suggest, focusing solely on caloric intake ignores the vital nutrients necessary for optimal health, positioning nutritionists as essential players in combating hunger through eco-nutrition initiatives (Hruby et al., 1978). Significantly, food security plays a crucial role in determining nutritional outcomes within households. The definition of food security encompasses not only access to adequate calories but also the availability of safe and nutritious food to meet dietary needs (Aldeseit et al., 2016).

Understanding that household food insecurity is a multidimensional issue allows for a more effective approach to alleviate hunger. Prioritizing equitable food distribution among family members, ensuring adequate household production for subsistence, and incorporating sustainable practices are vital strategies. Addressing the complexities of food security requires an examination of socio-demographic factors and the localized perceptions of hunger, which shape the experiences of individuals facing food insecurity. To effectively tackle the nutritional aspects of food and hunger, it is imperative to leverage contemporary research and innovative methods. As highlighted in recent studies, there is a pressing need for qualitative measures that resonate with those affected by food insecurity (Aldeseit et al., 2016). These approaches go beyond traditional quantitative assessments, tapping into the personal experiences and perceptions that underlie the food security narrative. By developing methodologies that consider the unique contexts of the poor, interventions can be more precisely tailored to meet the needs and preferences of vulnerable populations. In doing so, we can foster an environment where nutritional health flourishes, thereby reducing the prevalence of hunger and its associated impacts on society.

8.6.1. Importance of Balanced Nutrition for Health

Optimal health hinges on the consumption of balanced nutrition, which plays a vital role in physical growth, cognitive development, and overall well-being. A diet rich in fruits, vegetables, whole grains, and lean proteins provides essential nutrients that promote bodily functions and support the immune system. When the necessary nutrients are absent, as seen in communities grappling with food insecurity, children can suffer from significant developmental delays and health issues. For instance, in studies conducted on Mexican-origin households in Texas border colonias, a staggering 51% of families reported children facing hunger, linking inadequate nutrition directly with detrimental health outcomes for young individuals (Dean et al., 2013). This alarming situation underscores the crucial need for access to nutritious foods, which can prevent the onset of chronic health conditions that often stem from poor dietary habits established in childhood.

In addition to physical health, balanced nutrition significantly impacts mental and emotional well-being. The brain requires an array of nutrients to function optimally, including omega-3 fatty acids, vitamins, and minerals. Deficiencies in these nutrients can lead to mood disorders and cognitive decline, highlighting the intimate relationship between diet and mental health. The data collected from the aforementioned studies reveal that low participation rates in school-based nutrition programs exacerbate food insecurity, leaving many children without the necessary resources to develop both physically and mentally.

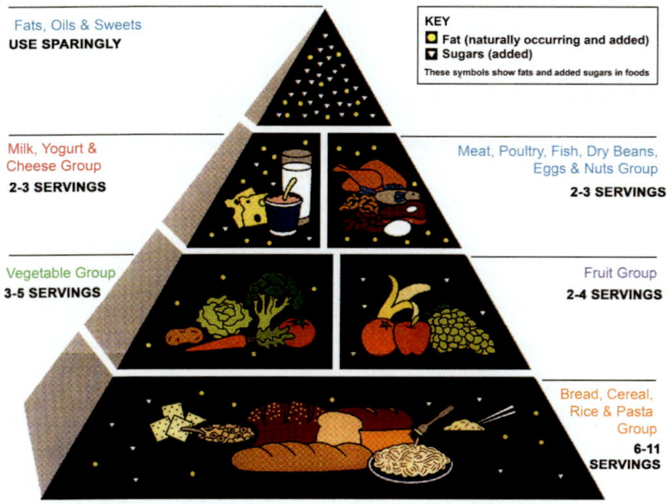

Figure 8.8. *Fundamentals of a balanced diet.*

Source: By The food pyramid, from http://www.nal.usda.gov/fnic/Fpyr/pyramid.gif, Public Domain, https://commons.wikimedia.org/w/index.php?curid=680809.

Consequently, ensuring that children have access to balanced meals is not merely a question of physical nourishment; it is a matter of fostering a supportive environment for mental resilience and academic success. Acknowledging this connection is vital in promoting programs that advocate for nutrition education and healthy food availability. Addressing the importance of balanced nutrition also stresses the need for systemic changes in food distribution and accessibility. Food insecurity, which affects a significant portion of the population, often prevents families from obtaining a diverse and healthy diet. Initiatives such as increasing access to federal nutrition assistance programs, like the Supplemental Nutrition Assistance Program (SNAP), are essential to combat the alarming rates of child hunger (Dean et al., 2013). A multi-faceted approach—including community resources, education on the importance of nutrition, and targeted food programs—could aid in bridging the gap for families in need. Ultimately, fostering an environment where all individuals can access balanced nutrition is crucial for ensuring that after-school meals and food programs do not just serve as temporary relief but as stepping stones towards long-term health equity and improved quality of life for future generations.

8.6.2. The Relationship Between Hunger and Malnutrition

The interplay between hunger and malnutrition is profoundly complex, and influenced by social, economic, and environmental factors. Food insecurity, a significant predictor of malnutrition, manifests when communities lack reliable access to a sufficient quantity of affordable, nutritious food. This lack of access can lead not only to undernutrition, as seen in stunted growth or low body mass index (BMI) among adolescents but also to overnutrition, where poor-quality diets result in obesity. As highlighted in recent studies, malnutrition rates diverge widely across different regions, demonstrating that food insecurity does not uniformly translate into malnutrition outcomes, particularly among youth (Balcombe et al. (n.d)). Hence, understanding this relationship requires a nuanced examination of how hunger differs across demographic and socio-economic groups and how it contributes to a spectrum of nutritional deficiencies and excesses. Moreover, the concept of income plays a crucial role in the relationship between hunger and malnutrition. Research shows that as income levels rise, calorie consumption also increases, but this increase is often accompanied by a shift toward higher-fat food options, which can exacerbate obesity issues in wealthier nations (Balcombe et al. (n.d)).

In lower-income settings, however, the primary concern often remains unmet caloric needs and micronutrient deficiencies. The disparity in consumption patterns reveals how hunger manifests differently across economic landscapes, leading to varied health consequences. As communities become wealthier, the challenge shifts; while poverty alleviation may initially improve hunger, it may inadvertently promote dietary patterns contributing to malnutrition in the form of obesity. This paradox complicates the pursuit of effective nutritional policies that aim to address both sides of malnutrition. Ultimately, addressing hunger must go hand in hand with comprehensive approaches to combat malnutrition. Public health interventions should prioritize improving food security by promoting access to diverse and nutritious food sources while also educating communities about healthy dietary choices. The findings from studies on adolescents indicate that food insecurity can negatively affect BMI-for-age and height-for-age metrics (Aiman et al., 2020), which emphasizes the critical need for tailored nutritional programs. Furthermore, longitudinal research is essential to establish clearer causal relationships between food insecurity and nutritional status, which can inform policy decisions. By comprehensively addressing systemic issues contributing to hunger and malnutrition, stakeholders can work towards fostering healthier populations capable of thriving in all aspects of life.

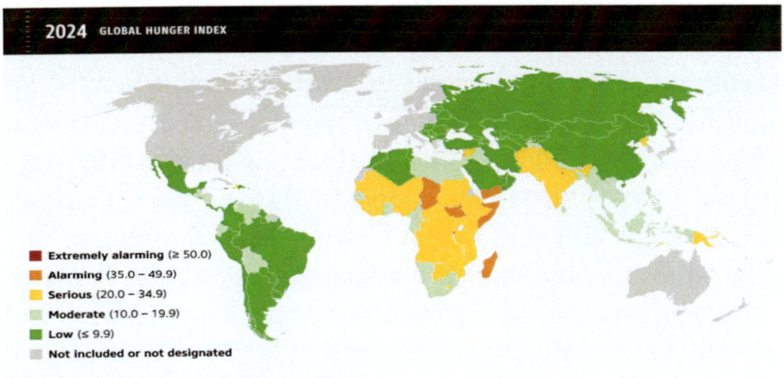

Figure 8.9. *Hunger and malnutrition.*

Source: By Unknown author – X (formerly Twitter), CC BY-SA 4.0, https://commons.wikimedia.org/w/index.php?curid=155813649.

8.6.3. Strategies for Improving Nutritional Outcomes

Effective strategies for enhancing nutritional outcomes must prioritize the integration of agricultural practices with public health initiatives. Increasing food production among smallholder farmers bolsters local economies and addresses food security by making nutrient-dense foods more accessible. It is essential to encourage smallholders to grow diverse crops that yield not just high quantities but also high nutritional value. This approach aligns with the need for enhancing the intake of essential vitamins and minerals, particularly in communities facing the dual challenges of food scarcity and malnutrition. Research shows that engaging smallholder farmers in local and regional procurement programs can further ensure a stable supply of nutritious foods, thereby improving local dietary patterns and overall nutrition (Donini et al., 2018). Community-based programs that focus on nutrition education also play a critical role in improving dietary practices among populations. By informing individuals about the importance of balanced diets and optimal food choices, these initiatives can address misconceptions and promote healthier eating habits. Additionally, training community members to utilize local agricultural products can enhance food preparation methods, making meals not only healthier but also culturally relevant. Targeted education can particularly benefit vulnerable groups, such as children and the elderly, who are often at higher risk of malnutrition.

As indicated by the systematic review on anorexia of aging, understanding the specific nutritional needs of different age groups provides a nuanced approach to implementing effective dietary interventions, ultimately fostering better health outcomes across diverse populations (Donini et al., 2018). Lastly, the collaboration between various stakeholders—governments, NGOs, healthcare providers, and the agricultural sector—is crucial in creating a multifaceted approach to improving nutritional outcomes. Policies designed to support agricultural production should also incorporate nutrition-focused frameworks, ensuring that food systems are not only sustainable but equitable. Further research in the area of agriculture-nutrition linkages is essential to identify effective interventions that can translate agricultural success into improved health benefits (Wiegers et al., 2011). Strengthening the connections between producers and consumers through equitable supply chains can lead not only to increased income for farmers but also to enhanced nutrition for families. By harnessing resources and expertise across sectors, such strategies can lead to lasting improvements in food security and nutritional health, ultimately contributing to a more resilient and healthy society.

8.6.4. Solutions to Hunger

Addressing the complex issue of hunger requires a multifaceted approach that encompasses food production, distribution, and access. Improving agricultural productivity through sustainable farming practices is critical. We can increase crop yields while minimizing environmental impact by investing in technology and education for farmers, especially in developing nations. Additionally, enhancing infrastructure for transportation and storage will reduce food loss and ensure that fresh produce is accessible to communities in need. These initiatives will support the growing global population and ensure that food availability meets demand, as current statistics indicate that worldwide, 815 million people still suffer from hunger, despite an overall increase in food production (Hoque, 2022). Moreover, tackling the socioeconomic factors contributing to hunger is essential. Addressing poverty through targeted economic policies can increase income levels, thereby improving access to nutritious food. Implementing social safety nets, such as food assistance programs and nutrition education, can directly alleviate food insecurity at the household level. We can combat the high rates of malnutrition and diet-related diseases that often accompany food insecurity by focusing on the availability of affordable, healthy options. Recognizing that food insecurity is both a cause and effect of poverty allows us to propose solutions that are not just temporary fixes but long-term strategies for systemic change (Adkins et al., 2019). Community-driven initiatives, such as food cooperatives and urban agriculture projects, have also shown promise in addressing hunger by involving local populations in food production.

These grassroots efforts empower communities to take control of their food systems and make fresh, healthy food more accessible. Educational programs that promote cooking skills and nutrition awareness can further enhance these initiatives, allowing families to maximize their resources and improve their diets. Collaborating with local governments and organizations is vital to ensure that these models are sustainable and scalable. As we consider potential solutions to hunger, it is crucial to recognize that systemic change must include diverse perspectives and collaborative action to create lasting impacts on food security (Adkins et al., 2019; Hoque, 2022).

8.6.5. Community-Based Initiatives and Programs

Community-based initiatives and programs play a crucial role in addressing food insecurity and promoting sustainable agricultural practices. By engaging local stakeholders, these programs foster a sense of ownership and accountability within the community, facilitating more effective solutions tailored to specific needs. For example, grassroots organizations often collaborate with local farmers, food banks, and community members to create networks that enhance food access and education. This participatory approach not only empowers individuals but also cultivates trust among community members, leading to lasting relationships that can mitigate the issues surrounding hunger. The integration of community voices into policymaking also amplifies the importance of local knowledge, ensuring that initiatives prioritize the unique characteristics of the area. Efforts to bolster food security are often embedded in broader legislative frameworks. Programs established through initiatives like the 2002 Farm Bill illustrate how policy can support community-based efforts. These initiatives create a structural foundation for enhancing food access, improving health outcomes, and fostering economic resilience within communities.

The commitment to healthy farms and communities is necessary for the current population and for future generations. Legislative support enables organizations to secure funding and resources essential for implementing innovative strategies that address both immediate and long-term food security challenges. By prioritizing these community-driven programs within larger legislative efforts, policymakers can ensure comprehensive and sustainable solutions to hunger. The interchange between food banks and community food security exemplifies the importance of collaborative programs in addressing hunger. Research indicates that food banks, while traditionally seen as emergency assistance, can play a transformative role in building resilience within local food systems (Fisher, 2005). By facilitating partnerships that emphasize education and resource sharing, food banks can contribute to a more integrated approach to food justice. The development of programs that connect food banks with local farmers, schools, and nutrition education initiatives can help create a sustainable ecosystem that empowers communities to take charge of their own food systems and combat hunger effectively. Ultimately, community-based initiatives not only address immediate needs but also lay the groundwork for systemic changes that promote equitable access to healthy food, reinforcing the intrinsic link between food, wellness, and community sustainability.

8.6.6. Technological Advancements in Food Production

Technological advancements have revolutionized food production, allowing farmers to increase yield and efficiency significantly. Innovations such as genetically modified organisms (GMOs), precision agriculture, and vertical farming have emerged as pivotal elements in addressing hunger and food scarcity. GMOs have been engineered to resist pests and withstand harsh environmental conditions, leading to higher productivity. Moreover, precision agriculture employs data analytics and satellite imaging to optimize planting patterns and resource usage, which conserves water and reduces the over-application of fertilizers. These technologies collectively contribute to greater food availability, supporting the notion that it is possible to produce enough calories to feed the growing global population while minimizing environmental impact. Despite these advancements, the paradox of hunger persists. While food production has increased substantially—rising from 2220 kilocalories per person per day in the 1960s to 2790 calories in 2008 (Adkins et al., 2019)—a staggering 815 million people still experience undernourishment. This statistic highlights that increased food availability does not equate to improved access for all individuals. Many factors contribute to this disparity, including economic inequality, food distribution inefficiencies, and inadequate infrastructure, particularly in developing regions. Thus, while technological progression has made it feasible to cultivate enough food, systemic issues must be addressed to ensure equitable distribution and consumption. Water resources play a crucial role in the effectiveness of modern agricultural techniques. Innovations such as drip irrigation, water-efficient crop selections, and rainwater harvesting have emerged to manage this vital resource more sustainably (Kumar et al.(n.d).). These methods contribute to improving crop resilience against drought while maximizing productivity. Nonetheless, ensuring access to water for irrigation remains a challenge, especially in areas experiencing severe climate change impacts. As technological advancements continue to push the boundaries of food production, it is essential to consider the interplay between these innovations and sustainable water management. The future of global food security hinges not only on producing more food but also on ensuring that resources like water are adequately managed and distributed.

8.6.7. Policy Recommendations for Reducing Hunger

Implementing comprehensive policies to enhance food security requires a multifaceted approach that addresses both immediate needs and systemic deficiencies in the agricultural sector. Investment in sustainable agricultural practices is crucial; this includes improving crop yields through research and development, as well as providing farmers with better access to technology and resources. Such investments can help mitigate the risks posed by climate change, which is expected to significantly reduce food crop yields in developing regions by 2050, exacerbating existing disparities in food access (Bäckman et al., 2009). Additionally, increasing funding for local agricultural research can help tailor solutions to specific regional challenges, thereby empowering farmers with the knowledge necessary to improve their production methods and adapt to changing environmental conditions. Equitable access to credit and financial resources is another essential policy recommendation to combat hunger. Rural farmers often struggle to secure financing for inputs or infrastructure improvements, which stifles their ability to enhance productivity. Initiatives like microfinance programs can provide the necessary capital for small-scale farmers, allowing them to invest in better equipment and adaptive practices.

Fair Trade certification is also noteworthy, as it guarantees minimum prices for farmers, providing a safety net that can help stabilize their incomes and livelihoods (Bäckman et al., 2009). Policymakers can create an environment conducive to sustainable agricultural development that ultimately contributes to hunger reduction by reforming financial systems to prioritize the needs of rural communities. Finally, collaboration between governments, NGOs, and international organizations can enhance the effectiveness of hunger reduction strategies. Expanding social safety nets is vital for supporting vulnerable populations who are most affected by food insecurity. Targeted food assistance programs can provide immediate relief while longer-term strategies are developed. Efforts to mitigate the adverse effects of climate change on agriculture are equally important, as these changes threaten to increase food insecurity in developing countries (Kahiluoto et al., 2009). Comprehensive climate policies should not only focus on adaptation strategies but also promote sustainable practices that empower local communities and contribute to resilient agricultural systems. By fostering cooperation and innovative policy solutions, we can create a platform for meaningful progress toward eradicating hunger globally.

8.7. CONCLUSION

In addressing the multifaceted challenges of food security and hunger, it is imperative to recognize the interconnectedness of local and global efforts.

The progression towards sustainable food systems necessitates a nuanced understanding of the balance between national policies and international obligations. While countries strive to attain self-sufficiency, as emphasized in the discussion on national food security, they must also comply with global initiatives aimed at eradicating poverty and hunger. The dual focus ensures that efforts to improve local livelihoods do not compromise the collective goals necessary to secure food for vulnerable populations worldwide. By fostering cooperation and coordination between nations, a more robust and resilient framework for food security can be established, ultimately reducing the prevalence of hunger on a global scale. Moreover, the impact of fisheries and aquaculture is pivotal in shaping solutions to food insecurity. As highlighted in recent studies, these sectors provide vital nutritional resources and economic opportunities for millions, particularly in developing regions facing significant challenges. Investing in sustainable practices within fisheries and aquaculture not only addresses hunger but also contributes to ecological balance and resource preservation. The findings from the detailed review of the role of these sectors emphasize the necessity of cohesive policy and governance frameworks to enhance their effectiveness.

Achieving significant improvements in food security will require harnessing technological advancements and ensuring equitable market access, allowing local communities to thrive while actively participating in the global economy. Ultimately, tackling the crisis of food insecurity involves a commitment to collaborative action at both the national and international levels. The establishment of comprehensive legal frameworks, such as an international agreement on world food security, is fundamental to promoting responsibility and accountability among participating nations. Such measures would foster a shared obligation to combat hunger and uphold the right to food for all individuals. By reconciling the tensions between national food policies and global food security initiatives, a pathway toward sustainable food systems can be devised. It is through this focused effort on collaboration and accountability that humanity can hope to achieve the ambitious goal of eliminating hunger and ensuring that everyone has access to sufficient, safe, and nutritious food.

8.7.1. Summary of Key Points Discussed

In the discourse surrounding food and hunger, one critical aspect highlighted is the vulnerability of food systems, particularly during disasters. Recent analyses have underscored the fragility of the disaster food pipeline in urban areas such as San Francisco. These reports emphasize that low-income and vulnerable populations are disproportionately affected during crises, revealing a significant inequity in access to essential resources. The need for a robust infrastructure that ensures food resiliency in times of disaster is paramount.

Thus, the identification of weaknesses in current food distribution mechanisms leads to tailored solutions that can mitigate the impact of emergencies, thereby safeguarding the nutritional needs of those most at risk (Bonini et al., 2014). Moreover, the integration of Community Food Assessments serves as a valuable tool for examining local food landscapes. By facilitating a platform for collaboration, these assessments empower communities to map out their food resources and articulate specific needs. Through case studies, we see effective strategies employed to foster community involvement, ensuring that voices from diverse demographics are included in the conversation surrounding food access. This participatory approach catalyzes localized change, allowing communities not only to identify issues but also to implement practical solutions tailored to their unique circumstances. Effective research methods and engagement strategies highlighted in this process further contribute to the overall objective of enhancing food security and addressing hunger (Fisher et al., 2002). In essence, the discussion on food and hunger brings to light the intersection of disaster preparedness and community empowerment. By acknowledging the systemic fragilities in food distribution networks and actively engaging communities in data collection and solution-building, a more resilient food system can emerge. These conversations foster an understanding of the multifaceted nature of hunger, revealing it as not merely a consequence of scarcity but as a result of complex socio-economic factors. The call for philanthropy to invest in enhancing disaster food systems signals a collective responsibility to ensure food security, particularly for those at the highest risk, thus shaping a more equitable future where hunger does not impede the well-being of communities.

8.7.2. The Importance of Collective Action

In addressing complex societal issues such as food security and hunger, the necessity for collective action becomes increasingly evident. No single entity can tackle the multifaceted challenges inherent in food systems alone; it requires a concerted effort among various stakeholders, including government entities, non-profit organizations, and local communities. Collaborative approaches not only maximize resource sharing and expertise but also promote a sense of shared responsibility. For instance, the New England Food Vision 2060 highlights how collaborative frameworks can spur innovative solutions and policy advocacy aimed at developing a resilient regional food system that emphasizes sustainable practices to mitigate hunger (Burke et al., 2012). When individuals and organizations unite for a common cause, they amplify their voices, creating a more substantial demand for policy changes that can lead to food justice and accessibility. Moreover, collective action fosters inclusivity, bringing together diverse perspectives that enrich the conversation surrounding food and hunger. By engaging a wide range of community members in the decision-making process, initiatives can address specific local needs and foster a sense of ownership among participants. This approach enhances the likelihood of long-term success in food system initiatives. As noted in discussions about food policy access and system sustainability, collective action allows for adaptive learning and resilience-building within the community (Burke et al., 2012).

When communities work together, they not only create plans but also continuously refine them based on shared experiences and outcomes. This dynamic process is crucial for addressing the fluid nature of hunger, which often fluctuates with economic, environmental, and social factors. Ultimately, the potential for transformative change lies in the power of collective action to unify efforts across various sectors. As issues such as food price volatility become increasingly intertwined with other global trends, coordinated responses become vital to ensure equitable distribution of resources. Collaboration across governmental, nonprofit, and private sectors can lead to innovative solutions that address the root causes of food insecurity, as highlighted by ongoing dialogues within regional food systems frameworks. Stakeholders can ultimately work towards creating sustainable and just food systems that alleviate hunger and foster a thriving community by focusing on collective impact. The transformation of these systems demands synchronized efforts, as only through unity can societies aspire to meaningful progress in combating the persistent issue of food insecurity.

8.7.3. Call to Action for Addressing Food and Hunger Issues

Many communities across the United States are grappling with food insecurity, a pressing issue that calls for immediate and collective action. Food deserts and inadequate access to nutritious foods exacerbate this problem, leaving millions vulnerable to health issues that stem from poor diets. To combat these challenges, individuals, organizations, and government bodies must collaborate in developing sustainable solutions. Initiatives such as community gardens, food banks, and local farmers markets can offer accessible, fresh produce to those in need, fostering both nutrition and community engagement. By pooling resources and knowledge, we can create a network that not only alleviates hunger but empowers communities to take charge of their food systems. Advocacy plays a critical role in addressing food and hunger issues, as systemic change is often necessary to ensure long-term solutions. Individuals must become informed about local policies and engage with representatives to push for more robust food assistance programs. This includes supporting legislation aimed at increasing funding for nutrition programs, and initiatives designed to mitigate food waste. Collaboration with local and national organizations can amplify efforts toward food justice, ensuring that marginalized communities have a voice in the dialogue surrounding hunger. By fostering a collective grassroots movement, we can create a powerful force that holds those in power accountable and drives progressive change in food policy. The future of our society hinges on the actions we take today regarding food security. It is not only an ethical obligation to ensure that all individuals have access to enough nutritious food, but it is also essential for the health and well-being of our nation. Education plays a significant role in this endeavor; by raising awareness about food insecurity and encouraging sustainable practices, we can cultivate a generation that prioritizes food access for all. Additionally, supporting local agricultural initiatives can bolster the economy while addressing hunger. Now is the time to act decisively, Advocate passionately, and unite purposefully. Together, we can create a more equitable food system that eradicates hunger, nourishes communities, and fosters resilience for future generations.

REFERENCES

1. Adkins, M. (2019). The Global Food Security Act: America's strategic approach to combating world hunger. *ScholarWorks@UARK*. Retrieved from: https://core.ac.uk/download/223094003.pdf (accessed on 5 December 2024).
2. Aggarwal, P. K., Baethegan, W. E., Cooper, P., Gommes, R., Lee, B., Meinke, H. B., Rathore, L. S., & Sivakumar, M. V. K. (2010). Managing climatic risks to combat land degradation and enhance food security: Key information needs. *Elsevier B.V.* Retrieved from: https://core.ac.uk/download/pdf/29240278.pdf (accessed on 5 December 2024).
3. Aiman, U., Bohari, B., Dewi, N. U., Hartini, D. A., Pradana, F., & Nurulfuadi. (2020). Food insecurity and anthropometry in adolescents: A literature review. *ID Design 2012/DOOEL Skopje*. Retrieved from: https://core.ac.uk/download/359939362.pdf (accessed on 5 December 2024).
4. Aldeseit, B., Hamad, H., & Khashroum, A. (2016). Nutritional and health outcomes of household food insecurity (HFIS). *Journal of Biology, Agriculture and Healthcare, 6*(4). Retrieved from: https://core.ac.uk/download/33892325.pdf (accessed on 5 December 2024).
5. Andrews, C. (n.d.). European external assistance & food security: The end of the line? Retrieved from: https://core.ac.uk/download/pdf/7032935.pdf (accessed on 5 December 2024).
6. Bäckman, S., Kahiluoto, H., Rötter, R., & Sumelius, J. (2009). Sustainable rural development with emphasis on agriculture and food security within the climate change setting: SARD-climate final report. *University of Helsinki Libraries*. Retrieved from: https://core.ac.uk/download/10929656.pdf (accessed on 5 December 2024).
7. Badiane, O., Benson, T., Castro de la Mata, G., Fanzo, J., Guinto, R. R., Hendriks, S., Montgomery, H., & Soussana, J. F. (2022). Climate, COVID-19 and conflict threaten health, food security and nutrition. *BMJ*. Retrieved from: https://core.ac.uk/download/566570095.pdf (accessed on 5 December 2024).
8. Balcombe, K. G., Salois, M. J., & Tiffin, J. R. (n.d.). Impact of income on calorie and nutrient intakes: A cross-country analysis. Retrieved from: https://core.ac.uk/download/pdf/6615136.pdf (accessed on 5 December 2024).
9. Bonini, C., & Rapp, S. (2014). San Francisco disaster food system report. *Walter & Elise Haas Fund*. Retrieved from: https://core.ac.uk/download/75782845.pdf (accessed on 5 December 2024).
10. Bos, M. S., Brown, M. A., Cline, S. A., Cohen, M. J., Pandya-Lorch, R.,

Rosegrant, M. W., & von Braun, J. (n.d.). New risks and opportunities for food security: Scenario analyses for 2015 and 2050. Retrieved from: https://core.ac.uk/download/pdf/6388619.pdf (accessed on 5 December 2024).

11. Burke, J. D. (2008). Poverty and plenty: The divided American plate. *University of New Hampshire Scholars' Repository*. Retrieved from: https://core.ac.uk/download/72048593.pdf (accessed on 5 December 2024).
12. Burke, J., & McCabe, M. S. (2012). The New England food system in 2060: Envisioning tomorrow's policy through today's assessments. *University of New Hampshire Scholars' Repository*.
13. Capron, G., Castro, J., Diez, A. L., Esquivel, M. T., Flores, R., Gonzalez Arellano, S., Hidalgo, H., Huarte T., M. C., Monterrubio, A., Morales, J., Sánchez-Mejorada, C., & Wigle, J. (2017). No. 07: The urban food system of Mexico City, Mexico. *Scholars Commons @ Laurier*. Retrieved from: https://core.ac.uk/download/87085613.pdf (accessed on 5 December 2024).
14. Chand, R. (n.d.). SAARC Agricultural Vision 2020. Retrieved from: https://core.ac.uk/download/pdf/6455745.pdf (accessed on 5 December 2024).
15. Cisneros-Montemayor, A., Singh, G., Cheung, W., & Ota, Y. (2017). Oceans and the sustainable development goals: Co-benefits, climate change & social equity. *The Nippon Foundation-UBC Nereus Program*. Retrieved from: https://core.ac.uk/download/87085613.pdf (accessed on 5 December 2024).
16. Crush, J. (2016). No. 01: Hungry cities of the global south. *Scholars Commons @ Laurier*. Retrieved from: https://core.ac.uk/download/157834980.pdf (accessed on 5 December 2024).
17. Davila, F., & Dyball, R. (2015). Transforming food systems through food sovereignty: An Australian urban context. *Cambridge University Press (CUP)*. Retrieved from: https://core.ac.uk/download/156624615.pdf (accessed on 5 December 2024).
18. Dean, W. R., Nalty, C. C., & Sharkey, J. R. (2013). Child hunger and the protective effects of supplemental nutrition assistance program (SNAP) and alternative food sources among Mexican-origin families in Texas border colonias. *Springer Science and Business Media LLC*. Retrieved from: https://core.ac.uk/download/231871516.pdf (accessed on 5 December 2024).
19. Djekic, I., Batlle-Bayer, L., Bala, A., Fullana-i-Palmer, P., & Jambrak, A. R. (2021). Role of the food supply chain stakeholders in achieving UN SDGs. *Sustainability, 13*(16), 9095. https://doi.org/10.3390/su13169095.

20. Donini, L. M., Lenzi, A., Perna, S., Poggiogalle, E., Rondanelli, M., & Spadaccini, D. (2018). Are the therapeutic strategies in anorexia of ageing effective on nutritional status? A systematic review with meta-analysis. *Wiley*. Retrieved from: https://core.ac.uk/download/188830393.pdf (accessed on 5 December 2024).
21. Dunyak, E. M. (2018). The end of the ramen diet: Higher education students and SNAP benefits. *ScholarWorks@UARK*. Retrieved from: https://core.ac.uk/download/159787585.pdf (accessed on 5 December 2024).
22. Ekwall, B. (2014). Global food loss and food waste and the environmental footprint. *Penn Collection: The Last Food Mile Conference*. Retrieved from: https://repository.upenn.edu/handle/20.500.14332/48549 (accessed on 5 December 2024).
23. Fahim, M. A. (n.d.). Impact of water scarcity on food security at micro level in Pakistan. Retrieved from: https://core.ac.uk/download/pdf/6681161.pdf (accessed on 5 December 2024).
24. Fan, S., You, L., & Yu, B. (n.d.). Toward a typology of food security in developing countries. Retrieved from: https://core.ac.uk/download/pdf/6290611.pdf (accessed on 5 December 2024).
25. Fisher, A. (2005). Building the bridge: Linking food banking and community food security. *Community Food Security Coalition*. Retrieved from: https://core.ac.uk/download/71338807.pdf (accessed on 5 December 2024).
26. Fisher, A., Burton, H., Joseph, H., & Pothukuchi, K. (2002). What's cooking in your food system? A guide to community food assessment. *Community Food Security Coalition*. Retrieved from: https://core.ac.uk/download/71342208.pdf (accessed on 5 December 2024).
27. Fisher, A., Gottlieb, B., Winne, M., & Forster, T. (2001). The healthy farms, food and communities act: Policy initiatives for the 2002 farm bill and the first decade of the 21st century. *Community Food Security Coalition*. Retrieved from: https://core.ac.uk/download/71342204.pdf (accessed on 5 December 2024).
28. Frayne, B., & Ziervogel, G. (2011). No. 08: Climate change and food security in southern African cities. *Scholars Commons @ Laurier*. Retrieved from: https://core.ac.uk/download/159478643.pdf (accessed on 5 December 2024).
29. Gloor, L., & Sotala, K. (2017). Superintelligence as a cause or cure for risks of astronomical suffering. *Informatica, 41*, 389–402.
30. Gul, S., Shafiq, U., Mir, S. A., Iqbal, G., & Lone, H. Q. (2024). Enhancing global food security through sustainable fisheries and aquaculture:

A comprehensive review. *Asian Journal of Agricultural Extension, Economics & Sociology, 42*(10), 60–70. https://doi.org/10.9734/ajaees/2024/v42i102563.

31. Harris, C., Agnew, D., Leotte, F., Arthur, R., Walmsley, S., Davies, T., & Peatman, T. (2013). Fisheries and aquaculture and their potential roles in development: An assessment of the current evidence. *University of Stirling (UoS)*. Retrieved from: https://core.ac.uk/download/75777502.pdf (accessed on 5 December 2024).
32. Heim, A., & Pyhälä, A. (2020). Changing food preferences among a former hunter-gatherer group in Namibia. Retrieved from: https://core.ac.uk/download/333882776.pdf (accessed on 5 December 2024).
33. Hoque, F. (2022). Food insecurity in Bangladesh: Causes and effects. *Global Journals Inc. (US)*. Retrieved from: https://core.ac.uk/download/581122530.pdf (accessed on 5 December 2024).
34. Hruby, J. (1978). About the problems in the field of nutrition: Remarks for research activities. *WP-78-066*. Retrieved from: https://core.ac.uk/download/33892325.pdf (accessed on 5 December 024).
35. IFPRI. (n.d.). *IFPRI Annual Report 2006–2007*. Retrieved from: https://core.ac.uk/download/pdf/6388533.pdf (accessed on 5 December 2024).
36. Ignatenko, I., Krechyk, A., & Perepelytsia, K. (2024). Food security: Legal problems and solutions. *Uzhhorod National University Herald. Series: Law, 2*, 64–69. https://doi.org/10.24144/2307-3322.2024.82.2.10.
37. Kahiluoto, H., & Rötter, R. (2009). Implications of and possible responses to climate change. *University of Helsinki Libraries*. Retrieved from: https://core.ac.uk/download/10929655.pdf (accessed on 5 December 2024).
38. Khanal, A. R., Neupane, R. C., Paudel, D., Poudel, P., & Sigdel, S. (2023). COVID-19 pandemic, climate change, and conflicts on agriculture: A trio of challenges to global food security. *Digital Scholarship @ Tennessee State University*. Retrieved from: https://core.ac.uk/download/588211576.pdf (accessed on 5 December 2024).
39. Kim, C. G., Koo, W. W., & Moon, W. (n.d.). World Agriculture Organization (WAO): New global governance for agricultural trade. Retrieved from: https://core.ac.uk/download/pdf/6672600.pdf (accessed on 5 December 2024).
40. Kulick, R. (2002). A measure for ending hunger in the United States. *Partners in Ending Hunger*. Retrieved from: https://core.ac.uk/download/71339029.pdf (accessed on 5 December 2024).
41. Kumar, M. D., Mudgerikar, A., Mukherjee, S., & Shah, Z. (n.d.). Water, human development, and economic growth: Some international

perspectives. Retrieved from: https://core.ac.uk/download/pdf/6764723.pdf (accessed on 5 December 2024).

42. Pandya-Lorch, R., & Spielman, D. J. (n.d.). Highlights from millions fed: Proven successes in agricultural development. Retrieved from: https://core.ac.uk/download/pdf/6329337.pdf (accessed on 5 December 2024).
43. Potorti, M. E. (2016). Food for freedom: The black freedom struggle and the politics of food. Retrieved from: https://open.bu.edu/bitstream/2144/15656/1/Potorti_bu_0017E_10953.pdf (accessed on 5 December 2024).
44. Retrieved from: https://core.ac.uk/download/72052332.pdf (accessed on 5 December 2024).
45. Sharkey, J. R., Dean, W. R., & Nalty, C. C. (2013). Child hunger and the protective effects of supplemental nutrition assistance program (SNAP) and alternative food sources among Mexican-origin families in Texas border colonias. *BMC Pediatrics, 13*, 143. https://doi.org/10.1186/1471-2431-13-143.
46. Timmer, P. (2004). Food security and economic growth: An Asian perspective. Retrieved from: https://core.ac.uk/download/pdf/7089736.pdf (accessed on 5 December 2024).
47. Timmermann, C. (2018). Food security as a global public good. Retrieved from: https://core.ac.uk/download/305120899.pdf (accessed on 5 December 2024).
48. Vanhaute, E. (2009). From famine to food crisis: What history can teach us about local and global subsistence crises. *MPRA Paper No. 17630*. Retrieved from: https://core.ac.uk/download/pdf/7105762.pdf (accessed on 5 December 2024).
49. Vrvic, M. M., Miletic, S., & Lugonja, N. (2023). How will chemistry help solve world hunger? *Proceedings, 91*(1), 235. https://doi.org/10.3390/proceedings2023091235.
50. Wiegers, E., Dorp, M., & Torgerson, S. (2011). Improving nutrition through agriculture: Viewing agriculture-nutrition linkages along the smallholder value chain. *Centre for Development Innovation, Wageningen University & Research Centre / REACH*. Retrieved from: https://core.ac.uk/download/pdf/29235287.pdf (accessed on 5 December 2024).

CHAPTER 9

AIR POLLUTION, CLIMATE CHANGE, AND OZONE DEPLETION

Contents

INTRODUCTION

The intricate web of environmental challenges we face today is primarily underscored by the symbiotic relationship between air pollution, climate change, and ozone depletion. Each of these issues poses distinct threats and interacts with the others, creating a complex matrix of environmental degradation. Air pollution, driven by industrial emissions, vehicle exhaust, and deforestation, directly impacts air quality and human health. Furthermore, its contribution to greenhouse gases exacerbates climate change, leading to more severe weather patterns, rising sea levels, and disrupted ecosystems. Understanding the interconnectedness of these phenomena is essential for formulating effective responses that can address multiple issues simultaneously. The rise in global temperatures can be significantly attributed to anthropogenic activities that release greenhouse gases into the atmosphere. As industrialization flourished in the 20th century, so did the emission of carbon dioxide and methane, further intensifying the greenhouse effect. This imbalance alters climatic patterns and impacts air quality, as higher temperatures can enhance the formation of ground-level ozone—one of the most dangerous air pollutants. Consequently, this climatic shift can lead to a spiral of worsening air quality and increased respiratory health issues. The repercussions of climate change extend beyond environmental alteration; they threaten food security and biodiversity, making it imperative to understand how these changes influence both the Earth's climate and its inhabitants. In addition to the immediate effects of climate change and air pollution, ozone depletion represents a critical layer of our planet's environmental crisis. The thinning of the ozone layer, primarily caused by human-made chemicals such as CFCs, leads to an increase in harmful ultraviolet radiation reaching the Earth's surface. This exacerbates health issues, including skin cancer and cataracts and impacts biodiversity by altering ecosystems. As the interplay of these issues continues to unfold, it is clear that the ramifications of air pollution, climate change, and ozone depletion are deeply intertwined. Addressing these challenges requires comprehensive policies and global cooperation to not only mitigate damages but also prioritize sustainable practices that safeguard the environment for future generations.

9.1. DEFINITION OF AIR POLLUTION

Air pollution is defined as the presence of harmful substances in the atmosphere, which can have detrimental effects on human health, ecosystems, and climate. These substances, referred to as pollutants, can be either gaseous, liquid, or solid and originate from a variety of sources, including industrial emissions, vehicle exhaust, and natural events like wildfires. The World Health Organization has identified several key pollutants, such as particulate matter, nitrogen oxides, sulfur dioxide, and volatile organic compounds, which contribute to various air quality issues. Understanding the definition of air pollution is crucial as it highlights the immediate threats posed by these contaminants and emphasizes the complexity of regulating and mitigating their impact in our interconnected environment. The scope of air pollution extends beyond local geographical boundaries, creating a significant challenge for international environmental law. As outlined in the literature, the regulatory framework governing the atmosphere encompasses a multitude of legal instruments aimed at addressing the diverse sources and effects of these pollutants. This complex regime complex, characterized by various treaties and agreements, has evolved to tackle the various economic activities that contribute to atmospheric degradation, including industrial processes and transportation.

Figure 9.1. *Sources of air pollution.*

Source: By (Uxud) – Own work, CC BY-SA 3.0, https://commons.wikimedia.org/w/index.php?curid=1955961.

The United Nations International Law Commission's inclusion of the topic of atmospheric protection in its work signals a growing recognition of the need for cohesive legal measures to combat air pollution and its associated threats, such as climate change and ozone depletion (Sand et al., 2016). Experts argue that empowering international cooperation and enhancing regulatory frameworks is vital to mitigate the impacts of air pollution on public health and the environment. In light of the increasing number of environmental agreements—now nearing

900—there is a pressing need to assess their effectiveness and identify ways to improve them to ensure they respond to contemporary challenges (Weiss et al., 1993). By establishing clearer definitions and targets, nations can collaborate more effectively in their efforts to reduce air pollution and adapt to its consequences. This collaboration is not just about reducing pollutants but about maintaining the integrity of air quality, which is intrinsically linked to broader environmental health and sustainability initiatives. Through such concerted efforts, a balanced approach can be achieved that reflects the complexities of air pollution and its relationship to climate change and ozone depletion (Figure 9.1).

9.1.1. Overview of Climate Change

Rising global temperatures have established climate change as one of the foremost challenges of our time, prompting urgent discussions surrounding environmental sustainability. The increase in greenhouse gases, primarily due to human activities such as fossil fuel consumption and deforestation, disrupts the natural balance of Earth's ecosystems. This imbalance results in extreme weather events, including floods, droughts, and hurricanes, exacerbating the vulnerabilities of both developed and developing nations. Although some regions may experience temporary benefits, such as extended growing seasons, the long-term repercussions of a warming planet are dire, revealing an urgent need for comprehensive policy reforms and global cooperation.

The consequences of climate change extend far beyond environmental degradation. As ecosystems shift, many plant and animal species face extinction, leading to the loss of biodiversity that is crucial for maintaining functional ecosystems. Additionally, those most affected often belong to marginalized communities, whose traditional lifestyles and food security are increasingly jeopardized. The ecological impacts, such as the effects of decreasing stratospheric ozone and pollution entering Arctic food chains, serve as stark reminders of the interconnectedness of climate science and social justice. Addressing climate change requires a nuanced understanding of both its ecological ramifications and its socio-economic implications, emphasizing the need for interdisciplinary approaches. International efforts to combat climate change have grown increasingly robust, particularly since the establishment of key legal instruments aimed at fostering environmental cooperation. The increase in international legal agreements highlights the global recognition of climate change as a pressing issue demanding collaborative action. National policies alone are insufficient; countries must engage in transnational strategies that address pollution and promote sustainable practices across borders. Continued dialogue and cooperation are imperative, as climate change knows no borders and its effects can reverberate worldwide. Ultimately, the global community must rally to address the multifaceted aspects of climate change, ensuring a sustainable future for generations to come (Weiss et al., 1993) (Figure 9.2).

Figure 9.2. *Climate change and global warming.*

Source: By Fredericknoronha – Own work, CC BY-SA 3.0, https://commons.wikimedia.org/w/index.php?curid=23249478.

9.1.2. Explanation of Ozone Depletion

The degradation of ozone in the stratosphere is a complex phenomenon primarily driven by human activities. Ozone, a molecule composed of three oxygen atoms, forms a protective layer that absorbs the sun's harmful ultraviolet (UV) radiation. In the mid-20th century, scientists began to notice alarming trends in ozone depletion, correlating with the release of chlorofluorocarbons (CFCs) and other ozone-depleting substances (ODS) into the atmosphere. These chemicals, once widely used in refrigeration, aerosol sprays, and foam manufacturing, break down in the stratosphere and release chlorine atoms, which can destroy thousands of ozone molecules. Consequently, the depletion of ozone has accelerated the rate of harm from UV radiation, posing risks to human health, ecosystems, and climate systems alike. The impact of ozone depletion extends beyond public health concerns; it also affects global environmental systems. Ollinger et al (1997) reported that the introduction of higher UV radiation levels disrupts the delicate balance of terrestrial ecosystems, notably impacting vegetation. Ozone pollution, particularly in lower atmospheric layers, has been shown to adversely affect mature forests, reducing leaf carbon gain and overall productivity. This damage compromises forest ecosystems' ability to sequester carbon, exacerbating climate change impacts.

Such consequences are particularly pronounced in regions experiencing high ozone levels, where reduced net primary production can lead to declines in biodiversity and habitat stability. All of these factors intertwine, illustrating how ozone depletion catalyzes wider environmental crises. Mitigating ozone depletion requires global cooperation and stringent regulatory measures. The 1987 Montreal Protocol marked a pivotal moment, as nations recognized the severe implications of ODS and committed to phasing out their production and use. This collective effort led to significant reductions in the atmospheric concentration of CFCs and enabled some recovery of the ozone layer, showcasing the potential efficacy of policy-driven action. Continuous monitoring and support for alternative substances are necessary to safeguard the progress made thus far. Additionally, public awareness and education on ozone-related issues are vital to fostering individual and community-driven initiatives that would further support the recovery of this crucial component of Earth's atmosphere. The success of our environmental policies will depend on collaborative action to ensure that both human health and ecosystems benefit from a restored ozone layer (Figure 9.3).

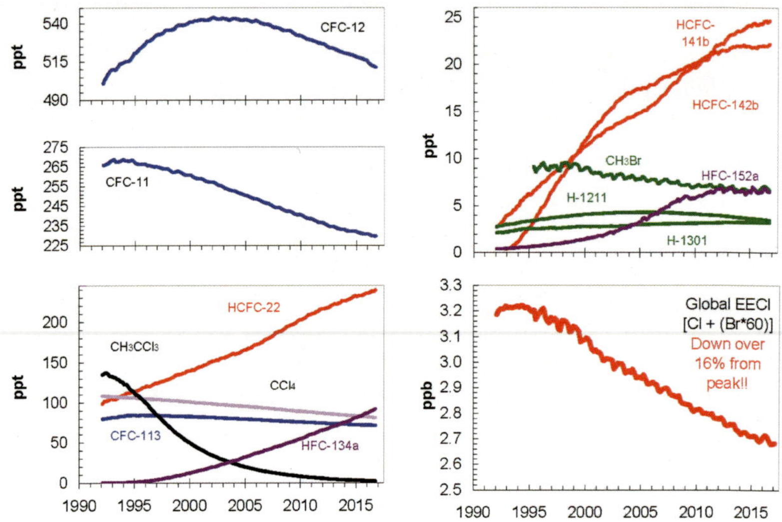

Figure 9.3. *Ozone layer depletion.*

Source: By https://www.esrl.noaa.gov/gmd/hats/graphs/graphs.html, Public Domain, https://commons.wikimedia.org/w/index.php?curid=126002612.

9.2. SOURCES OF AIR POLLUTION

A significant contributor to air pollution stems from transportation emissions, which include pollutants released from vehicles such as cars, trucks, and airplanes. As urban areas continue to expand, the reliance on motorized transport increases, exacerbating the levels of harmful gases entering the atmosphere. These emissions primarily consist of nitrogen oxides, carbon monoxide, and volatile organic compounds, which can lead to deteriorating air quality and health concerns for urban populations. As cities grow, so does the necessity for efficient transportation systems. This interplay often results in inadequate infrastructure, leading to congestion and longer commute times, further amplifying the pollution problem. The actions taken to mitigate these emissions, such as promoting electric vehicle technology and enhancing public transportation infrastructure, are critical steps toward improving air quality and minimizing environmental impacts. Industrial activities also play a crucial role in air pollution, emitting significant quantities of pollutants into the atmosphere. Factories releasing particulate matter, sulfur dioxide, and chemical byproducts contribute to an environment laden with harmful substances. For example, industries engaged in mining, manufacturing, and energy production often rely on fossil fuels, resulting in substantial emissions that affect local ecosystems and have far-reaching consequences on air quality, even in remote regions like the Arctic. The presence of black carbon and other pollutants in these areas emphasizes the transboundary nature of air pollution, showing that sources far from the Arctic can influence the climate and health of its inhabitants (Arnold et al., 2016).

Furthermore, these industrial emissions can intensify climate change impacts, leading to adverse effects on both ecosystems and human societies. Lastly, agricultural practices also represent a significant source of air pollution, particularly through the release of ammonia and other byproducts from livestock and fertilizers. Intensive farming methods generate large quantities of ammonia, which can combine with other pollutants to form particulate matter and contribute to poor air quality. The environmental consequences of these practices can lead to not only localized air pollution but can also have more substantial implications, affecting public health and climate patterns globally. This situation is further compounded by the increasing demand for food as the global population grows. Innovations in sustainable agriculture must be embraced to reduce these emissions and mitigate the harmful effects they bear. Research initiatives such as those highlighted in the PACES program aim to address the complex interactions between agriculture, air pollution, and climate change, opening pathways to a more sustainable future (Arnold et al., 2016).

9.2.1. Industrial Emissions

Industrial emissions significantly contribute to the ongoing challenges of air pollution and climate change. Many industrial activities release a variety of pollutants, including volatile organic compounds (VOCs), nitrogen oxides (NOx), and particulate matter. These emissions can lead to detrimental effects on air quality and public health, facilitating the formation of ground-level ozone, which poses serious respiratory risks. Beyond immediate health concerns, these emissions can disrupt local ecosystems and contribute to larger climate patterns. As industries continue to expand their operations, the need for effective regulatory measures and innovative emission reduction technologies becomes paramount to address the growing environmental crisis. The interconnectedness of industrial emissions and climate change is further exemplified by the feedback mechanisms that emerge from these activities. For instance, increasing emissions of ozone-depleting substances and greenhouse gases can disrupt natural processes within ecosystems, such as those occurring in the Arctic. The decline in stratospheric ozone levels, exacerbated by industrial pollutants, affects global temperatures and the delicate balance of natural habitats. Cuevas et al (2015) point out that emissions can influence atmospheric chemistry, leading to harmful interactions that result in negative geochemical feedback loops. This underscores the complexity of the issue, as industrial practices contribute directly to pollution and induce broader climatic changes that can further impair ecological health. Mitigation strategies aimed at reducing industrial emissions are crucial for promoting environmental resilience. Policymakers must prioritize the adoption of cleaner technologies and sustainable practices to limit the extent of air pollution. As Educational initiatives that raise awareness about the consequences of decreasing ozone levels and eco-toxic substances entering food chains can empower communities to advocate for better industrial regulations. Collaborative efforts between industries, governments, and scientists can pave the way for innovative approaches to emissions management, ensuring that economic growth does not come at the expense of environmental well-being. The path forward requires a concerted focus on transitioning toward greener alternatives, protecting both human health and the fragile ecosystems that sustain life on Earth.

9.2.2. Transportation and Vehicular Emissions

The impact of transportation on air pollution is significant, as it remains one of the largest contributors to greenhouse gas emissions globally. Vehicles powered by fossil fuels release a variety of pollutants, such as carbon monoxide (CO) and volatile organic compounds (VOCs), which hinder air quality and exacerbate health issues. Particularly evident during large events, like the Hajj pilgrimage in Saudi Arabia, studies have shown elevated levels of CO and VOCs along pilgrimage routes due to the influx of millions of vehicles and people. For instance, a study conducted during the 2012 Hajj found CO levels reaching up to 57 ppmv, underscoring the urgent need to address transportation-related emissions and improve urban air quality (Aburizaiza et al., 2014). As urban populations grow, the challenge of managing pollution from transportation becomes even more pressing, necessitating innovative solutions and regulations. The approaches to mitigating vehicular emissions involve a combination of technological advancements and policy implementation. The transition toward electric and hydrogen-powered vehicles can drastically reduce reliance on fossil fuels and lower overall emissions. Policies promoting public transportation, carpooling, and biking can also decrease the number of vehicles on the road, further minimizing the air pollution caused by transportation.

Additionally, modernizing existing transportation infrastructure to incorporate green technologies can enhance efficiency. The biogas generated from animal waste, as evaluated in various studies, illustrates the importance of seeking renewable energy sources (Hollas et al., 2024). Such initiatives combat the negative environmental effects of conventional transportation and contribute to the circular economy by harnessing waste materials for energy production. Ultimately, addressing transportation and vehicular emissions requires a multifaceted approach that considers environmental, social, and economic factors. The need for collaboration among stakeholders—including government agencies, businesses, and communities—is essential to create cohesive strategies that promote cleaner transport options. Public awareness and education campaigns informing individuals about the health impacts of air pollution can also drive behavioral changes, fostering community support for sustainable practices. Because of impending climate challenges and the urgency of mitigating air pollution, it is crucial to implement effective solutions and promote advocacy for cleaner transportation systems. Through these concerted efforts, we can work toward a healthier environment and a more sustainable future, ultimately reducing the contributions of transportation to climate change and ozone depletion (Figure 9.4).

Figure 9.4. *Vehicle emissions.*

Source: By Ruben de Rijcke – Own work, CC BY-SA 3.0, https://commons.wikimedia.org/w/index.php?curid=8735834.

9.2.3. Agricultural Practices

Agricultural practices play a significant role in influencing environmental quality, particularly regarding air pollution and climate change. Conventional farming techniques often rely heavily on chemical fertilizers and pesticides, which can lead to the release of VOCs and other harmful pollutants into the atmosphere. These emissions contribute to local air quality degradation and to larger issues, such as the formation of ground-level ozone, which poses health risks to communities. Transitioning to sustainable agricultural strategies, such as organic farming and integrated pest management, can mitigate these negative impacts.

By reducing chemical inputs, sustainable practices help lower emissions and promote healthier ecosystems, thereby contributing to the reduction of air pollution and supporting overall climate resilience. Furthermore, the globalization of agricultural markets has significant implications for environmental governance and agricultural sustainability. As noted, globalization could lead to a phenomenon where the "greening" of the global North is accompanied by the "browning" of the global South (James Boyce). This illustrates the potential disparity in environmental practices, where developed nations adopt eco-friendly techniques while developing countries may prioritize productivity over sustainability for economic gain. Such a divide exacerbates environmental challenges in vulnerable regions and highlights the necessity of equitable policies that promote sustainable agriculture across borders.

Only through collaborative efforts and shared knowledge can we ensure that agricultural practices do not disproportionately harm economically disadvantaged areas while still addressing global food demands. Finally, adapting agricultural practices in response to climate change is crucial for both environmental sustainability and food security. Techniques such as agroecology, which emphasize ecological principles, can enhance resilience to climate impacts while promoting soil health and biodiversity. Moreover, incorporating traditional knowledge and local practices can empower communities and foster adaptive capacity.

The interplay of globalization and local practices presents both challenges and opportunities for addressing climate issues. As the interconnected nature of global markets can lead to both higher or lower environmental quality (James Boyce), it is essential to prioritize strategies that harmonize agricultural productivity with responsible stewardship of natural resources. In doing so, we can create a future where agricultural practices contribute positively to air quality, climate stability, and sustainable development.

9.3. EFFECTS OF AIR POLLUTION ON HUMAN HEALTH

The consequences of air pollution on human health are both profound and far-reaching, significantly impacting individuals across various demographics. Exposure to pollutants such as particulate matter, nitrogen oxides, and volatile organic compounds can lead to a range of respiratory diseases, including asthma and chronic obstructive pulmonary disease (COPD). As noted in the special edition of ChemMatters focusing on atmospheric chemistry, air pollution not only exacerbates existing health conditions but also contributes to premature deaths from cardiovascular disease. These health ramifications are not confined to urban centers; rural and sub-urban areas also experience deteriorating air quality, as pollutants can be transported over long distances. Children, the elderly, and individuals with pre-existing health conditions are particularly vulnerable, experiencing more severe effects than their healthier counterparts. The correlation between air quality and human health emphasizes the critical need for policies addressing air pollution. A growing body of research underscores the link between air pollution and mental health disorders, another dimension of its impact on human well-being. Emerging studies indicate that long-term exposure to polluted air increases the risk of developing conditions such as anxiety and depression. This troubling trend is compounded by the fact that various air pollutants can induce inflammatory responses in the brain, exacerbating neurodegenerative disorders. Given that mental health is an integral component of overall health, these findings prompt urgent considerations for public health strategies. The UN International Law Commission has recognized the necessity of international frameworks that address atmospheric protection beyond traditional air law, as this is crucial in combating the multifaceted implications of air pollution on mental and physical health (Sand et al., 2016). Creating effective policies will require a comprehensive understanding of such health impacts and the formulation of sustainable solutions. Addressing air pollution necessitates a collaborative approach that involves government agencies, health organizations, and community stakeholders. Public health campaigns aimed at raising awareness can play a pivotal role in educating individuals about the risks associated with air pollution and promoting measures to mitigate exposure. A ChemMatters (2001) publication encourages engagement with scientific principles to demystify the consequences of atmospheric changes, emphasizing the importance of understanding chemical reactions in pollution formation. Additionally, implementing regulations aimed at reducing industrial emissions and promoting cleaner energy sources could drastically improve air quality and, by extension, public health. Advocacy for cleaner transportation options, such as electric vehicles, can further contribute to this effort. Ultimately, the fight against air pollution will require concerted efforts to develop innovative policies and raise awareness, fostering a healthier environment for current and future generations (Figure 9.5).

Figure 9.5. *The effects of air pollution on human health.*

Source: By Simons/Staff Sgt. – http://www.af.mil/shared/media/photodb/photos/021205-O-9999G-021.jpg, Public Domain, https://commons.wikimedia.org/w/index.php?curid=320999.

9.3.1. Respiratory Diseases

Respiratory diseases represent a significant public health concern, exacerbated by the impacts of air pollution and climate change. Conditions such as asthma,

COPD, and lung cancer have been increasingly linked to harmful pollutants released into the atmosphere. The deterioration of air quality often leads to heightened irritability of the respiratory system, causing exacerbations in those already afflicted by chronic conditions. Moreover, as urban areas continue to grow and industrial activities expand, the concentration of harmful particles in the air rises, increasing the risk of respiratory diseases among populations. This escalating health crisis underscores the need for comprehensive regulatory measures aimed at reducing emissions and improving air quality to protect vulnerable groups, including children and the elderly. In addition to direct health impacts, the interconnectivity between environmental factors and respiratory health highlights important considerations for policy and planning. The phenomenon of climate change contributes to the exacerbation of respiratory illnesses through increasing temperatures, which can lead to the elevation of ground-level ozone, a harmful air pollutant. According to findings presented in the EOS-Aura special edition of ChemMatters (2001), good, and bad ozone levels fluctuate with changing climatic patterns Consequently, the respiratory health of millions is jeopardized as hotter temperatures foster conditions ideal for ozone formation, leading to more frequent and severe smog episodes.

Addressing these challenges requires a multifaceted approach that encompasses environmental protection and health education and community engagement in recognizing and mitigating the risks associated with air pollution. Effective management of respiratory diseases necessitates a thorough understanding of the intricate relationships between natural and anthropogenic factors. Recent observations have indicated that environmental crises, such as air pollution, carry consequences that extend beyond immediate health effects. The enduring nature of these crises implies that respiratory diseases influenced by pollution cannot be viewed in isolation but must be integrated into broader discussions about sustainability and public health (Danu Marcela-Cornelia, 2010). Combatting the rising tide of respiratory illnesses will require proactive engagement from policymakers, healthcare providers, and communities. As efficient strategies are developed to address both air pollution and its health impacts, a comprehensive framework will be essential for fostering resilient populations capable of adapting to ongoing environmental changes.

9.3.2. Cardiovascular Issues

Air pollution has emerged as a significant contributor to cardiovascular issues, posing a serious threat to public health globally. Exposure to airborne pollutants, including particulate matter and toxic gases, has been linked to increased rates of heart disease, hypertension, and stroke. The mechanisms by which air pollution affects cardiovascular health are multifaceted, involving inflammation, oxidative stress, and endothelial dysfunction. As pollutants enter the bloodstream, they can trigger inflammatory responses that lead to atherosclerosis—an accumulation of plaque in the arteries that can obstruct blood flow, ultimately resulting in severe cardiovascular events. Moreover, vulnerable populations, such as the elderly and those with pre-existing conditions, are particularly susceptible to these adverse effects, highlighting the urgent need for public health interventions aimed at reducing exposure to air pollution. The implications of climate change further exacerbate cardiovascular issues through increased heat exposure and environmental degradation. Rising temperatures and their associated health risks contribute to cardiovascular strain, particularly among those with heart diseases. Extreme heat is known to elevate the likelihood of heat-related illnesses and can exacerbate existing heart conditions, leading to higher mortality rates. Additionally, extreme weather events, such as hurricanes and floods, can disrupt healthcare services and access, complicating the management of cardiovascular diseases. These challenges underscore the interconnected nature of environmental factors and their impact on human health.

A synthesis of the effects of climate change and air pollution is critical to develop strategies that prioritize heart health while addressing the root environmental causes. Furthermore, the depletion of the ozone layer plays a crucial role in cardiovascular health, as the increased exposure to UV radiation can have indirect effects on heart function. Although primarily associated with skin cancer, heightened levels of UV radiation can provoke systemic inflammation and contribute to the exacerbation of existing cardiovascular diseases. The link between ozone depletion and cardiovascular effects is often overlooked; however, research emphasizes the intricate relationship between environmental changes and human health outcomes, noting potential impacts on diseases influenced by climate and pollution (Martens et al., 1998). These health challenges demand comprehensive policies and proactive approaches that prioritize intervention in air quality and climate action to safeguard cardiovascular health while addressing the broader implications of environmental degradation. Efforts to mitigate air pollution, combat climate change, and protect the ozone layer are not merely environmental imperatives—they are essential measures for preserving public health and ensuring a healthier future.

9.3.3. Long-Term Health Impacts

The intricate relationship between air pollution and long-term health impacts cannot be overstated, particularly in the context of global environmental challenges such as climate change and ozone depletion. Exposure to pollutants, including harmful trace gases like tropospheric ozone and particulates such as black carbon, has been shown to have deleterious effects on respiratory and cardiovascular health. Individuals residing in areas with significant air pollution experience higher rates of chronic respiratory diseases, decreased lung function, and increased mortality. Furthermore, as air pollution amplifies due to climate change, the population's vulnerability increases, presenting a dual challenge for public health. Addressing these intertwined issues necessitates a comprehensive understanding of the sources and long-term consequences of air pollution on human health, leading to the urgent need for proactive public health strategies. The cycle of air pollution and its health impacts disproportionately affects vulnerable populations, particularly in the Arctic regions where local and distant emissions intersect. Air pollution is not only a regional concern but also a driver of global change that has cascading effects on ecosystems and human health. This is particularly alarming given the increasing shipping and industrial activities in the Arctic, which contribute to the release of toxic substances (Arnold et al., 2016). The presence of pollutants like polycyclic aromatic hydrocarbons in this fragile environment raises the stakes for indigenous communities whose livelihoods and health depend on the land and resources. Understanding these health impacts holistically requires a trans-disciplinary approach, integrating social and economic factors with the chemical and physical aspects of air pollution, as highlighted in current international initiatives advocating for coordinated research and monitoring. Efforts to mitigate the long-term health impacts of air pollution must also consider the broader regulatory and legal frameworks governing atmospheric protection. The evolving landscape of international environmental law hints at the necessity for a regime that comprehensively addresses air quality and its effects on global health (Sand et al., 2016). This includes recognizing the atmospheric commons as a shared resource influenced by multiple jurisdictions and interests. Future initiatives should focus on establishing robust guidelines that reduce emissions at the source and take into account the far-reaching consequences of air pollution on health. By fostering international cooperation and innovative solutions, stakeholders can work towards creating healthier environments that mitigate the profound long-term health impacts projected due to the interplay of pollution, climate change, and ozone depletion.

9.4. RELATIONSHIP BETWEEN AIR POLLUTION AND CLIMATE CHANGE

The complexities underlying air pollution and climate change are deeply intertwined, manifesting in both environmental degradation and public health crises. Air pollutants, primarily from fossil fuel combustion, contribute significantly to atmospheric warming. These pollutants, such as carbon dioxide and methane, not only pose direct threats to human health but also exacerbate climatic shifts. The relationship extends beyond mere causation; for instance, increased temperatures can lead to higher levels of ozone formation in the lower atmosphere. This dual effect underscores the urgent need for addressing air pollution as part of the broader climate change agenda. Addressing air pollution could yield immediate health benefits and contribute substantially to climate change mitigation efforts. Furthermore, the ramifications of air pollution on ecosystems highlight another layer of this relationship. Mature forests, for example, are adversely affected by ozone pollution, which can impede their growth and reduce overall ecosystem productivity. A study indicated that reductions in leaf carbon gain correlated with ozone flux levels, leading to declines in net primary production by up to 16% in polluted areas (Ollinger et al., 1997). As trees struggle to grow in compromised air quality, their ability to sequester carbon diminishes, creating a reinforcing cycle of climate change.

Figure 9.6. *Air pollution and climate change.*

Source: https://www.rifs-potsdam.de/sites/default/files/styles/cinematic_xxl/public/istockAntonio-BanderAS-.jpg?itok=nTRp5DJV.

As forests fail to perform their role as carbon sinks effectively, the increasing levels of atmospheric carbon exacerbate global warming. Thus, mitigating air pollution is essential for preserving vital ecosystems that act as natural buffers against climate change. The ethical implications of air pollution and climate change further complicate the interaction between these phenomena. Climate change is not only an environmental issue but also a question of equity between generations and communities, as highlighted by the profound consequences of current practices on future generations (Weiss et al., 2008). As air pollution disproportionately affects marginalized communities—those with the least capacity to adapt and respond—addressing air quality becomes a matter of social justice. Effective climate policy must therefore include stringent regulations on pollutants that protect the environment and ensure equitable health outcomes. Ultimately, recognizing the interconnectedness of air pollution and climate change can steer us toward comprehensive solutions that promote sustainability while fostering equity among all communities (Figure 9.6).

9.4.1. Greenhouse Gases and Their Sources

An understanding of greenhouse gases is crucial to grasping the complexities of climate change. These gases include carbon dioxide (CO_2), methane (CH_4), nitrous oxide (N_2O), and fluorinated gases, each playing a significant role in trapping heat within the Earth's atmosphere. CO_2, primarily released through fossil fuel combustion for electricity and transportation, is the most abundant anthropogenic greenhouse gas. Its persistent presence is exacerbated by deforestation, which reduces the number of trees available to absorb CO_2. CH_4, although released in smaller quantities, is more potent as a heat-trapping agent. It originates from livestock digestion, agricultural practices, and the decay of organic waste in landfills. This multifaceted nature of greenhouse gas sources necessitates a comprehensive approach to mitigation, underscoring the inherent interconnectedness between human activity and environmental health. Agricultural practices are a key contributor to greenhouse gas emissions, underscoring the role of food production in this global challenge. The use of fertilizers in agriculture leads to increased N_2O emissions, a gas that is known to have a warming potential substantially higher than that of CO_2. Moreover, livestock farming, particularly cattle, significantly contributes to methane emissions through enteric fermentation and manure management.

As global populations continue to rise, the demand for food increases, often prompting agricultural expansion into sensitive ecosystems. Such practices enhance greenhouse gas emissions and lead to biodiversity loss and soil degradation. Therefore, shifting towards sustainable agricultural practices becomes paramount in effectively addressing the adverse effects of climate change while ensuring food security. Addressing greenhouse gas emissions involves collaborative efforts spanning multiple sectors, including energy production, transportation, and waste management. Policymakers play a critical role by establishing regulations and incentives aimed at reducing emissions and promoting renewable energy sources. Public awareness campaigns are also essential, educating individuals about their carbon footprints and encouraging behavioral changes such as reducing waste, utilizing public transportation, and supporting eco-friendly products. Furthermore, technological advancements in carbon capture and storage are now being explored as viable pathways for mitigating emissions from fossil fuel sources. The urgency of the climate crisis demands immediate action, as cited in educational materials focused on atmospheric chemistry, which highlight the science of ozone depletion and its interplay with greenhouse gases (University of the Arctic, 2001). Prioritizing holistic, integrated approaches will ultimately pave the way for meaningful progress in combating climate change and protecting our planet's future (Figure 9.7).

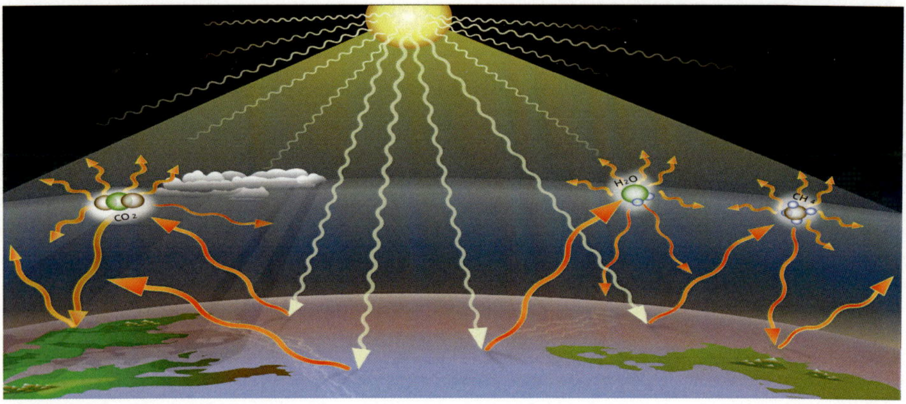

Figure 9.7. *Greenhouse gases.*

Source: By A loose necktie – Own work, CC BY-SA 4.0, https://commons.wikimedia.org/w/index.php?curid=80356809.

9.4.2. Feedback Loops in Climate Systems

Understanding feedback loops in climate systems is essential to grasping the broader implications of environmental change. These loops occur when an initial change in the climate system triggers processes that either amplify or diminish the effects of that change. For instance, as global temperatures rise due to increasing greenhouse gases, polar ice melts, which reduces the albedo effect—its ability to reflect sunlight. This reduction leads to greater absorption of solar energy, further exacerbating warming. Thus, the process becomes self-reinforcing, indicating how interconnected are the components of climate system. The relationship between these feedback loops and atmospheric changes reveals much about our vulnerabilities in the face of ongoing climate shifts. Another critical dimension of climate feedback loops revolves around human activities, particularly those contributing to air pollution. Pollutants such as aerosols can induce cooling by reflecting sunlight, thereby masking some of the warming effects of greenhouse gases. However, these aerosols also pose severe health risks and can alter weather patterns, leading to unintended consequences. For instance, the reduction of air pollutants in efforts to improve air quality may inadvertently accelerate warming if other greenhouse gas emissions go unchecked. As stated in (Perelet and Renat, 2007), development, and environmental challenges are ultimately intertwined, emphasizing the need for a nuanced understanding of how air pollution and climate dynamics interact. Thus, addressing feedback mechanisms requires a multi-faceted strategy that considers environmental and human factors collectively. Lastly, the implications of feedback loops extend beyond immediate climate changes to long-term sustainability and resilience. The Planetary Accounting Framework discussed aims to quantify these interconnections and inform better decision-making at multiple levels (Meyer et al., 2018). Policymakers can better understand how feedback loops operate and develop strategies to mitigate their risks by managing the Earth system using scientific limits. Effective resource management and emissions reductions must align with scientific insights about these feedback cycles to create a sustainable future. Without this integrated approach, attempts to combat climate change could inadvertently destabilize other aspects of the climate system, leading to outcomes that may be negative for both the environment and human societies. In a world where environmental and developmental challenges are distinctly interlinked, a comprehensive understanding of feedback loops becomes essential for meaningful action.

9.4.3. Impact on Global Temperatures

The gradual rise in global temperatures is intricately linked to factors such as air pollution, the increase of greenhouse gases, and ozone depletion. The burning of fossil fuels, industrial processes, and deforestation all contribute to a rise in carbon dioxide and other greenhouse gas concentrations in the atmosphere. This leads to an enhanced greenhouse effect, which traps heat and raises average global temperatures. The interconnection between these pollutants and climate systems is crucial; as temperatures rise, feedback mechanisms may further amplify warming. For instance, as ice and snow cover decrease, less solar energy is reflected back into space (a phenomenon known as albedo), and more is absorbed by the Earth, exacerbating warming trends. In the Arctic particularly, the effects of climate change manifest strikingly. The decline in stratospheric ozone, alongside climate change, has led to shifts in global weather patterns and regional temperatures (University of the Arctic).

As the ozone layer thins, it allows greater amounts of ultraviolet radiation to penetrate, which can further affect local ecosystems and weather variations. The warming Arctic is causing unprecedented ecological shifts, influencing the biodiversity of the region and the lifestyles of indigenous communities. Species once adapted to colder conditions are struggling to survive as their habitats undergo rapid transformation, illustrating a clear link between global temperature increases and localized ecological impacts. Moreover, air pollution has complex interactions with climate dynamics that severely threaten global temperatures. Pollutants such as particulate matter and aerosols can have both warming and cooling effects, altering cloud formation and longevity. The culminating effects of air pollution complicate temperature regulation and climate predictions, posing challenges for global climate strategies. Programs targeting air quality improvements can also yield significant climate benefits, as cleaner air often correlates with reduced greenhouse gas emissions. This intricate relationship highlights the importance of addressing air pollution from a health perspective and in the broader context of climate change and its impact on global temperatures.

9.5. OZONE LAYER AND ITS IMPORTANCE

The ozone layer plays a crucial role in shielding the Earth from harmful UV radiation emitted by the sun. By absorbing the majority of UV radiation, the ozone layer acts as a protective barrier, reducing the incidence of skin cancers, cataracts, and other health issues among living organisms. Its presence directly influences ecological stability, particularly in marine and terrestrial ecosystems, where increased UV exposure can disrupt photosynthesis, affect plankton populations, and ultimately alter food chains. As we confront the escalating challenges of air pollution and climate change, understanding the functions of the ozone layer becomes increasingly vital to maintaining public health and environmental balance, highlighting the urgent need for effective policies and conservation measures. The degradation of the ozone layer is primarily attributed to anthropogenic chemicals, such as CFCs, which have been extensively used in industrial applications. This depletion poses a significant threat not only to human health but also to biodiversity. For instance, studies have shown that regions with higher ozone levels experience greater primary productivity, crucial for supporting diverse ecosystems and sustaining food supplies (Ollinger et al., 1997). Furthermore, the integrity is intertwined with various climate processes, as it influences temperatures and weather patterns. Therefore, the impacts of ozone depletion extend beyond local environments, contributing to global climate change phenomena that further exacerbate environmental degradation. Efforts to mitigate ozone layer depletion have led to significant international cooperation, exemplified by the Montreal Protocol, which successfully phased out the use of many ozone-depleting substances (Weiss et al., 1993). This landmark agreement demonstrates how collective action can yield positive outcomes for both the environment and humanity. Observations indicate that the ozone layer is gradually recovering, which could, in turn, help stabilize climate conditions and reduce negative health outcomes associated with increased UV radiation. However, continued vigilance and commitment to environmental agreements are essential to ensure the long-term health of the ozone layer. As the world confronts the intertwined crises of air pollution and climate change, recognizing the crucial function of the ozone layer importance is crucial for safeguarding our planet's future.

9.5.1. Role of the Ozone Layer in Protecting Life

The ozone layer plays a crucial role in maintaining the delicate balance of life on Earth. Located in the stratosphere, this layer absorbs a significant portion of the sun's harmful UV radiation. If the ozone layer were to shrink or deplete, the increase in UV radiation would pose serious threats to all forms of life, ranging from phytoplankton in the ocean to humans on land. An increase in UV exposure can lead to higher instances of skin cancers, cataracts, and immune system suppression in humans, while also severely disrupting the growth and photosynthesis of plants. Ultimately, the ozone layer's protective shield is essential for sustaining ecosystems, securing food chains, and promoting biodiversity, illustrating its necessity not only for individual species but for the planet as a whole. The degradation of the ozone layer is a harrowing consequence of human activity, primarily attributed to the release of CFCs and other ozone-depleting substances. These compounds, released from products like aerosols and refrigerants, have led to significant thinning of the ozone layer, especially over polar regions. International efforts, such as the Montreal Protocol established in 1987, have made strides in limiting the production and use of these harmful chemicals. However, continued vigilance is necessary in enforcing and expanding such treaties to combat both existing and emerging threats to the ozone layer. As the global community grapples with climate change and air pollution, it must also remain committed to protecting the ozone layer, emphasizing the interconnectedness of environmental issues. As research on atmospheric science evolves, it becomes increasingly clear that the ozone layer is not just a static barrier but a dynamic component of Earth's environment. New regulations and international collaborations, as discussed in studies by the UN International Law Commission, are essential for addressing the complex network of atmospheric protections that extend beyond mere regulation of air quality (Sand et al., 2016). The ozone layer's significance is also underscored by the diverse array of international legal instruments aimed at safeguarding our planet's atmosphere, highlighting the need for a comprehensive approach to environmental law (Weiss et al., 1993). By recognizing the vital role of the ozone layer in protecting life, we can inspire collective action, ensuring a sustainable future where both humanity and nature thrive, undeterred by the ramifications of pollution and climate change (Figure 9.8).

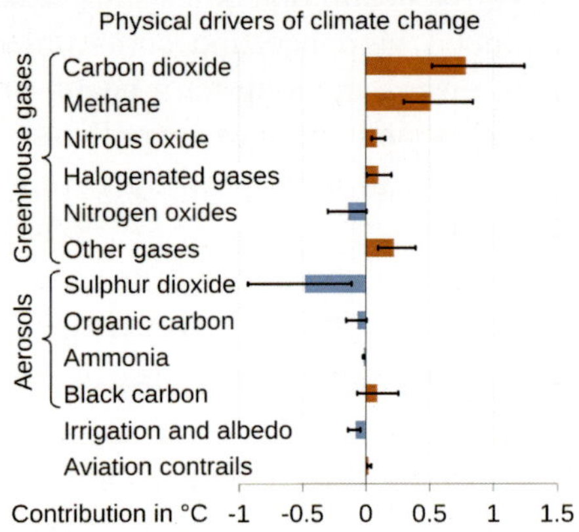

Figure 9.8. *Health benefits of protecting the ozone layer.*

Source: By Eric Fisk – Own work, CC BY-SA 4.0, https://commons.wikimedia.org/w/index.php?curid=81034563.

9.5.2. Natural Versus Anthropogenic Ozone Depletion

Ozone depletion occurs through both natural processes and human activities, each contributing significantly to the thinning of the ozone layer. Natural ozone depletion primarily results from stratospheric reactions involving solar radiation and varied trace gases. For example, volcanic eruptions can release significant amounts of sulfur dioxide and chlorine compounds, which photovoltaic reactions transform into ozone-depleting substances. These natural phenomena tend to occur over longer timescales and can affect global ozone concentrations. Their impact is typically more localized and less predictable, but they still underscore the natural variability of the Earth's atmosphere and its inherent processes that contribute to ozone dynamics. In contrast, anthropogenic factors have dramatically accelerated ozone depletion rates, primarily through the release of CFCs and other synthetic chemicals. These compounds, used in refrigeration, aerosols, and solvents, have been shown to break down ozone molecules significantly when they reach the stratosphere. As the global community witnessed, the use of these substances led to alarming ozone hole phenomena, particularly over Antarctica, demonstrating a clear human impact on the atmospheric environment.

Research indicates that anthropogenic emissions have driven a substantial increase in tropospheric ozone levels, creating a harmful feedback loop where increased surface ozone enhances emissions of reactive natural compounds, thus further depleting ozone. The interaction between natural and anthropogenic ozone depletion provides critical insights into the complexities of atmospheric chemistry and climate change. For instance, while oceanic emissions of iodine compounds can help mitigate ozone loss by destroying harmful agents, the presence of elevated anthropogenic tropospheric ozone exacerbates the overall problem. This dynamic illustrates that human activities have introduced significant stressors to natural systems, triggering processes such as enhanced inorganic iodine emissions, which ultimately work to counterbalance some of the detrimental effects of increased ozone exposure (Arnold et al., 2016). As climate change progresses, understanding these interactions will be vital, emphasizing the necessity for continued research and international cooperation, as initiated by initiatives like PACES, to address the multifaceted challenges posed by both natural and anthropogenic influences on ozone depletion (Arnold et al., 2016).

9.5.3. Consequences of Ozone Layer Depletion

The depletion of the ozone layer has profound implications for both human health and ecological stability. As the ozone layer thins, it allows increased UV radiation from the sun to penetrate the Earth's atmosphere, which can lead to a marked increase in skin cancers, cataracts, and other health issues among humans. Increased UV exposure can also compromise immune system functioning, further exacerbating health risks. Beyond impacts on human health, elevated levels of UV radiation can severely affect aquatic ecosystems, disrupting the growth and development of phytoplankton, which serve as the foundation of oceanic food chains. These biological disruptions underscore the urgency of addressing ozone depletion, as the consequences ripple outwards, affecting not only individual health but also broader environmental systems. Further complicating the issue, the interaction between ozone depletion and climate change presents additional challenges. Ozone-depleting substances, such as CFCs, are greenhouse gases that contribute to global warming alongside their destructive effects on the ozone layer. As the climate warms, researchers note potential feedback loops that could intensify both ozone depletion and climate change effects. For instance, the loss of stratospheric ozone could alter weather patterns, which then may lead to shifts in carbon storage capacities in forests and oceans. These entangled processes underscore the complexity of environmental systems and illustrate how ozone depletion not only threatens human health but also disrupts ecological balance and climate stability (Danu Marcela-Cornelia, 2010). Finally, the cultural and societal implications of ozone layer depletion cannot be overlooked. Communities that rely on traditional lifestyles, particularly those in vulnerable regions like the Arctic, face significant challenges as environmental changes disrupt their longstanding practices. Increased UV radiation can impact local wildlife, further threatening the food security of indigenous peoples who depend on these species for sustenance. Furthermore, as these communities grapple with the ecological impacts of ozone depletion, their cultural identities are put at risk, emphasizing the interconnection between environmental health and cultural preservation. Addressing the consequences of ozone layer depletion requires a concerted effort from environmental and health sectors and from policymakers who recognize the human dimensions of these crises (University of the Arctic).

9.6. INTERNATIONAL EFFORTS TO COMBAT AIR POLLUTION

International efforts to combat air pollution have increasingly prioritized the development and implementation of comprehensive agreements and protocols that reflect a shared responsibility among nations. One significant instance of this is the Kigali Amendment to the Montreal Protocol, which addresses hydrofluorocarbons (HFCs)—potent greenhouse gases that contribute to air pollution and climate change. By incorporating HFCs into the original framework, the amendment exemplifies how international cooperation can effectively target specific harmful substances. This model has succeeded in reducing the emissions of ozone-depleting chemicals, demonstrating that with shared commitment, nations can work towards mitigating air pollution. However, challenges remain, particularly with the potential for illicit trade in banned substances, which calls for vigilant enforcement and collaboration among member states (Welch et al., 2017). Litigation has emerged as a powerful tool to enhance international efforts against air pollution, particularly in nations where governmental action may be lacking. In the United States, environmental NGOs and state governments have initiated legal challenges to compel compliance with emissions reduction commitments, such as those outlined in the Kyoto Protocol. These lawsuits aim to hold corporations and regulatory bodies accountable for their contributions to greenhouse gas emissions, effectively leveraging the judicial system as a means to influence policy change. This strategy seeks to address the immediate impacts of air pollution and serves as a platform for raising public awareness and fostering a culture of accountability regarding environmental degradation (McGee et al., 2005).

By employing litigation, advocates can reinforce the importance of adhering to international agreements, ultimately galvanizing further action against air pollution. A comprehensive approach to combating air pollution necessitates the integration of scientific research, technology, and public policy. Innovations in clean energy, transportation, and waste management can significantly reduce emissions and improve air quality, but these solutions require international collaboration to maximize effectiveness. Nations can share best practices and technologies through frameworks like the United Nations Framework Convention on Climate Change (UNFCCC), fostering a spirit of collective responsibility. The interplay between domestic initiatives and global commitments creates a multifaceted strategy that addresses air pollution on various fronts. Ultimately, while international agreements provide the groundwork for action, the active engagement of governments, citizens, and organizations is crucial to realizing the full potential of these efforts and creating a cleaner, healthier environment for future generations (Figure 9.9).

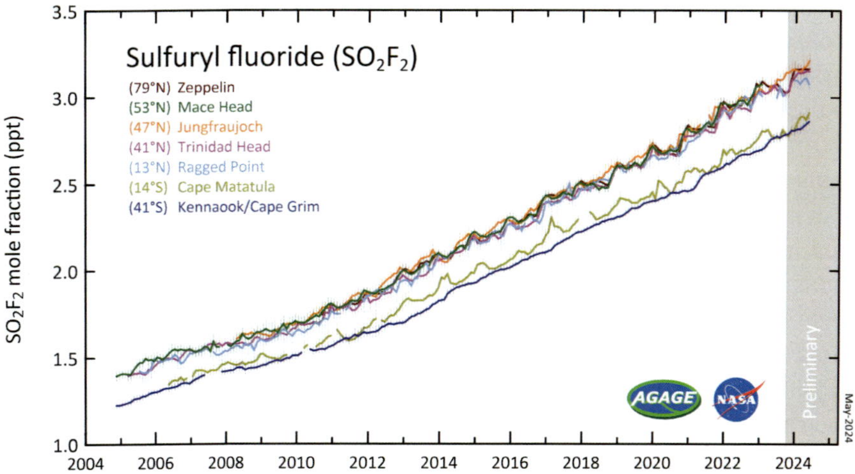

Figure 9.9. *Effective air pollution prevention strategies.*

Source: By Jens Mühle – Own work, CC BY-SA 4.0, https://commons.wikimedia.org/w/index.php?curid=96687369.

9.6.1. The Paris Agreement

Global climate change remains one of the most pressing challenges of our time, necessitating a concerted international response. The Paris Agreement, adopted in 2015, signifies a landmark commitment by nations to collectively curb greenhouse gas emissions and limit global warming to well below 2 , aiming for 1.5 if possible. This agreement reflects an understanding that actions taken in one part of the world can have far-reaching implications for air quality and climate stability elsewhere. Its success relies on the cooperative will of countries to implement their Nationally Determined Contributions (NDCs), which outline individual climate action plans. The framework encourages transparency and accountability, thus fostering a sense of shared responsibility among nations, and inspiring more ambitious targets over time, ultimately fostering a global culture of climate mitigation. The manifestations of air pollution, climate change, and ozone depletion are interconnected, and the Paris Agreement acknowledges this complexity. The agreement aims to tackle the root causes of these environmental threats by addressing emissions from various sectors, including energy, transportation, and agriculture. For instance, as reported, industrial air pollution costs European Union citizens hundreds of euros annually, impacting health and the quality of life (Sand et al., 2016). This illustrates that effective climate policy, as envisioned by the Paris Agreement, must also focus on reducing air pollution to protect public health.

The interplay between greenhouse gas emissions and local air pollutants emphasizes the necessity for holistic climate action, which the agreement seeks to promote through comprehensive national strategies that factor in the socio-economic implications of environmental policies. The ongoing challenge lies in the implementation and adaptation of the Paris Agreement amid varying national interests and economic conditions. For developing nations, the pathway to reducing emissions often intersects with issues of poverty and economic development, complicating their commitments to the agreement. Furthermore, the recognition of "regime complexes" managing atmospheric commons, as highlighted in the discussed international law frameworks (Sand et al., 2016), indicates that multiple legal, economic, and political avenues must be navigated. To maximize efficacy, nations must align their climate strategies with proven methodologies for assessing air quality and health impacts, as suggested by the Uniform World Model (Spadaro, 2011). This integration of climate commitments with effective pollution control initiatives is crucial for synergistic progress, highlighting the Paris Agreement not only as an environmental treaty but as a foundation for sustainable global development.

9.6.2. The Montreal Protocol

An agreement that emerged in the late 20th century, the Montreal Protocol represents a significant international effort to combat substances harmful to the ozone layer. Established in 1987, it aimed to phase out the production and consumption of ozone-depleting chemicals, primarily CFCs. As countries recognized the critical role of the ozone layer in shielding the earth from harmful ultraviolet radiation, this treaty garnered unprecedented global support. Its innovative mechanisms, including flexible implementation schedules for developing nations, encouraged widespread participation. The success is evident in the substantial reduction of CFC emissions, fostering a growing optimism about international cooperation in environmental preservation. However, the journey is not yet complete; as new challenges arise, including the need to manage HFCs, an alternative with high global warming potential, vigilance, and adaptation in policy is essential to sustain the health of our atmosphere.

The unique structure of the Montreal Protocol, emphasizing both binding commitments and compliance flexibility, has facilitated effective collaboration among diverse nations. Developing countries, granted grace periods for implementation, have received ongoing technological and financial support to transition away from harmful substances. This inclusion underscores the commitment to equity and shared responsibility in tackling global environmental issues. Moreover, the evolution of the Protocol to address emerging threats, such as climate-forcing chemical substances, showcases its adaptability to changing scientific and environmental landscapes. As nations advance in industrialization and global trade, the risk of environmental dumping poses a growing concern, complicating the objectives of the Protocol. (Andersen et al., 2018) highlight this issue, emphasizing that while developed nations work to reduce HFCs, developing countries must simultaneously navigate the challenges posed by increased industrial waste and pollution. Moving forward, the Montreal Protocol serves as a vital framework for addressing the interconnected crises of air pollution, climate change, and ozone depletion. Its foundational principles can guide future international agreements, fostering cooperation on reducing greenhouse gas emissions and promoting clean technologies. However, the potential for environmental dumping complicates these efforts, illustrating a critical need for robust legal and policy tools to prevent the exploitation of nations with less stringent regulatory frameworks. Analyzing the implications of such practices has the necessity for developing countries to champion proactive measures against unwanted waste, as they transition towards sustainable development. The path forward hinges on collective action that integrates environmental justice, equitable resource management, and continuous investment in clean technologies. Ultimately, the success of the Montreal Protocol and future agreements will depend on the global community's commitment to environmental stewardship and responsible governance (Andersen et al., 2018).

9.6.3. National Regulations and Policies

The development of national regulations and policies is essential in combating pressing environmental issues, such as air pollution and climate change. These regulations serve as a framework within which industries must operate, ensuring that harmful emissions are minimized and public health is protected.

In many cases, governments have been reluctant to impose stringent regulations until significant damage has occurred, often resulting in a reactive rather than proactive approach. For instance, as environmental degradation intensifies, nations gradually move to implement measures like cap-and-trade systems or emissions standards. The reliance on uniform rules can provide some relief, yet it often fails to address the underlying externalities associated with pollution and resource exploitation, which frequently leads to overexploitation and environmental degradation over time (Libecap, 2007). While national policies are crucial, they also need to reflect an understanding of intergenerational equity, particularly concerning climate change—a challenge that disproportionately impacts future generations. Climate change, as highlighted in previous analysis, raises significant ethical concerns regarding how current decisions affect not just contemporaneous society but also those who will inhabit the planet in years to come (Brown Weiss et al., 2008).

As environmental costs rise, the need for comprehensive regulations that prioritize long-term sustainability becomes undeniable. The crafting of such regulations necessitates a careful balancing act, as policymakers must navigate the interests of various stakeholders, including industry leaders, environmental advocates, and the public. Thus, defining policies that aim to mitigate pollution and to promote sustainable development is vital in creating an equitable future. The effectiveness of national regulations often hinges on their enforcement and adaptability. Policies must evolve in response to emerging data and trends, as static regulations can become obsolete in the face of rapid technological advancement and changing environmental conditions. Effective governance involves setting ambitious targets and ensuring compliance through penalties for violations and incentives for adherence. Furthermore, engaging local communities in the regulatory process helps to tailor policies to specific environmental issues, fostering greater accountability and shared responsibility. In this regard, the relationship between national regulations and local practices is crucial in achieving collective goals related to air quality and climate stability, ultimately demonstrating the importance of a multifaceted approach to environmental governance that blends regulation, innovation, and community involvement (Libecap, 2007).

9.7. TECHNOLOGICAL INNOVATIONS FOR MITIGATING POLLUTION

Innovations in technology play a crucial role in addressing the persistent issue of pollution, particularly in urban areas where industrial activities and vehicle emissions contribute to degraded air quality. Modern advancements such as electric vehicles (EVs) offer a promising alternative to traditional fossil fuel-powered transportation, significantly reducing greenhouse gas emissions. Additionally, the implementation of smart grid technologies allows for more efficient energy consumption and management, decreasing reliance on coal and natural gas. As cities increasingly adopt these technologies, the combined effect can lead to improved public health, as cleaner air correlates with lower rates of respiratory diseases. Addressing the challenges related to air pollution is essential not only for the current population but also for the sustainability of future generations, as highlighted by Olewilers focus on maintaining non-decreasing per capita consumption for future societal welfare (Sharpe et al., 2002). Moreover, industrial processes have also benefitted significantly from technological advancements aimed at pollution reduction.

The adoption of cleaner production techniques allows manufacturers to minimize waste and decrease emissions at the source. Technologies such as carbon capture and storage (CCS) are being integrated into power plants and other heavy industries to mitigate the release of carbon dioxide into the atmosphere. Industries can transition to more sustainable practices while simultaneously complying with stringent environmental regulations by investing in these innovative solutions. Yet, it is important to acknowledge that while these technologies offer immense potential, they also necessitate a paradigm shift in how environmental management is perceived. The traditional view that pits economic development against environmental preservation is increasingly being challenged, as integrated approaches to resource management are gaining traction (Colby, 1989). Finally, information technology enhances public engagement and awareness regarding pollution issues, increasingly influencing consumer behavior and policy decisions. Digital platforms facilitate real-time data sharing and monitoring of air quality, empowering communities to take action in advocating for cleaner environments. Furthermore, innovative applications can analyze personal energy consumption patterns, helping individuals make informed choices that contribute to pollution mitigation. As the interplay between technological advancements and public policy evolves, the hope is that these innovations will bridge the gap between economic growth and environmental stewardship. This alignment between technological progress and sustainable practices underscores the critical importance of continual investment in research and development, demonstrating that effective solutions to air pollution and climate change are necessary and achievable.

9.7.1. Renewable Energy Sources

In the ongoing battle against climate change and pollution, renewable energy sources emerge as critical alternatives to fossil fuels. These sources, including solar, wind, hydroelectric, and geothermal energy, present significant advantages in reducing greenhouse gas emissions and improving air quality. By harnessing natural processes, renewable energy generates power without the harmful byproducts associated with burning coal or oil. For instance, solar panels convert sunlight directly into electricity, offering a clean and abundant energy solution. Transitioning to renewable energy sources not only mitigates the adverse effects of climate change but can also promote sustainable development in vulnerable regions, countering the impacts of human activities that contribute to environmental degradation (Leiserowitz, 2007). The scalability of renewable technologies is a crucial factor in their potential to reshape the global energy landscape. As advancements in technology continue to drive down costs, more communities become capable of adopting these clean energy solutions.

Figure 9.10. *Renewable energy.*

Source: By Thomas Lloyd Group – Own work, CC BY-SA 4.0, https://commons.wikimedia.org/w/index.php?curid=77438927.

Implementing distributed energy systems, such as rooftop solar installations, allows for localized energy production, reducing dependency on grid systems that often rely on fossil fuels. Projects like the FLOWS initiative exemplify the importance of integrating sustainable practices into public policy and spatial planning (Leiserowitz, 2007). Stakeholders can foster community engagement and encourage broader acceptance of a cleaner energy paradigm by raising awareness about the benefits of renewable sources through innovative methods and exhibitions. This shift is essential for achieving climate resilience and preserving our planet for future generations. Finally, the transition to renewable energy sources must be paired with comprehensive policies and education to maximize their effectiveness. Governments and organizations have a responsibility to promote and support renewable energy initiatives at local, regional, and national levels. This includes investing in research and development, creating financial incentives for adoption, and implementing regulations that favor clean energy solutions. Collaborative efforts, such as public-private partnerships, can further amplify the impact of these initiatives, ensuring that communities are equipped to tackle environmental challenges sustainably (Abeyasekera et al., 2007). As the world faces unprecedented environmental crises, the transition to renewable energy stands not only as a viable alternative but as a necessary response to air pollution, climate change, and ozone depletion, facilitating a healthier planet and society (Figure 9.10).

9.7.2. Air Purification Technologies

The rising levels of air pollution worldwide have necessitated the development and implementation of advanced air purification technologies. These innovations aim to mitigate harmful pollutants and improve indoor air quality, which is essential for human health and well-being. Various methods—including mechanical filters, chemical scrubbers, and biological air purification—serve distinct functions in removing airborne contaminants. Mechanical filters target particulate matter, while chemical scrubbers neutralize gases like volatile organic compounds (VOCs) and carbon dioxide. Biological approaches utilize living organisms, such as plants and microbes, to break down pollutants. As urban areas continue to expand and industrial activities increase, these technologies stand at the forefront of efforts to combat air pollution, aligning with the broader objectives of environmental movements and legislative initiatives aimed at preserving air quality and promoting sustainability (Winter et al., 2024). Moreover, the effectiveness of these air purification technologies is often contingent upon their integration with existing environmental management strategies. For instance, wastewater treatment, which plays a crucial role in reducing aquatic pollution, can also positively impact air quality by minimizing odors and harmful emissions associated with untreated effluents (Etsuyankpa et al., 2024). As such, a comprehensive approach to environmental management must consider how various technologies work in concert.

Emphasizing synergy between air purification and water treatment could lead to more robust solutions in the fight against pollution, offering dual benefits. Additionally, public awareness of the necessity of both air and water purification technologies is vital to encouraging broader adoption within communities, ultimately fostering healthier environments and improving quality of life. Looking ahead, ongoing research and technological advancements are essential to enhancing the efficacy of air purification systems. Innovations like photocatalytic oxidation and advanced ionization processes hold promise for transforming how pollutants are managed in urban settings. Furthermore, with increasing concerns about climate change and ozone depletion, these technologies must adapt to address the shifting landscape of air quality issues.

As governments and organizations alike recognize the urgency of pollution control, investment in research and development becomes critical. These efforts will not only enhance air purification technologies but will also pave the way toward sustainable solutions that align with global environmental goals, ultimately contributing to a cleaner, healthier planet for future generations (Etsuyankpa et al., 2024; Winter et al., 2024).

9.7.3. Sustainable Transportation Solutions

The integration of sustainable transportation solutions is crucial for addressing the pressing challenges posed by air pollution and climate change. Urban areas are often major contributors to greenhouse gas emissions, primarily due to conventional transportation methods reliant on fossil fuels. Transitioning to EVs offers a promising remedy, significantly reducing emissions while enhancing energy efficiency. The development of comprehensive charging infrastructure is essential, enabling widespread adoption and minimizing range anxiety among potential users. Moreover, public transportation systems, such as buses and trains, should be electrified or powered by renewable energy sources, further curtailing the carbon footprint of transit operations. By investing in sustainable transportation technologies, cities not only improve air quality but also foster greater public health and environmental sustainability. Enhanced mobility options, including cycling and walking, also play a pivotal role in sustainable transport solutions. Creating pedestrian-friendly environments and dedicated bike lanes can reduce reliance on cars, effectively decreasing traffic congestion and lowering emissions. Initiatives like bike-sharing programs and pedestrian pathways encourage healthy lifestyles while supporting environmental goals. Integrating these options with public transport ensures that all demographics have access to efficient and eco-friendly travel methods. As a result, urban planners must prioritize these alternatives in their designs to create more sustainable cities. local governments can encourage behavioral shifts towards greener choices and significantly mitigate the impacts of air pollution by facilitating diverse transportation modes, thus supporting broader climate action objectives.

Community engagement and education are necessary to foster a culture of sustainability surrounding transportation choices. Research highlights that public interactions with the scientific community influence attitudes and behaviors related to climate issues, underscoring the need for grassroots movements to advocate for cleaner transit options (Kuo et al., 2010). By raising awareness about the environmental and health benefits of sustainable transportation, individuals may become more inclined to embrace these solutions. Furthermore, incorporating local concerns into transportation policies can lead to more effective and widely supported initiatives. Collaborative efforts between citizens, policymakers, and transportation experts are vital to creating a cohesive approach to sustainable transportation. This collective action not only addresses the urgent issues of air pollution and climate change but also cultivates a more resilient and health-conscious society, ready to tackle future environmental challenges (Figure 9.11).

Figure 9.11. *Sustainable transportation and mobility solutions.*

Source: By Michael Movchin / Felix Müller, CC BY-SA 3.0, https://commons.wikimedia.org/w/index.php?curid=27783265.

9.8. CONCLUSION

Addressing the interconnected issues of air pollution, climate change, and ozone depletion reveals a pressing need for global cooperation and strict regulatory measures. As discussed throughout this chapter, the exponential growth of international environmental agreements highlights the urgency of collective action. In a mere few decades, the number of multilateral agreements has surged to nearly hundreds, signaling a shift in how nations perceive their environmental responsibilities (Weiss et al., 1993). This growing legal framework has the potential to foster collaboration and encourage countries to adopt sustainable practices. However, the mere establishment of these agreements is not sufficient; they require earnest implementation and compliance to be effective in combating the environmental crises we face. Recent scientific findings underscore the severity of the challenges posed by climate change and its associated effects. The exploration of celestial phenomena, such as the supernova 2006gy, illustrates the immense energy outputs that can arise from astronomical events, akin to the extensive impacts of climate change on our planet (Melott et al., 2007). Such cosmic events can serve as metaphors for understanding the far-reaching consequences of human-induced environmental degradation. Just as the light and energy from a supernova can disrupt galactic surroundings, the atmospheric changes resulting from air pollution and greenhouse gas emissions can irreversibly alter ecosystems and weather patterns. This parallel reinforces the necessity for immediate action to mitigate further damage and adapt to ongoing environmental changes.

Ultimately, the conclusion drawn from examining air pollution, climate change, and ozone depletion is that urgent action is essential for the preservation of the planet and for the survival of future generations. As noted, the proliferation of international environmental law indicates an awakening to these global issues, yet the implementation of effective policies remains a challenge (Weiss et al., 1993). Nations must come together to prioritize the protection of the environment while also realizing the socio-economic benefits that accompany a transition to greener policies. Failure to act decisively can result in catastrophic implications, aligning with the consequences illustrated by cosmic events like supernovae (Melott et al., 2007). The time for complacency has passed; individuals, governments, and organizations must collaborate to forge a sustainable future for all.

9.8.1. Summary of Key Points

The intricate relationship between air pollution and climate change has profound implications for ecosystems and human health. As pollutants such as black carbon and ozone emerge as significant contributors to climate change, they also exacerbate health problems and environmental degradation. The Arctic region, in particular, is vulnerable due to shifted climate patterns influenced by both local and remote pollution sources. In this context, understanding the dynamics of air pollution becomes crucial. For instance, the PACES international initiative seeks to enhance predictive capabilities regarding the sources and impacts of Arctic air pollutants, revealing the pressing need for interdisciplinary collaboration in climate science. The effectiveness of these initiatives may hinge on developing localized strategies that address broader systemic issues in pollution management and climate resilience.

Equally important is the need to evaluate economic implications related to air pollution and climate change. The Regional Index of Sustainable Economic Wellbeing (RISEW) highlights the challenges in quantifying economic progress that accounts for environmental health and social welfare. As outlined, traditional metrics often fail to encompass the broader societal impacts of pollution. This oversight can lead to misguided policy decisions that prioritize short-term economic growth over long-term sustainability. By integrating environmental quality into economic frameworks, policymakers can better understand the trade-offs and long-term implications of their decisions. This holistic approach can ultimately align economic systems with sustainable development goals, ensuring the well-being of current and future generations. Finally, a comprehensive understanding of the effects of air pollution on climate change underscores the need for urgent action and policymaking at various levels. The intricate connections highlighted in the literature emphasize that addressing air pollution should not be viewed in isolation, as its effects ripple through various environmental and socio-economic systems. The qualitative analysis from recent studies calls for transdisciplinary approaches, merging social, economic, and environmental sciences to craft effective solutions in combating pollution. Additionally, increasing the quality and quantity of monitoring efforts can significantly aid in forming robust data sets to inform policy changes. As nations navigate the challenges of air pollution, coordinated efforts will be critical to mitigate its dual impact on climate change and public health, thereby fostering a healthier planet for all.

9.8.2. Importance of Collective Action

Collective action serves as a crucial mechanism for addressing pressing global issues such as air pollution, climate change, and ozone depletion. When individuals, organizations, and governments unite in purpose, they amplify their impact and drive meaningful change. For instance, environmental NGOs and state governments in the United States have recently pursued legal avenues to compel compliance with international agreements like the Kyoto Protocol, showcasing how collective efforts can challenge governmental inertia and promote accountability (McGee et al., 2005). Such collaboration not only raises awareness but also engages multiple stakeholders, ensuring that various perspectives and expertise contribute to finding sustainable solutions. The urgency of environmental crises necessitates a unified approach so that actions taken at local, national, and international levels can effectively align and support one another.

The shared responsibility of protecting the environment underscores the significance of collective action. In the face of complex challenges, individual efforts often fall short, while coordinated initiatives can leverage resources, knowledge, and technology for greater efficacy. For example, the development of innovative materials, such as copper-infused fabric aimed at preventing bacterial growth in clinical settings, exemplifies how interdisciplinary collaboration can yield products that enhance public health and safety (Victor et al., 1999). Similarly, addressing climate change requires a concerted effort across various sectors including industry, policy, and science. Collaborations among disparate groups can utilize their unique strengths, ultimately leading to more impactful solutions that tackle the root causes of environmental degradation rather than merely addressing the symptoms. The success of collective action relies, in part, on strong communication and trust among participants. Establishing networks that foster transparency and collaboration enables stakeholders to share best practices, overcome challenges, and celebrate successes together. In the realm of environmental protection, building these connections is essential for ensuring that communities are engaged and mobilized to tackle local and global issues. As seen in the legal battles fought by American groups against regulatory inertia, these united efforts are often necessary to challenge powerful interests and advocate for systemic changes (McGee et al., 2005). Only through a coordinated approach can society hope to effectively combat air pollution, climate change, and ozone depletion, paving the way for a healthier planet for future generations.

9.8.3. Call to Action for Future Generations

To cultivate a sustainable future, individuals of tomorrow must embrace proactive measures to combat the environmental crises that loom large today. This commitment involves advocating for policy changes and adopting lifestyle choices that minimize ecological impacts. Education plays a pivotal role in this journey; by integrating environmental science into school curricula, future generations can become informed citizens equipped with the knowledge to challenge harmful practices and promote sustainability. Initiatives such as tree planting, recycling drives, and community clean-up projects can foster a sense of stewardship toward the planet, galvanizing young people to take ownership of their surroundings. With a dedicated effort to create a culture of environmental awareness, today's youth can emerge as the catalysts for change, inspiring their peers and future adults to prioritize the health of our planet above all else. Furthermore, collective action rooted in a sense of responsibility toward Earth can amplify individual efforts, leading to substantive change. Social media, technology, and community organizations provide platforms for young advocates to voice concerns regarding air pollution, climate change, and ozone depletion. These tools can be mobilized to create campaigns, share impactful stories, and engage diverse audiences, thereby transforming public discourse on climate issues. Future generations must not only participate in these movements but also spearhead new initiatives that challenge the status quo. By fostering collaborative dialogues among scientists, policymakers, and activists, they can elevate climate action to the forefront of societal priorities. This collective resolve will empower them to form alliances that bridge gaps across various disciplines, ultimately resulting in innovative solutions to the global environmental challenges we face. In conclusion, the onus rests heavily on the shoulders of future generations to act decisively and collectively against environmental degradation. To effect lasting change, they must cultivate resilience, remain critical thinkers, and drive systemic transformation across all sectors of society. The legacy of our present choices will greatly impact the sustainability of future environments, making it crucial for young leaders to engage in advocacy and policy-making efforts. By leveraging their passions and skills, future generations can forge a new path where clean air, a stable climate, and healthy ecosystems become the norm rather than the exception. Transitioning to a more sustainable future is not just a hopeful aspiration; it is an obligation that young people must champion, ensuring that they leave behind a planet capable of sustaining life for generations to come.

REFERENCES

1. Abeyasekera, K., & Matthews, G. (2007). *Sustainable Exhibit Design: Guidelines for Designers of Small Scale Interactive and Travelling Exhibits*. University of Lincoln, School of Film and Media and Changer Agency. Retrieved from: https://core.ac.uk/download/56852.pdf(accessed on 5 December 2024).
2. Aburizaiza, O. S., Barletta, B., Blake, D. R., Blake, N. J., Gartner, A., Khwaja, H., Meinardi, S., Siddique, A., Simpson, I. J., & Zeb, J. (2014). Air quality in Mecca and surrounding holy places in Saudi Arabia during Hajj: Initial survey. *eScholarship, University of California*. Retrieved from: https://core.ac.uk/download/287624244.pdf (accessed on 5 December 2024).
3. Andersen, S. O., Carvalho, S., Ferris, R., Gonzalez, M., Picolotti, R., & Zaelke, D. (2018). Defining the legal and policy framework to stop the dumping of environmentally harmful products. *Duke University School of Law*. Retrieved from: https://core.ac.uk/download/213020936.pdf (accessed on 5 December 2024).
4. Arnold, S. R., Baklanov, A., Brock, C. A., Dibb, J. E., Eckhardt, S., Fidel, M., et al., (2016). *Arctic Air Pollution: Challenges and Opportunities for the Next Decade*. University of New Hampshire Scholars' Repository. *Elementa, 4*(104). https://doi.org/10.12952/journal.elementa.000104. Retrieved from: https://core.ac.uk/download/215538408.pdf (accessed on 5 December 2024).
5. Boyce, J. (n.d.). Green and brown? Globalization and the environment. Retrieved from: https://core.ac.uk/download/pdf/6548979.pdf (accessed on 5 December 2024).
6. Brown Weiss, E. (1993). *International Environmental Law: Contemporary Issues and the Emergence of a New World Order*. Scholarship @ Georgetown Law. Retrieved from: https://core.ac.uk/download/70375508.pdf (accessed on 5 December 2024).
7. *ChemMatters: Special Issue on NASA's EOS Aura Mission (Sept. 2001)*. Retrieved from: https://core.ac.uk/download/pdf/5221085.pdf (accessed on 5 December 2024).
8. Colby, M. E. (1989). *Environmental Management in Development: The Evolution of Paradigms*. The World Bank Strategic Planning and Review Department Policy Planning and Research Staff.
9. Cuevas, C. A., Fernandez, R. P., Kinnison, D. E., Lamarque, J. F., Prados Roman, C., & Saiz-Lopez, A. (2015). A negative feedback between anthropogenic ozone pollution and enhanced ocean emissions

of iodine. *Copernicus GmbH*. Retrieved from: https://core.ac.uk/download/288802299.pdf (accessed on 5 December 2024).

10. Danu, M. C. (2010). Causal relations between natural risk and business risk. *Annals of Faculty of Economics, University of Oradea, Faculty of Economics, 1*(2), 242–248.

11. *Environment and Global Climate Change*. University of the Arctic. Retrieved from: https://core.ac.uk/download/pdf/5210824.pdf (accessed on 5 December 2024).

12. Etsuyankpa, B., Augustine, A., Tanko, M., Mathew, J., Haruna, I., Adamu Makanta, S., Salihu, M., & Mamman, A. (2024). An overview of wastewater characteristics, treatment, and disposal: A review. *Journal of Applied Sciences and Environmental Management, 28*, 1553–1572. https://doi.org/10.4314/jasem.v28i5.28.

13. Hollas, C. E., do Amaral, K. G. C., Lange, M. V., Higarashi, M. M., Steinmetz, R. L. R., Mariani, L. F., Nakano, V., Sanches-Pereira, A., de Martino Jannuzzi, G., & Kunz, A. (2024). Livestock waste management for energy recovery in Brazil: A life cycle assessment approach. *Environmental Science and Pollution Research International, 31*(3), 4705–4720. https://doi.org/10.1007/s11356-023-31452-1.

14. Khallaf, M. (2011). *The Impact of Air Pollution on Health, Economy, Environment, and Agricultural Sources*. InTech. https://doi.org/10.5772/1000.

15. Kuo, S. Y. (2010). *The Public Understanding of Climate Change: A Case Study of Taiwanese Youth*. Digital Commons @ NJIT. Retrieved from: https://core.ac.uk/download/232274196.pdf (accessed on 5 December 2024).

16. Leiserowitz, A. (2007). *International Public Opinion, Perception, and Understanding of Global Climate Change*. Human Development Occasional Papers (1992–2007) HDOCPA-2007-31, Human Development Report Office (HDRO), United Nations Development Programme (UNDP).

17. Libecap, G. (2007). *Open-Access Losses and Delay in the Assignment of Property Rights*. National Bureau of Economic Research, Inc, NBER Working Papers, 50.

18. Martens, W. J. (1998). *Health Impacts of Climate Change and Ozone Depletion: An Ecoepidemiologic Modeling Approach*. Retrieved from: https://core.ac.uk/download/pdf/4017579.pdf (accessed on 5 December 2024).

19. McGee, H. (2005). *Litigating Global Warming: Substantive Law in Search of a Forum*. Seattle University School of Law Digital Commons. Retrieved from: https://core.ac.uk/download/235983186.pdf (accessed on 5 December 2024).

20. Melott, A. L., Anthony-Twarog, B. J., Thomas, B. C., Fields, B. D., & Predehl, P. (2007). Superluminous supernovae: No threat from Eta Carinae. *Mary Ann Liebert Inc*. Retrieved from: http://arxiv.org/abs/0705.4274 (accessed on 5 December 2024).

21. New Economics Foundation. (2005). *Interim Report on the Feasibility of Constructing a Regional Measure of Domestic Progress (MDP) for East Midlands Development Agency*. East Midlands Development Agency. Retrieved from: https://core.ac.uk/download/30624484.pdf (accessed on 5 December 2024).

22. Ollinger, S. V., Aber, J. D., & Reich, P. B. (1997). Simulating ozone effects on forest productivity: Interactions among leaf-, canopy-, and stand-level processes. *Ecological Applications, 7*(6), 1237–1251. https://doi.org/10.1890/1051-0761(1997)007[1237:SOEOFP]2.0.CO;2 (accessed on 5 December 2024).

23. Perelet, R. (2007). Interlinkages: Governance for sustainability. In K. Meyer (Ed.), *Global Environmental Outlook (GEO-4)*. UNEP. Retrieved from: https://core.ac.uk/download/195696507.pdf (accessed on 5 December 2024).

24. Prados-Roman, C., Cuevas, C. A., Fernandez, R. P., Kinnison, D. E., Lamarque, J. F., & Saiz-Lopez, A. (2015). A negative feedback between anthropogenic ozone pollution and enhanced ocean emissions of iodine. *Atmospheric Chemistry and Physics, 15*(4), 2215–2224. https://doi.org/10.5194/acp-15-2215-2015.

25. Sand, P. H., & Wiener, J. B. (2016). *Towards a New International Law of the Atmosphere?* Duke University School of Law. Retrieved from: https://core.ac.uk/download/62571126.pdf (accessed on 5 December 2024).

26. Sharpe, A., St-Hilaire, F., Banting, K., & Olewiler, N. (2002). *The Review of Economic Performance and Social Progress 2002: Towards a Social Understanding of Productivity*. Retrieved from: https://core.ac.uk/download/pdf/7033280.pdf (accessed on 5 December 2024).

27. Spadaro, J. V. (2011). *The Uniform World Model: A Methodology for Predicting the Health Impacts of Air Pollution*. BC3 Working Paper Series 2011-12.

28. Uz, S. S., Ruane, A. C., Duncan, B. N., Tucker, C. J., Huffman, G. J., Mladenova, I. E., Osmanoglu, B., Holmes, T. R. H., McNally, A., Peters-Lidard, C., Bolten, J. D., Das, N., Rodell, M., McCartney, S., Anderson,

M. C., & Doorn, B. (2019). Earth observations and integrative models in support of food and water security. *Remote Sensing Earth Systems Science, 2*(1), 18–38. https://doi.org/10.1007/s41976-019-0008-6.

29. Victor, D. G. (1999). *Enforcing International Law: Implications for an Effective Global Warming Regime*. Duke University School of Law. Retrieved from: https://core.ac.uk/download/62547055.pdf (accessed on 5 December 2024).
30. Welch, G. D. (2017). HFC smuggling: Preventing the illicit (and lucrative) sale of greenhouse gases. *Digital Commons @ Boston College Law School*. Retrieved from: https://core.ac.uk/download/83100966.pdf (accessed on 5 December 2024).
31. Winter, J. (2024). *A Historical Perspective Review of the Environmental Pollution*. Insights Publisher. Retrieved from: https://core.ac.uk/download/618266480.pdf (accessed on 5 December 2024).

CHAPTER 10

CONVENTIONAL AND SUSTAINABLE ENERGY

Contents

INTRODUCTION

The debate surrounding energy sources has never been more urgent, as societies grapple with the complex interdependencies between environmental sustainability and economic growth. The conventional forms of energy, primarily fossil fuels such as coal, oil, and natural gas, have powered industrial progress for centuries, fostering unprecedented development and convenience. However, the environmental consequences of these energy sources have prompted widespread concern, particularly in light of climate change. As emissions from burning fossil fuels continue to rise, alternative solutions must be pursued to mitigate the adverse impacts on the planet and future generations. By examining both conventional and sustainable energy sources, we can better understand their respective roles in our energy landscape and the challenges we face in transitioning to a more sustainable future. Conventional energy production is characterized by its extensive infrastructure and established technologies that harness finite resources, often leading to economic stability and job creation. These energy industries are deeply entrenched in global markets, providing a reliable supply to meet growing demand. Yet, the volatility of fossil fuel markets and geopolitical tensions underscore the vulnerabilities associated with this energy dependence.

Additionally, environmental degradation, health impacts, and social inequities further complicate the narrative surrounding conventional energy. As awareness of these issues grows, stakeholders—from policymakers to citizens—are increasingly calling for a shift towards renewable energy sources that promise reduced environmental footprints and a more sustainable energy future. Understanding the inherent challenges of conventional energy allows for a more informed discussion on how to integrate more sustainable practices effectively. The transition to sustainable energy is not merely a technological challenge; it also encompasses economic, social, and political dimensions. Renewable energy sources like solar, wind, and hydroelectric power offer promising alternatives with significantly lower greenhouse gas emissions. However, scaling these technologies requires not only substantial investment but also supportive policies and public buy-in. While the potential benefits—such as job creation in new sectors and energy independence—are substantial, the path is fraught with obstacles, including infrastructure requirements and energy storage challenges. Ultimately, the balancing act between maintaining reliable energy supplies and advancing sustainability necessitates a nuanced approach that embraces innovation, interdisciplinary collaboration, and adaptive policies. The pursuit of a more sustainable energy framework invites critical engagement, demanding that we weigh the short-term conveniences of conventional energy against the long-term health of our planet and society.

10.1. DEFINITION OF CONVENTIONAL ENERGY

Conventional energy primarily encompasses the traditional sources of energy that have dominated global energy production for decades. These sources include fossil fuels such as coal, oil, and natural gas, which are extracted from the Earth and utilized for a variety of purposes, including electricity generation, heating, and transportation. The extraction and combustion of these fuels release significant amounts of greenhouse gases, contributing to climate change and various environmental degradation issues. As a result, while conventional energy has played a crucial role in industrial development and economic growth, its long-term viability is increasingly questioned due to its environmental impact and the finite nature of available resources. This brings into stark focus the need for a shift towards more sustainable energy practices, driven by growing public concern over climate issues and energy security. The dependence on conventional energy sources poses systemic vulnerabilities that can undermine energy security. As highlighted by many studies, these interconnected systems often lead to increased vulnerability during disruptions caused by geopolitical tensions, resource depletion, or natural disasters (Adger et al., 2010).

Furthermore, the fossil fuel industry is subject to market fluctuations that can destabilize economies reliant on these energy sources. In contrast, adopting localized and decentralized energy approaches may enhance resilience against such vulnerabilities by diversifying energy sources and minimizing dependency on fossil fuels. Innovative development driven by user-focused approaches, such as the energy aspects of Private Finance Initiative (PFI) projects, reflects an emerging trend to address these vulnerabilities. Thus, understanding conventional energy within this context reveals the urgent need for a paradigm shift to mitigate inherent risks. The conversation surrounding conventional energy necessitates a deeper examination of its life cycle impacts, particularly about sustainable alternatives. The life cycle assessment of buildings utilizing sustainable and renewable technologies indicates that the long-term benefits often outweigh the initial costs associated with transitioning from conventional energy. Costs incurred during the construction phase can be recouped through significant energy savings, improved environmental quality, and lower overall operational expenses. By critically analyzing the implications of continuing reliance on conventional energy versus adopting innovative, sustainable solutions, stakeholders can better appreciate the financial, environmental, and social advantages inherent in repositioning their energy strategies. Ultimately, the transition from conventional to sustainable energy is not merely an environmental imperative but a necessary economic evolution for a resilient, future-minded society (Hartungi et al., 2009).

10.2. DEFINITION OF SUSTAINABLE ENERGY

The concept of sustainable energy revolves around harnessing natural resources in a manner that balances ecological preservation with human energy needs. It seeks to ensure that energy production and consumption do not deplete resources or cause environmental harm, thereby safeguarding these resources for future generations. Sustainable energy aims to reduce our reliance on fossil fuels by integrating renewable energy sources—such as solar, wind, and hydropower—into the energy mix, which contribute significantly to greenhouse gas emissions and climate change. This shift addresses the urgent need for energy transition and promotes energy security, minimizing vulnerabilities associated with fossil fuel dependency. As tensions between centralized energy systems and localized approaches become evident, it is crucial for energy policy to prioritize sustainable practices to achieve long-term resilience and environmental stewardship (Adger et al., 2010).

Examining the economic value of sustainable energy systems reveals that these approaches can often yield significant long-term savings. Life Cycle Costing (LCC) provides a framework for assessing the total cost of energy technologies over their lifetime, including initial investment, operational, and maintenance costs. This analysis has shown that while the upfront costs of implementing sustainable technologies may initially seem high, the subsequent savings from energy efficiency and conservation often outweigh these expenses (Hartungi et al., 2009). Furthermore, enhanced indoor environmental quality and improved worker productivity can emerge as additional benefits, creating a compelling case for sustainable energy adoption. Thus, understanding the broader economic implications of sustainable energy is crucial in making informed decisions about resource allocation and energy policy. Ultimately, the transition to sustainable energy is not just an environmental imperative; it is also a catalyst for social and economic development. As communities increasingly adopt renewable energy practices, they foster local job creation and energy independence, which can mitigate social issues such as fuel poverty. Furthermore, by investing in innovative energy solutions, cities, and regions can enhance their resilience against climate-related disruptions, thus ensuring a more secure and stable energy future. Integrating sustainable energy into urban planning and infrastructure development holds the potential to address pressing systemic vulnerabilities while engendering a collective responsibility toward environmental stewardship. The dual focus on sustainability and resilience not only supports individual and community well-being but also strengthens the overall economy, making the adoption of sustainable energy a viable and necessary pathway forward.

10.3. IMPORTANCE OF THE ENERGY DEBATE IN CONTEMPORARY SOCIETY

Contemporary society is experiencing an unprecedented shift in its energy landscape, driven by pressing environmental concerns and technological advancements. The debate surrounding energy sources and policies is crucial as it shapes not only national strategies but also global dynamics. The divergence in civil nuclear policies between Germany and the UK exemplifies how political decisions impact energy frameworks (Aliseda et al., 2015). The commitment to phase out nuclear power by 2022 in Germany reflects a broader societal consensus on prioritizing sustainability, whereas the UK's push for a nuclear renaissance invokes discussions about energy security and environmental implications. This contrast underscores the necessity of ongoing dialogues about energy choices, which are fundamental to addressing climate change and promoting a transition towards sustainable energy practices. Furthermore, the energy debate also encompasses the complex interplay between economic, political, and social factors, influencing public policies and corporate strategies. A pivotal aspect of this discussion is what researchers call flex crops and commodities, which represents a growing recognition of the adaptability required in agriculture and energy sectors amid crises (Borras et al., 2014). The dual concepts of multiple-ness and flexible-ness beckon critical examination of how current practices can evolve to meet future challenges. Understanding these dynamics is essential, as they reveal how energy systems can respond to fluctuations in demand, environmental policies, and international markets. The implications of this flexibility are significant, as they can lead to more resilient energy frameworks capable of addressing the complexities of contemporary society. Lastly, exploring the energy debate illuminates a path toward innovative solutions and collaborative approaches to energy challenges. Societies can create frameworks that reflect diverse values and priorities by emphasizing the importance of democratic engagement in energy policymaking (Aliseda et al., 2015). This engagement is imperative for fostering public trust and accountability, especially as decisions about energy infrastructure resonate with wider societal impacts. The evolving narrative around energy, including the discourse surrounding flex policy narratives calls for transparency and inclusivity within strategic planning. The role of social movements and community involvement in shaping energy policy is vital, driving forward the collective aspiration for a sustainable and just energy future, demonstrating that energy debates are not merely technical discussions but fundamental to social equity and environmental stewardship (Borras et al., 2014).

10.4. OVERVIEW OF CONVENTIONAL ENERGY SOURCES

The prevailing conventional energy sources play a pivotal role in powering the global economy. Primarily, fossil fuels, including coal, oil, and natural gas, constitute the backbone of energy production worldwide. These resources are favored for their high energy density and established extraction and distribution infrastructures. However, the environmental repercussions of fossil fuel consumption are increasingly drawing scrutiny. Combustion of these fuels contributes to greenhouse gas emissions, significantly impacting climate change and air quality. As concerns about sustainability intensify, there is a vital need for a comprehensive evaluation of how conventional energy sources influence not only economic growth but also environmental health, underscoring the importance of integrating cleaner alternatives into our energy mix.

In addition to fossil fuels, nuclear energy represents another significant conventional energy source, lauded for its capacity to generate large amounts of electricity with relatively low greenhouse gas emissions. Nuclear power plants use uranium as fuel, harnessing the energy released during nuclear fission to produce electricity. This source is often praised for its reliability and efficiency, capable of providing baseload power that can stabilize the grid. However, issues surrounding nuclear waste disposal, potential accidents, and public perception continue to challenge its acceptance. The long-term implications of nuclear energy, alongside fossil fuels, prompt critical discussions on sustainable energy management. As highlighted in existing literature, understanding the life cycle and ecological impacts of different energy sources, including conventional ones, is crucial for developing strategies that balance energy needs with environmental stewardship (Ekbladh et al., 2002). While conventional energy sources have historically dominated the energy landscape, the growing emphasis on sustainability and renewable alternatives poses a pressing challenge to their future viability. As awareness of climate change grows, many nations are transitioning toward cleaner energy systems. This shift is not merely about reducing carbon footprints; it also involves recognizing the long-term costs associated with conventional energy reliance, including health impacts and ecological degradation. For instance, the life cycle costing of conventional versus sustainable technologies reveals stark contrasts in potential savings and benefits, particularly in energy conservation and environmental quality (Hartungi et al., 2009). Ultimately, the transition from conventional to sustainable energy sources will require collaborative efforts across sectors to create policies that promote innovation and ecological responsibility, ensuring a more sustainable energy future for generations to come.

10.4.1. Fossil Fuels: Coal, Oil, and Natural Gas

The reliance on fossil fuels—coal, oil, and natural gas—has long been a cornerstone of global energy production, but the implications of this dependence create a complex narrative. As nations industrialize, they tend to increase their consumption of fossil fuels to meet energy demands. However, this reliance poses significant environmental challenges due to the release of greenhouse gases that contribute to climate change.

The need for a shift towards energy sources that yield lower carbon emissions is becoming increasingly urgent. In the context of this crisis, the conventional significance of fossil fuels cannot overshadow the pressing necessity for sustainable alternatives that promote long-term ecological health, fostering a balance between energy needs and environmental stewardship. The projected production peaks of fossil fuels suggest an impending transition that may not only reshape energy landscapes but also influence economic stability.

For instance, recent analyses indicate that coal production in China is expected to peak around 2020 at significant levels, while oil is anticipated to reach its zenith shortly after. As noted in research, these declines in production correlate with diminishing energy return on investment (EROI), particularly in coal and conventional oil and gas sectors (Gupta et al., 2017). This trend signals that continued reliance on these fuels may soon become untenable, highlighting an urgent need for diversification in energy sourcing.

The anticipated gaps between supply and demand underscore the necessity for robust policymaking focused on energy efficiency and alternative resources, as dependence on imported fossil fuels could introduce risks of volatility in energy security and economic fluctuations. Efforts to transition to renewable energy sources thus gain importance as we grapple with the realities of fossil fuel constraints. The exploration of alternatives such as wind, solar, and biomass aims to reduce environmental impact and offers potential for sustainable economic growth. According to studies, this transition is not merely technical; it involves navigating political, social, and economic landscapes to foster broad acceptance and integration of cleaner technologies.

The potential for innovation within the renewable energy sector provides opportunities for job creation and technological advancement, equipping societies to meet both their energy needs and climate goals. A strategic shift towards these sustainable alternatives can pave the way for cleaner air and a more resilient planet, underscoring the critical need to reevaluate our energy paradigms in the face of finite fossil fuel resources (Elliott et al., 2003) (Figure 10.1).

Figure 10.1. *Natural gas processing.*

Source: By C.Stadler/Bwag – Own work, CC BY-SA 4.0, https://commons.wikimedia.org/w/index.php?curid=109396940.

10.4.2. Nuclear Energy and Its Role

Nuclear energy continues to play a pivotal role in the global energy landscape, particularly in the transition toward a low-carbon future. As nations grapple with the pressing need to reduce greenhouse gas emissions, nuclear power emerges as a reliable alternative to fossil fuels, providing a significant amount of baseload electricity with minimal carbon output. Research suggests that a balanced energy portfolio, including nuclear alongside renewable sources, is essential for meeting stringent climate targets. For example, the Hydrogen Council has set ambitious goals, predicting a potential shift in the energy system where nuclear technology can facilitate hydrogen production—an area that is crucial for decarbonizing sectors such as transportation and industry (Yue et al., 2021). This dual capability not only underscores nuclear energy's versatility but also highlights its contribution to the sustainability agenda. In addition to its environmental advantages, the economic viability of nuclear energy offers compelling reasons for its continued utilization. Despite challenges related to public perception and safety concerns, advancements in reactor technology and regulatory frameworks have made nuclear plants safer and more efficient.

Countries exploring hydrogen production through nuclear energy demonstrate its growing importance in the energy market. As noted, hydrogen can significantly assist in balancing the intermittency of renewable sources like wind and solar, thereby enhancing overall energy security (Yue et al., 2021). The ability of nuclear power to operate at high capacity factors ensures a consistent energy supply, which is essential for economic stability and energy affordability in a rapidly changing market. Energy policies around the world are increasingly recognizing the necessity of incorporating nuclear energy into their strategies. The shift toward cleaner energy sources demands a comprehensive approach, balancing the needs for energy security, economic growth, and environmental sustainability. As fluctuating prices of renewable technologies create uncertainty, the stability offered by nuclear power serves as an anchor for energy policy (Chapman et al., 2020). Furthermore, collaboration between governments, energy producers, and technology developers can foster innovations that enhance the safety and efficiency of nuclear power plants. By embracing nuclear energy's potential, countries can position themselves at the forefront of the clean energy transition, ensuring a more sustainable and resilient energy future that aligns with global climate goals (Figure 10.2).

Figure 10.2. *Nuclear energy.*

Source: By Hansueli Krapf – Own work: Hansueli Krapf (User Simisa (talk · contribs)), CC BY-SA 3.0, https://commons.wikimedia.org/w/index.php?curid=110649554.

10.4.3. Environmental Impacts of Conventional Energy

The reliance on conventional energy sources, primarily fossil fuels, has profound environmental repercussions that manifest at multiple levels. These energy sources contribute significantly to greenhouse gas emissions, leading to climate change and its associated adverse effects on ecosystems and human societies. The combustion of coal, oil, and natural gas not only releases carbon dioxide but also emits pollutants such as sulfur dioxide and nitrogen oxides, which are responsible for acid rain and respiratory diseases.

Additionally, the extraction processes for these fuels often result in habitat destruction, water contamination, and soil degradation, further exacerbating biodiversity loss. In essence, the environmental footprint of conventional energy extends beyond just emissions, as it encompasses a wide range of ecological impacts that threaten planetary health and sustainability. A crucial concern in the discourse on conventional energy is its role in resource depletion and land use conflicts. The extraction of fossil fuels, such as through drilling and mining, often requires substantial land and water resources, leading to significant ecological disruption.

For instance, a report underscores the environmental hotspots associated with agricultural practices that depend on conventional energy, particularly in irrigation systems that draw heavily on groundwater (Penning Rowsell et al., 2021). This strains local water supplies and impacts agricultural sustainability, creating a paradox of intensifying energy use while degrading environmental conditions. Furthermore, the competing interests between energy production and other land uses, such as agriculture or conservation, can lead to conflicts and further environmental degradation, emphasizing the need for a balanced approach to resource management.

Addressing the environmental impacts of conventional energy necessitates a multifaceted strategy that encourages the transition to sustainable alternatives. While immediate shifts may pose challenges, integrating renewable energy sources can significantly mitigate the adverse effects associated with fossil fuels. For example, Zira et al (2022) highlight how changing energy systems and practices in livestock production can lower negative impacts, exemplifying the potential of innovative approaches to improve environmental sustainability. By investing in renewable technologies, promoting energy efficiency, and rethinking consumption patterns, it is possible to break the cycle of dependency on conventional energy sources. Ultimately, transitioning to sustainable energy not only responds to the environmental crises generated by conventional energy systems but also revitalizes communities and economies poised for a healthier, more sustainable future.

10.5. OVERVIEW OF SUSTAINABLE ENERGY SOURCES

The rising concern over environmental sustainability has shifted the focus toward renewable energy sources, which offer a viable solution to the challenges posed by conventional energy systems. Not only do these sources—such as solar, wind, and hydroelectric power—significantly reduce carbon emissions, but they also provide a sustainable alternative to fossil fuels. As the electricity demand continues to grow, particularly with the proliferation of technologies like the Internet of Things (IoT), exploring sustainable energy solutions becomes critical. Innovations in energy capture and storage are essential for maximizing the potential of renewable sources, ensuring a reliable supply of electricity without compromising environmental health. Efforts such as the deployment of ultra-low power IoT devices powered by sustainable radio frequency wireless energy transfer highlight the intersection of technology and sustainable practices, indicating a shift toward a more integrated approach to energy generation and usage (Alves et al., 2023). Furthermore, the integration of sustainable energy sources involves not just the technology itself but the socio-economic factors that dictate their successful implementation. Achieving a transition to renewable energy systems requires a balance between economic prosperity, social equity, and environmental health. While renewable technologies are increasingly economically viable, policy frameworks and community engagement play vital roles in their adoption and success.

The move toward sustainable energy sources is as much about societal buy-in and legislative support as it is about technical advancements. As discussed in the introductory chapter of the book "Future Electricity Technologies and Systems," understanding these interconnections is essential for formulating comprehensive policies that promote sustainable energy development (Jamasb et al., 2005). Only through a concerted effort to harmonize these elements can we ensure a sustainable energy future.

The exploration of sustainable energy sources also poses unique challenges that need to be addressed to maximize their potential. One major concern is the integration of renewable resources into existing energy grids, which can be complex due to their variable nature. Energy storage technologies, such as batteries and pumped hydroelectric storage, are crucial to mitigating these challenges by providing backup power and stabilizing supply. However, the economic implications of these technologies must be carefully analyzed to ensure that they contribute to the overall efficiency of the energy system. As highlighted in discussions surrounding current energy-saving technologies, merely shifting to renewable energy is not enough; we also need to adapt our consumption patterns and develop infrastructure that can accommodate these changes (Alves et al., 2023). We can pave the way for a robust and resilient sustainable energy future by addressing these issues head-on.

10.5.1. Solar Energy and Its Applications

Harnessing solar energy is essential for transitioning to sustainable energy systems, as it provides a clean and renewable source of power. With the advancement of technology, solar panels have become more efficient and accessible, allowing homes and businesses to generate their own electricity. This distributed energy model reduces reliance on fossil fuels and minimizes greenhouse gas emissions, contributing to environmental conservation. Moreover, solar energy systems can be easily integrated into various applications, from residential rooftops to large-scale solar farms, illustrating their versatility. As communities become more aware of energy consumption and environmental impacts, the adoption of solar technology represents a significant shift toward sustainable practices that are crucial for addressing climate change. In recent years, solar energy applications have expanded into agriculture, enhancing food security and economic sustainability.

Innovative solutions, such as IoT-based solar-powered systems for mushroom cultivation, exemplify how renewable energy can be integrated into farming practices. These systems monitor environmental conditions to optimize crop growth, demonstrating the potential of solar power to increase agricultural productivity while minimizing resource use. The integration of technology, such as sensors connected to the Blynk IoT application, allows farmers to manage their operations efficiently, ensuring optimal conditions are maintained for the growth of mushrooms and other crops. This approach supports farmers in achieving higher yields and provides an alternative source of income, thereby strengthening rural economies and promoting sustainable development. Solar energy has that reliability extends beyond agricultural applications, influencing how we approach urban infrastructure and smart cities. Cities can reduce their carbon footprint and enhance energy efficiency by incorporating solar technologies into transportation systems, such as solar-powered electric vehicles and charging stations. Additionally, implementing solar energy in building designs can significantly lower energy costs while improving residents' quality of life. The increasing use of renewable energy sources, particularly solar energy, is crucial in efforts to create interconnected and automated urban environments. As demonstrated in various studies, the synergy between energy and technology opens up new possibilities for sustainable development, as communities embrace innovative solutions to climate challenges (Mahvish Konain et al., 2023; Olubunmi Adeolu Adenekan et al., 2024). We take an important step toward a more sustainable, equitable, and resilient future by investing in solar energy applications (Figure 10.3).

Figure 10.3. *Solar panels: An array of solar panels convert sunlight to electricity.*

Source: By AleSpa – Own work, CC BY-SA 3.0, https://commons.wikimedia.org/w/index.php?curid=29290121.

10.5.2. Wind Energy and Its Potential

Utilizing wind energy presents a transformative opportunity for both energy production and environmental sustainability. As one of the most abundant renewable resources, wind can be harnessed to generate electricity on a scale capable of meeting the needs of growing populations while significantly reducing greenhouse gas emissions. Wind turbines convert kinetic energy from wind into mechanical power, which is then transformed into electricity. The scalability of wind energy systems, ranging from small home installations to extensive wind farms, allows diverse applications across urban and rural settings. This versatility enhances its accessibility, making it a practical solution for energy demands globally. The financial viability of wind energy has increased notably due to advancements in technology and a decrease in capital costs. As the report by the Green Growth Action Alliance indicates, a significant gap exists in green investment, which presents a challenge for broader adoption. Nevertheless, wind energy continues to demonstrate compelling returns on investment, leading to increased interest from policymakers and financial institutions. (World Economic Forum, 2013).

Furthermore, as clean-energy technologies mature, wind energy becomes more competitive with traditional fossil fuels, contributing to energy independence and security for nations increasingly reliant on imported oil or gas. Public and private investments in wind energy infrastructure can stimulate local economies while fostering a sustainable energy future. Moreover, the global implications of widespread wind energy adoption extend to social and environmental benefits that are essential for achieving sustainability goals. It holds the promise not only to reduce fossil fuel dependency but also to create jobs and enhance energy access in underserved communities. Renewable energy options, including wind, are observed to be viable for poverty alleviation in developing regions, according to various studies (Worldwatch Institute, 2005). With careful planning and integration into energy systems, wind energy can help mitigate climate change and bolster resilience against its effects. The collective engagement of stakeholders—governments, financial bodies, and communities—is crucial to realizing the full potential of wind energy and transitioning towards a more sustainable energy paradigm (Figure 10.4).

Figure 10.4. *Brazos wind farm in Texas.*

Source: https://upload.wikimedia.org/wikipedia/commons/8/8b/GreenMountain-WindFarm_Fluvanna_2004.jpg.

10.5.3. Biomass and Geothermal Energy

The potential of biomass energy as a renewable resource has gained significant attention in the context of sustainable energy solutions. Biomass, which includes organic materials such as plant matter, agricultural residues, and waste, can be converted into biofuels, electricity, and heat. Utilizing biomass reduces reliance on fossil fuels and minimizes waste, effectively transforming unusable materials into energy. This conversion process can be executed through various methods, including combustion, fermentation, and anaerobic digestion. The versatility of biomass stands out, as it can contribute to energy needs while also engaging in environmental stewardship—addressing both energy consumption and waste management challenges. However, it is crucial to develop sustainable practices for biomass production to prevent negative impacts on food supply and biodiversity, ensuring that the pursuit of greener energy sources does not inadvertently exacerbate other ecological issues.

Geothermal energy presents another viable alternative in the realm of renewable energy sources, capturing the Earth's internal heat for a range of applications. By tapping into geothermal reservoirs located beneath the Earth's surface, this energy source offers a consistent and reliable power supply that is not subject to the fluctuations experienced by solar and wind energy. Enhanced geothermal systems (EGS) further optimize this potential by facilitating the extraction of heat from hot, dry rocks where water can be injected and vaporized. The stability of geothermal energy makes it particularly appealing for baseload power generation, which is essential for meeting continuous electricity demands in urban centers.

As countries strive to meet targets outlined in global climate agreements, such as the Paris Agreement, expanding geothermal energy infrastructure can play a pivotal role in reducing greenhouse gas emissions while providing sustainable energy solutions (Samer Fawzy et al., 2020). Further integrating biomass and geothermal energy into conventional energy frameworks can facilitate a smoother transition towards a more sustainable energy landscape. Both of these resources address the critical challenge of reducing carbon emissions while ensuring energy security. As methane, the second most significant greenhouse gas, shores up substantial contributions to climate change, understanding, and managing sources of emissions becomes paramount (Marielle Saunois et al., 2020). Biomass can help sequester carbon, while geothermal power operates with minimal emissions.

By synergizing these renewable sources, policymakers, and energy planners can diversify energy portfolios, enhance public acceptance, and mobilize technological advancements to optimize energy generation. Ultimately, the blend of biomass and geothermal technology can play a crucial role in combating climate change and moving towards a carbon-neutral future, especially when bolstered by supportive legislation and community engagement (Figure 10.5, 10.6).

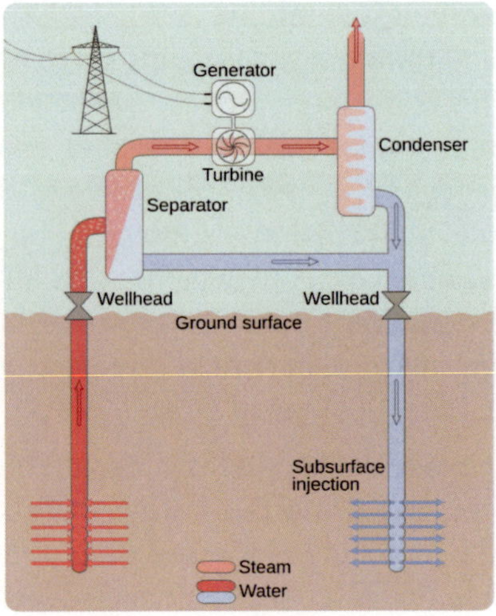

Figure 10.5. *Geothermal energy.*

Source: By Goran tek-en, CC BY-SA 4.0, https://commons.wikimedia.org/w/index.php?curid=31077638.

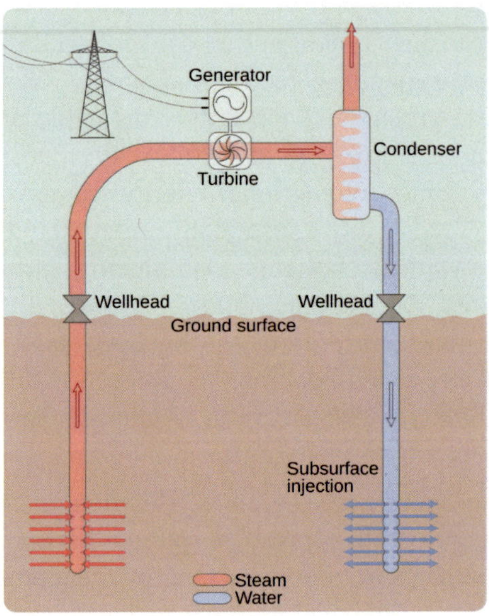

Figure 10.6. *Enhanced geothermal systems.*

Source: By Goran tek-en, CC BY-SA 4.0, https://commons.wikimedia.org/w/index.php?curid=31077637.

10.6. ECONOMIC IMPLICATIONS OF ENERGY SOURCES

The diversification of energy sources is critical for promoting economic stability and resilience. Traditional reliance on fossil fuels can lead to severe economic vulnerabilities, particularly for net energy exporting countries like Nigeria, which depend heavily on oil and gas revenue for government funding. As renewable energy consumption rises in developed net energy importing countries, it negatively impacts oil exports from NEECs, as evidenced by the findings of recent studies, which illustrate a significant reduction in revenue from these traditional sources (Hassan et al., 2018).

This evolving landscape compels exporting nations to explore alternative economic avenues, emphasizing the necessity for a versatile energy approach that fosters both growth and security in an increasingly competitive global market. Transitioning to renewable energy sources not only enhances ecological sustainability but also presents significant economic prospects. The exploration and deployment of renewable technologies such as wind, solar, and biomass can stimulate job creation and economic diversification. Investment in these sectors often leads to reduced energy costs over time, making businesses more competitive and improving overall economic output.

Additionally, sustainable energy resources contribute less to pollution and greenhouse gas emissions, fostering healthier communities and potentially lowering healthcare costs associated with pollution-related illnesses (Maghear Diana, 2011). As countries endeavor to meet global environmental standards and commitments, the economic implications of this shift will be crucial in ensuring long-term viability and conformity with sustainable development goals. In conclusion, the economic implications of energy sources are multifaceted and deeply interconnected with both local and global markets.

As the balance shifts from conventional fossil fuels to more sustainable energy options, the necessity for strategic policy formulations becomes evident. Policymakers need to develop frameworks that support renewable energy investments and innovations, ensuring that these transitions do not come at the cost of economic stability. By harnessing the vast potential of renewable resources while simultaneously addressing the challenges posed by diminishing reliance on fossil fuels, countries can create robust economies poised to thrive in a sustainable future (Hassan et al., 2018; Maghear Diana, 2011). Thus, embracing these alternative energy strategies is essential for environmental integrity and fostering resilient economies that can weather the uncertainties of a changing energy landscape.

10.6.1. Cost of Extraction and Production of Conventional Energy

The extraction and production of conventional energy, such as coal, oil, and natural gas, entails significant financial investment and operational costs. This includes not only the expenses associated with drilling and mining techniques but also the environmental compliance measures and regulation adherence necessary in many regions. The technological advances in extraction methods have led to more efficient production; however, these innovations often require substantial initial capital. Moreover, as easily accessible reserves deplete, companies are increasingly compelled to engage in more challenging and expensive extraction projects, such as offshore drilling or hydraulic fracturing. Thus, the economic viability of these endeavors can fluctuate dramatically based on market prices, operational efficiency, and regulatory environments, creating a precarious balance that companies must navigate to maintain profitability in the volatile energy market. In analyzing the broader implications of conventional energy extraction costs, one must consider the resultant environmental impact and the associated social costs. The United Nations argues that water access is a human right essential to the enjoyment of life and health (Auch et al., 2018). This assertion becomes particularly pertinent when discussing the water-intensive nature of certain extraction methods, such as fracking, which can significantly disrupt local water supplies.

The withdrawal of water resources from communities in favor of industrial use shifts the supply dynamics, potentially undermining the rights of local residents and causing long-term ecological damage. As energy companies weigh these social responsibilities against their profit margins, the reality emerges that the actual cost of extraction extends beyond financial figures—it encapsulates the broader societal and environmental costs brought forth by conventional energy practices. Lastly, the financial burden linked to conventional energy extraction translates into long-term sustainability challenges. Although traditional energy sources have been integral to powering industrial growth and economic development, the escalating production costs call into question their future viability. The need for transition to more sustainable energy solutions grows increasingly urgent given the environmental degradation associated with fossil fuel extraction and production highlighted in recent studies (Elliott et al., 2003). As the global community acknowledges that the dependency on fossil fuels largely contributes to escalating greenhouse gas emissions, the economic rationale for investing in renewable alternatives becomes more compelling. Shifting investments towards renewable energy sources addresses environmental and social concerns and mitigates the economic risks associated with volatile fossil fuel markets, paving the way for a more resilient energy future.

10.6.2. Investment and Infrastructure for Sustainable Energy

A robust investment landscape is crucial for advancing sustainable energy infrastructure, as it fosters innovation and resilience while addressing systemic vulnerabilities within energy systems. Traditional approaches often create interconnected vulnerabilities that can undermine energy security and efficiency. In contrast, focusing on localized energy solutions enhances resilience and better meets the unique needs of communities. For instance, drawing from ongoing evaluations of a Private Finance Initiative (PFI) project, it appears that user-focused processes can motivate developers to adopt innovative approaches that align with environmental standards and mitigate vulnerabilities. By embedding open-source principles within energy system development, the potential for enhanced collaborative solutions increases, promoting sustainable practices that also address social issues such as fuel poverty and economic inequality, which are critical facets of energy policy. Government support plays an integral role in the transition toward sustainable energy markets. By encouraging international cooperation and creating favorable conditions for private sector investments, governments can stimulate technological advancements and infrastructure development. As noted, international markets enhance R&D opportunities by broadening the technology landscape. The existence of larger, interconnected markets incentivizes technology developers to innovate and refine production processes, thereby facilitating the integration of more efficient energy solutions. However, challenges remain in ensuring that national governments adequately align their interests to prevent free-riding on global initiatives while maximizing domestic advantages. Ultimately, a strategic approach to fostering international collaboration is essential, as it mitigates costs and accelerates the adoption of advanced energy technologies. Achieving effective investment in sustainable energy infrastructure demands a comprehensive understanding of both market dynamics and regulatory frameworks. Policymaking should aim to create environments that prioritize long-term investments into renewable technologies while ensuring equitable access to energy resources. By addressing both the benefits of a diverse energy portfolio and the costs associated with transitioning away from conventional energy, policymakers can outline clear pathways for sustainable development. Investments that emphasize resilience can help shield communities from energy vulnerabilities and environmental challenges, reinforcing the importance of diverse energy sources. As we move toward a more sustainable future, the integration of innovative funding mechanisms and cross-sector partnerships will be fundamental in overcoming existing barriers, ultimately paving the way for a more efficient, secure, and equitable energy landscape (Figure 10.7).

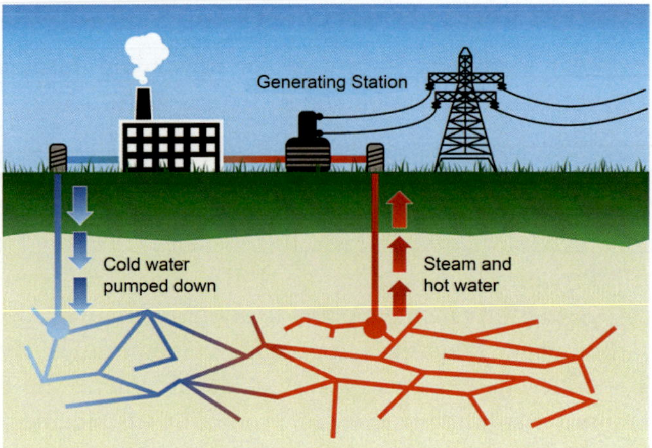

Figure 10.7. *Infrastructure is expanding for renewable energy.*

Source: https://commons.wikimedia.org/wiki/File:Geothermal-energy.png.

10.6.3. Job Creation in Renewable Energy Sectors

The shift toward renewable energy has emerged as a pivotal driver for job creation, reflecting broader economic trends and the urgent need for sustainable practices. As countries commit to reducing carbon emissions, investments in renewable sectors like solar, wind, and hydropower have surged. This transition not only addresses climate change but also fosters innovation and entrepreneurship within local communities. As renewable energy projects increase, they create a wide range of job opportunities, from research and development to installation and maintenance. These positions often require diverse skill sets, catering to both high-skilled engineers and on-the-ground technicians, thereby broadening the workforce participation in the energy sector. Moreover, the International Energy Agency (IEA) highlights the scale of necessary investments in the clean energy sector to combat climate change effectively. According to the IEA, an additional \$36 trillion in clean energy investment is needed through 2050 to adhere to global temperature goals (Mark Fulton et al., 2014). This funding translates into substantial job creation as these investments lead to the construction and operation of renewable energy facilities. In addition to direct employment in the renewable energy sector, indirect job creation in related industries, such as manufacturing, supply chain logistics, and research institutions also expands, creating a robust economic ecosystem. The ripple effect of these investments sustains long-term employment growth while simultaneously addressing environmental challenges.

Policymakers, investors, and businesses need to collaborate in promoting renewable energy initiatives to sustain this momentum. Engaging communities through education and training programs is crucial, as these efforts ensure a skilled workforce ready to meet the demands of an evolving energy landscape. Moreover, governments can create a favorable regulatory environment that incentivizes clean energy projects, thereby spurring investment and job creation. As the global community aims for ambitious climate targets, the renewable energy sector presents a solution for environmental sustainability and a pathway for resilient economic growth (Buckley et al., 2014). By prioritizing these efforts, societies can harness the potential of renewable energy, fostering job creation and a sustainable future for all (Figure 10.8).

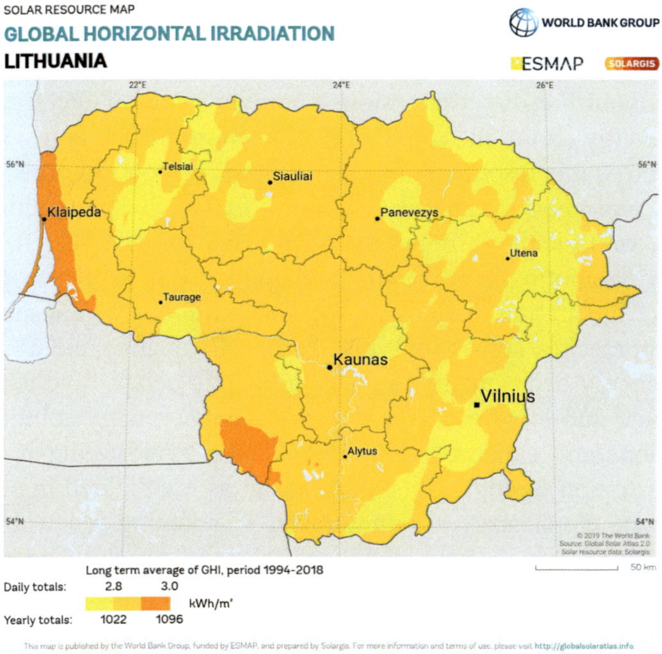

Figure 10.8. *Solar power potential.*

Source: By © 2019 The World Bank, Source: Global Solar Atlas 2.0, Solar resource data: Solargis. – https://solargis.com/maps-and-gis-data/download/lithuania, CC BY 4.0, https://commons.wikimedia.org/w/index.php?curid=86911845.

10.7. ENVIRONMENTAL IMPACT COMPARISON

The examination of environmental impacts between conventional and sustainable energy sources reveals a significant disparity in their influence on the planet. Conventional energy systems, primarily reliant on fossil fuels, contribute substantially to greenhouse gas emissions, which are the leading drivers of climate change. These sources not only release harmful pollutants into the atmosphere but also pose considerable risks to water quality and biodiversity through processes such as mining and drilling. In contrast, sustainable energy alternatives, particularly solar power, present a unique opportunity to mitigate these negative effects. Solar energy is an inexhaustible and CO_2-emission-free resource that generates clean electricity while minimizing air and water pollution (Ramrao et al., 2021). This comparison highlights the pressing need for a transition to greener energy solutions that prioritize environmental health. Investments in renewable energy technologies underscore the global shift toward more sustainable practices. Recent trends indicate that solar power, among other renewables, has received substantial financial support, reflecting its perceived potential to drastically reduce environmental impacts. These systems produce electricity without the direct emissions common in fossil fuel generation by harnessing energy from the sun. This transition trades conventional fuel reliance for renewable resources and reduces local carbon footprints, benefiting communities and ecosystems alike (Ramrao et al., 2021). Additionally, the development of organic farming practices aligns with sustainable energy principles by promoting biodiversity and reducing dependence on external inputs, further illustrating that environmental health can be enhanced through integrative approaches across multiple sectors. Evaluating the environmental outcome of energy decisions necessitates a holistic perspective that considers both immediate and long-term impacts. Sustainable energy solutions present lower emissions and support ecological balance by fostering practices like carbon sequestration through green vegetation, as highlighted in studies. This interdependence emphasizes the importance of adopting renewable methods like solar energy, which help stabilize ecosystems while addressing the pressing challenge of climate change. Furthermore, the principles of sustainable agriculture illustrate how energy practices intertwine with land management, indicating that efforts to promote ecological health must transcend purely energy-related metrics. In summary, the comparison between conventional and sustainable energy practices illustrates the critical nature of our choices today, as they hold the keys to a more sustainable and environmentally sound future (Ramrao et al., 2021).

10.7.1. Carbon Emissions From Conventional Energy

The reliance on conventional energy sources such as fossil fuels has significant implications for global carbon emissions. A large proportion of greenhouse gas emissions—including carbon dioxide—stem from the combustion of coal, oil, and natural gas, which are central to today's energy landscape. In fact, in many developed nations, these sources dominate energy production, contributing to an alarming increase in atmospheric carbon levels. The embodied emissions associated with energy extraction, transportation, and consumption further exacerbate climate change. As buildings and infrastructure continue to utilize materials like steel and concrete, largely produced from carbon-intensive processes, the challenge of mitigating these emissions becomes even more daunting. These materials not only require substantial energy inputs but also continue to emit carbon over their lifecycle, further entrenching the environmental crisis (Gallersdörfer et al., 2020). Transitioning away from conventional energy methods is imperative reducing emissions and for pursuing sustainability in the built environment. The use of alternative, renewable energy sources such as wind, solar, and hydroelectric power can significantly decrease carbon output. As public awareness of climate change grows, there is a renewed impetus to innovate and invest in clean energy technologies that minimize environmental damage. However, this transition must also be mindful of resource availability and the implications of materials used in renewable energy projects. Significant carbon emissions are associated with the manufacture of solar panels and wind turbines, indicating that every energy solution demands a critical examination of its entire life cycle (Gallersdörfer et al., 2020).

Ultimately, while conventional energy contributes significantly to carbon emissions, a concerted shift towards more sustainable practices holds promise for mitigating climate change. The future of energy stands at a crossroads where choices made today will dictate the well-being of the planet for generations to come. While conventional energy forms have fueled economic growth over the decades, their contribution to carbon emissions presents unfathomable risks to global health and environmental stability. Strategies such as enhancing energy efficiency and adopting carbon capture technologies are essential components in decreasing reliance on carbon-intensive energy. Furthermore, efforts to integrate bio-based materials in construction, which store carbon, highlight the potential synergistic benefits of sustainable practices leading to a more resilient energy paradigm. As urban centers expand and energy demands soar, policies promoting both renewable energy and sustainable materials are crucial. Building a cleaner energy future requires collaboration among governments, industries, and communities to achieve significant reductions in carbon emissions (Pomponi et al., 2020).

10.7.2. Land Use and Habitat Disruption From Energy Sources

The expansion of energy production, particularly through conventional means, has led to significant land use changes and habitat disruption. Fossil fuel extraction, including coal mining and oil drilling, directly alters landscapes, often leading to deforestation and the fragmentation of ecosystems. This land degradation eliminates essential habitats for countless species and disrupts local communities that depend on these environments for their livelihoods. The consequences are far-reaching: as habitats are destroyed, biodiversity declines and the delicate balance of ecosystems is thrown into disarray. Moreover, the infrastructure required for the transportation and processing of these energy sources further exacerbates land use issues, severing interconnected habitats and hindering the movement of wildlife, which is essential for their survival and genetic diversity. In contrast, sustainable energy sources offer a pathway to mitigate some of these habitat disruptions, but they are not without their own challenges. Renewable energy projects, such as wind farms or solar arrays, often necessitate large areas of land, raising concerns about their impact on local ecosystems and land use patterns. The siting of these facilities can lead to habitat loss and may encroach upon agricultural land, leading to new conflicts over resource allocation. However, the argument for sustainable energy hinges on its potential to reduce the overall environmental footprint. As highlighted by Braun et al. (2007), transitioning to renewable energy sources is critical in addressing climate change and its associated challenges. Thus, while renewable sources do impact land use, they may ultimately contribute to the conservation of natural habitats by reducing reliance on more harmful energy extraction methods. The interaction between land use, habitat disruption, and energy sources will be pivotal in shaping environmental policy and energy strategies moving forward. As we grapple with the challenges posed by climate change, it is essential to critically evaluate the long-term consequences of our energy choices. The urgent call for a shift toward non-GHG-emitting energy sources (Braun et al., 2007), emphasizes the need to prioritize ecological integrity while meeting human energy demands. Ultimately, the conversations surrounding energy production must encompass a holistic view of land use practices that respect both the environment and social frameworks. We can strive for solutions that promote sustainability while minimizing habitat disruption and fostering resilience in the face of climate change challenges by thoughtfully considering the implications of our energy decisions.

10.7.3. Long-Term Sustainability of Renewable Energy

The growing emphasis on renewable energy sources is primarily driven by the necessity to reduce carbon emissions and combat climate change. As traditional fuel stocks deplete, many energy providers face the challenge of maintaining grid reliability without over-reliance on fossil fuels. Transitioning to renewable sources like solar and wind energy is essential, but it requires efficient management and integration into existing energy systems. An effective strategy involves employing advanced scheduling models and forecasting techniques to optimize energy production and consumption. For instance, the implementation of Mixed Integer Programming (MIP) in energy allocation can significantly minimize costs while enhancing sustainability, as outlined in studies emphasizing the importance of balancing various energy sources (Fekih Hassen et al., 2024). However, the transition to renewable energy also necessitates innovative approaches to address intermittent generation issues. The integration of advanced technologies, such as long-short term memory (LSTM) networks for solar energy prediction, can facilitate effective load management and enhance the overall efficiency of renewable energy systems. These systems must account for fluctuations in energy production due to varying weather conditions while ensuring a stable energy supply for consumers. Energy providers can dynamically schedule resources, reducing reliance on traditional fuels, particularly during peak demand by using Teaching-Learning Optimization (TLBO) alongside predictive models (Uttamarani Pati et al., 2024). Ultimately, the long-term sustainability of renewable energy hinges on the ability to harness advanced analytics and optimization techniques to create a more resilient energy grid. In addition to technological advancements, the long-term sustainability of renewable energy is also contingent upon regulatory frameworks and consumer engagement. Governments can accelerate the transition by establishing supportive policies that incentivize renewable energy adoption and investment in infrastructure. Integrating blockchain technology into energy trading platforms enhances transparency and security, facilitating a more robust interaction between utilities and consumers. Such measures are vital for fostering an environment where individuals become active participants, referred to as prosumers, in the energy market. By empowering consumers to contribute to energy supply and demand dynamics, the overall resilience of the grid is strengthened, leading to a more sustainable and efficient energy ecosystem. Thus, a multifaceted approach that combines technology, policy, and community engagement is essential for achieving the long-term sustainability of renewable energy resources.

10.8. TECHNOLOGICAL ADVANCEMENTS

Within the realm of energy production, technological advancements have dramatically reshaped both conventional and sustainable practices. New methodologies in artificial intelligence and machine learning have optimized energy efficiency across various sectors. These advancements foster smarter grid systems capable of predicting consumption patterns and managing resources more effectively. As urban areas continue to grow, the integration of high-tech urban agriculture has emerged as a particularly relevant innovation. By leveraging technologies like LED lighting and controlled environment systems, cities can not only enhance food production but also promote sustainable practices within their ecosystems, as illustrated in the case of agricultural initiatives in Amsterdam (Carsjens et al., 2020). By aligning agricultural practices with energy efficiency, cities are paving the way for more sustainable urban environments. Moreover, the intersection of biotechnology and data management has opened new frontiers in energy sustainability. Research exploring DNA as a storage medium exemplifies how innovative thinking can address existing limitations in data handling and archival processes (Akhtar and Tajbuil Rawol, 2024). Such advancements command attention as they propose methods to store vast amounts of information compactly and sustainably. This is particularly crucial in an age defined by the digitization of energy management systems, where conventional practices need augmentation to meet ever-increasing demands. The embrace of DNA-based storage techniques could streamline operations in energy networks, offering scalable solutions that align with the principles of sustainability. As these technologies evolve, they represent a significant leap towards a more efficient and responsible future. Finally, the economic implications of technological advancements cannot be overstated. As new tools and systems are developed, they lead to cost savings for energy producers and stimulate job creation in burgeoning fields, such as renewable energy technology and smart grid management. The rise of alternative energy sources, supported by cutting-edge research like that in high-tech urban agriculture and DNA data storage, can facilitate a paradigm shift in energy consumption, reducing reliance on fossil fuels and lowering carbon emissions. With support from both governments and private sectors, continued investment in technological innovation is essential for achieving sustainability goals. The successful integration of these advancements into conventional energy frameworks ultimately enhances the resilience and sustainability of our energy systems, ensuring a more stable and environmentally conscious future.

10.8.1. Innovations in Fossil Fuel Extraction and Efficiency

Innovations in fossil fuel extraction and efficiency are critical to addressing both energy demands and environmental concerns. Techniques such as hydraulic fracturing and horizontal drilling have revolutionized the oil and gas industry, increasing the feasibility of accessing previously unreachable reserves. These advancements enable the extraction of fossil fuels from unconventional sources, thereby significantly enhancing output while lowering the cost of production. As global energy consumption continues to rise, particularly in emerging economies, such innovations facilitate a more sustainable approach to fossil fuel extraction, ensuring that energy needs can be met with greater efficiency. Ultimately, the adoption of these technologies boosts energy security and contributes to the economic viability of fossil fuel resources, which play a crucial role in the current energy landscape. Nonetheless, the environmental impacts of fossil fuel extraction cannot be ignored. Innovations aimed at improving efficiency must also encompass measures that mitigate adverse effects on ecosystems and communities. For instance, advancements in emission control technologies and carbon capture and storage (CCS) represent steps toward reducing the carbon footprint associated with fossil fuel utilization. These technologies allow for the continued use of fossil fuels while addressing greenhouse gas emissions, thus providing a bridge toward more sustainable energy systems. Additionally, the integration of waste management practices in extraction processes highlights the potential for reducing pollution and resource waste. By focusing on innovation that aligns economic objectives with environmental stewardship, the fossil fuel industry can play a transformative role in striving toward a more balanced energy strategy. While the drive for innovation in fossil fuel extraction offers both opportunities and challenges, the ultimate goal should be a seamless integration with sustainable practices. The evolution of the fossil fuel sector must consider the potential of biobased resources and their sustainability implications. The use of underutilized residues from agricultural processes, like palm oil production, encourages a more circular economy by valuing waste products as resources. This perspective can influence the fossil fuel industry by promoting hybrid systems that utilize both fossil fuels and sustainable bio-based alternatives. Moreover, ongoing research and development into cleaner extraction and processing technologies can significantly aid in reducing inefficiencies and the environmental impact of fossil fuel consumption. By marrying technological innovation with a commitment to sustainability, the fossil fuel sector can evolve in a manner that contributes positively to the larger energy ecosystem (Dam et al., 2009).

10.8.2. Breakthroughs in Renewable Energy Technologies

The evolution of renewable energy technologies has demonstrated significant promise in transforming our global energy landscape. Innovations such as advanced solar photovoltaics, offshore wind turbines, and novel energy storage solutions have dramatically improved efficiency and accessibility. For instance, the development of bifacial solar panels, which capture sunlight on both sides, has enhanced energy output compared to traditional designs. As these technologies become more cost-effective, they make it feasible for a broader range of consumers and industries to adopt renewable sources. Consequently, the increasing affordability of renewable solutions plays a critical role in mitigating the reliance on fossil fuels, signaling a gradual shift towards a more sustainable energy future. Technological advancements are not solely confined to solar and wind energy; breakthroughs in energy storage and grid management are equally pivotal in facilitating the transition to renewable energy. Effective management of supply and demand is essential for integrating intermittent renewable sources into the energy grid. Strategies such as grid-scale battery storage and smart grid technology allow for better balancing of energy loads and improved reliability. According to the ITIF, the advancement of technology is crucial in overcoming barriers to renewable energy adoption (Cohen et al., 2015). By enhancing grid infrastructure and developing innovative storage methods, we can ensure a seamless flow of clean energy, preparing the stage for a substantial move away from conventional energy sources. Moreover, the critical importance of government intervention in fostering renewable energy innovation cannot be overlooked. Effective public policy can accelerate the transition to a more sustainable energy system by supporting research and development initiatives. ITIF highlights the necessity for strategic actions and investment to address technological and political challenges inherent in the renewable energy sector (Atkinson and Hackler, 2010). Countries can collaboratively tackle pressing energy issues, navigating the complex web of market dynamics and regulatory frameworks by forging strong partnerships between the public and private sectors. Applying targeted policies and incentives, governments can stimulate innovation, create jobs, and drive economic growth, ultimately leading to a transformation in how we generate and consume energy, moving closer to a fully renewable-based economy.

10.8.3. The Role of Energy Storage Solutions

As the global energy landscape evolves, the necessity for effective energy storage solutions has become increasingly apparent. These systems play a crucial role in managing the intermittency associated with renewable energy sources such as solar and wind power. Energy storage technologies, including batteries, pumped storage hydroelectricity, and thermal storage, provide a means to store excess energy generated during peak production times. This stored energy can then be released during periods of high demand or low generation, enhancing grid reliability and overall efficiency. By ensuring a stable energy supply, these solutions maximize the use of renewable resources and contribute to a reduction in greenhouse gas emissions, thus aligning with sustainable energy goals. The integration of energy storage solutions is pivotal in facilitating the transition from conventional energy systems to greener alternatives. As identified in the literature, the efficiency of solar-to-fuel systems hinges on the ability to store energy effectively for later use (Smith et al., 2019). Technologies such as electrochemical storage and grid-scale batteries are essential for supporting the increasing share of variable renewable energy sources in power generation. Moreover, these storage systems enable more flexible energy management, allowing grid operators to respond swiftly to fluctuating demand and supply conditions. This adaptability is particularly vital as independent power producers increasingly generate energy from renewables, demonstrating how energy storage not only complements these systems but is integral to their success.

A comprehensive approach to energy storage can significantly enhance the resilience and sustainability of energy systems. As highlighted in various studies, the future of energy production lies in the synergy between diverse technologies, including traditional fossil fuels, renewables, and efficient energy storage (Intergovernmental Panel on Climate Change, 2023). With advances in battery technology, particularly in improving cycle life and reducing costs, there is potential for widespread adoption. Furthermore, integrating energy storage into existing infrastructure can alleviate stress on conventional systems during peak load periods, while fostering advancements in smart grid technology. By facilitating a more responsive energy grid that can adapt to the dynamics of renewable integration, energy storage solutions serve as the backbone for a sustainable energy future, illustrating their pivotal role in the transition toward a low-carbon economy.

10.9. POLICY AND REGULATION

Effective policy and regulation are fundamental to shaping the future of energy, especially as the global community grapples with the implications of climate change and the transition from conventional to sustainable energy sources. Governments must implement comprehensive policies that not only promote renewable energy development but also ensure that traditional fossil fuel industries adhere to stricter environmental standards. This balancing act requires a detailed understanding of socioeconomic factors and workforce transitions, emphasizing the need for policy frameworks that facilitate investment in clean technologies. By fostering a regulatory environment that encourages public and private stakeholders to invest in sustainable energy infrastructure, nations can accelerate their shift towards a low-carbon economy, ultimately benefiting both environmental and economic interests. Investment from Multilateral Development Banks (MDBs) can play a pivotal role in supporting such transitions. Ballesteros and Nakhooda (2010) highlight the importance of MDBs in financing electricity projects, stressing the need for robust policy reforms and institutional capacities that can attract both public and private investments. Regulation must target specific barriers to sustainable energy investment, such as outdated grid technologies or insufficient economic incentives for renewable energy development. Policymakers are tasked with creating a legal and regulatory landscape that supports innovation and competition, ensuring that the market does not favor entrenched interests within the conventional energy sector. By addressing these challenges, MDBs, and policymakers can collaboratively create an environment conducive to sustainable energy investments, facilitating more resilient and inclusive energy systems. Moreover, the interplay between trade and climate policy is becoming increasingly crucial as nations navigate their energy futures.

Ackerman et al (2016) report that there is a profound tension between existing trade agreements and climate objectives, which is often overlooked in policy discussions. Policymakers must revise international trade frameworks to ensure they support rather than hinder climate action. For instance, investing in green technologies can be stymied by trade rules that favor fossil fuel dependence. Governments can align economic policy with environmental objectives by updating trade and investment legislation, creating a cohesive approach to global energy governance. Addressing these regulatory intricacies is vital for achieving the ambitious targets outlined in international climate agreements, paving the way for a more sustainable energy landscape.

10.9.1. Government Incentives for Renewable Energy

Government incentives play a crucial role in the transition from conventional energy sources to renewable energy solutions. By providing financial support and subsidies, governments can lower the initial installation costs of renewable energy systems, making them more accessible to both businesses and consumers. These incentives may take various forms, including tax credits, grants, and feed-in tariffs that guarantee fixed payments for energy produced from renewable sources. As outlined by MDB analysis, successful implementation of these incentives hinges on complementary policy reforms and regulations that empower public and private investment in sustainable energy. This coordinated effort can lead to a significant increase in renewable energy adoption, mitigating the reliance on fossil fuels and fostering a more environmentally friendly energy landscape (Athena and Nakhooda, 2010). The effectiveness of these incentives in facilitating renewable energy deployment can be seen in the growing solar-thermal market in Europe. Despite the mature nature of solar technologies, many countries have struggled to meet EU targets for solar-thermal installations. By analyzing successful market development in certain EU countries, it becomes evident that obligation schemes are instrumental in maximizing installations. These schemes can drive local industry growth without imposing a direct burden on public budgets. When governments create a stable regulatory environment, it not only boosts market confidence but also stimulates innovations within the renewable sector, encouraging investments that aim to achieve long-term sustainability (Ramos et al., 2017). Moreover, the integration of hybrid technologies, such as photovoltaic-thermal (PV-T) collectors, further emphasizes the impact of government support on renewable energy innovation. Such PV-T technologies offer enhanced efficiency by simultaneously producing heat and electricity from a single system, particularly in areas where space for solar installations is limited. Government incentives can accelerate research and development in these hybrid systems, leading to broader adoption and a more significant share of renewable energy in the overall energy mix. As the transition to sustainable energy intensifies, sustained government support amplifies the capabilities of existing technologies and paves the way for new solutions that address the challenges of energy storage and efficiency, thereby laying a foundation for a cleaner, more sustainable energy future.

10.9.2. Regulations on Fossil Fuel Emissions

The need for stringent regulations on fossil fuel emissions has become increasingly evident as the evidence of climate change continues to mount. Regulations play a crucial role in setting limits on the amount of greenhouse gases that can be emitted by various industries, thereby fostering a cleaner environment. For instance, governmental policies can demand the adoption of cleaner technologies and fuel alternatives, incentivizing corporations to invest in sustainable practices. Authors of the Working Group on Trade, Investment, and Climate Policy highlight the potential of investment treaties to direct capital toward climate-positive initiatives. By creating an economic environment that favors sustainable energy investments, these regulations can mitigate the adverse effects of fossil fuel reliance and push society toward greener alternatives, ultimately contributing to global efforts aimed at reducing temperatures and preserving ecological balance (Ackerman et al., 2016).

The economic implications of regulating fossil fuel emissions are multifaceted, offering both challenges and opportunities for innovation. While strict regulations may initially appear burdensome to industries reliant on fossil fuels, they can drive technological advancements in renewable energy sources and increase energy efficiency. Economic evaluations of trade agreements, such as the Transatlantic Trade and Investment Partnership (TTIP), indicate that failing to incorporate environmental considerations could result in significant economic costs and lost opportunities for sustainability (Athena and Nakhooda, 2010). By establishing a framework for emissions reductions, governments encourage businesses to adapt, creating a competitive market for sustainable technologies and practices. As fossil fuel regulations unfold, they target existing emissions and shape the future landscape of energy production and consumption, urging industries to transition toward lower-carbon alternatives. As international pressure mounts concerning climate change, countries are increasingly recognizing the importance of collaborative emissions regulations. The global nature of the climate crisis necessitates a unified approach, where nations work together to set standards that limit fossil fuel emissions across borders. The role of Multilateral Development Banks (MDBs) becomes critical in this context, as they can facilitate funding for sustainable projects while ensuring compliance with international standards. MDBs can enable countries to implement necessary reforms, bolstering public and private investments in sustainable energy solutions. This proactive stance toward emissions regulation aligns with broader environmental goals and promotes a significant shift in how nations prioritize energy consumption and production. Ultimately, through coordinated regulations, the world can move closer to meeting climate targets and ensuring a sustainable future for subsequent generations.

10.9.3. International Agreements and Their Impact on Energy Policies

International agreements inherently shape national energy policies by establishing frameworks and commitments that countries must adhere to in their pursuit of sustainable development. These agreements often aim to mitigate environmental impacts and foster cooperation on climate change initiatives. For example, the Paris Agreement represents a significant step toward unifying global efforts to limit greenhouse gas emissions and limit global temperature rise. While such treaties provide foundational support for transitioning to renewable energy, they also compel nations to reevaluate their reliance on conventional energy sources. Consequently, countries are increasingly incorporating renewable energy goals and stringent emissions targets into their energy policies, which influences investment, innovation, and regulatory measures within the energy sector. Furthermore, international agreements frequently address the intersection of trade and energy policies, creating both opportunities and challenges for sustainable energy development. The negotiations surrounding trade agreements, as discussed in the context of the Trans Pacific Partnership (TPP) and the Transatlantic Trade and Investment Partnership (TTIP), reveal that environmental considerations are often secondary to economic objectives. The failure to adequately address the implications of shipping and maritime transport on climate change under these agreements serves as a cautionary example, highlighting that trade and investment rules must be realigned with climate commitments to achieve effective energy policy outcomes (Ackerman et al., 2016). Countries must prioritize sustainability and work collaboratively on frameworks that directly support the transition to cleaner energy sources to navigate the complex landscape of international trade. Finally, the shifting landscape of international agreements influences national policies through the promotion of collaborative technologies and investments in climate-friendly initiatives. Scholars and policymakers argue for the inclusion of provisions in international treaties that reroute investments toward renewable energy projects, thereby facilitating climate change mitigation and adaptation strategies relevant to both developed and developing nations (Ackerman et al., 2016). These agreements can significantly impact local energy landscapes, fostering innovation and economic resilience By redirecting global capital toward sustainable energy infrastructure. The potential of international treaties to create synergies between energy security and environmental responsibility underscores the importance of continued engagement and negotiation to ensure that energy policies remain aligned with global sustainability goals. This ongoing dialogue must prioritize a balance between economic growth, energy independence, and environmental preservation, ultimately shaping a cleaner energy future.

10.10. CONCLUSION

In summarizing the implications of our analysis on conventional and sustainable energy, it becomes apparent that a transition towards more sustainable practices is not merely beneficial but essential. The environmental impacts of agricultural activities, as highlighted by previous studies, reveal a pressing need to reconsider current energy practices within the agricultural sector. By adopting energy-efficient methods, particularly those observed in organic farming, we can mitigate the negative impacts of conventional practices. For example, research shows a significant improvement in energy efficiency in organic production methods—almost two-fold compared to conventional ones—indicating that a shift in approach can yield both ecological and economic benefits (Dessane, 2003). Therefore, addressing this transition requires a comprehensive understanding of sustainable practices and their potential for reducing the overall ecological footprint.

Moreover, integrating mission investments into philanthropic efforts can enhance sustainable energy initiatives significantly. According to findings from an FSG report funded by The Surdna Foundation, mission investments are not standalone tools; rather, when combined with traditional grants, they contribute to a more holistic approach to tackling pressing issues like climate change. This interconnectedness implies that foundations and organizations can generate a more substantial impact by diversifying their investment strategies to include sustainable projects. Such collaboration aligns with our findings, underscoring the necessity of strategic partnerships in promoting a shift towards renewable and sustainable energy sources, which in turn can contribute to broader environmental goals. Collectively, these efforts will facilitate a more sustainable energy landscape. In conclusion, the evidence presented in our discourse on conventional versus sustainable energy illustrates the urgent need for transformation. The contrasting impacts of agricultural practices on energy use, as discussed, highlight the imperative to adopt organic and other sustainable methods to mitigate climate change effectively. The dual approach of implementing energy-efficient agrarian practices alongside strategic investments can pave the way for a more sustainable future. As such, embracing these practices will not only enhance energy efficiency but also address pressing environmental issues emanating from traditional energy practices. Promoting a synergistic approach involving agricultural reform and mission investing is vital in moving society towards sustainable energy solutions, ensuring a balanced coexistence between ecological sustainability and economic growth (Kramer et al., 2010).

10.10.1. Summary of Key Points

In examining energy production, it is essential to highlight the contrasting approaches of conventional and sustainable energy systems. Conventional energy sources, such as fossil fuels, have powered industrial growth and economic expansion. However, reliance on these sources significantly contributes to environmental degradation and climate change, evidenced by the consensus among scientists that human activities are likely responsible for recent climate fluctuations (Losonc et al., 2012). As urban populations grow and energy demands increase, the negative externalities of fossil fuel usage—like air pollution and greenhouse gas emissions—become more pronounced, leading to harmful effects on both human health and ecosystem stability. In response to the pressing challenges posed by climate change and resource depletion, sustainable energy solutions have garnered increasing attention. These renewable energy sources, including wind, solar, and biomass, offer viable alternatives that can help mitigate environmental impacts while also promoting energy security.

Notably, initiatives aimed at enhancing organic farming practices demonstrate how sustainability can be seamlessly integrated into traditional methods. For instance, research has emphasized effective nutrient management and weed control strategies that improve agricultural productivity and minimize ecological footprints (Keatinge, 2000). This dual focus on productivity and sustainability underscores the potential of renewable energy to transform both the energy sector and agricultural practices. Ultimately, the choice between conventional and sustainable energy sources hinges on a complex interplay of environmental, economic, and social factors. Transitioning to sustainable energy addresses urgent environmental challenges and presents opportunities for innovation and job creation in emerging sectors. Furthermore, as traditional energy sources become increasingly volatile due to resource scarcity and geopolitical tensions, investing in sustainability aligns both with contemporary global imperatives and with the ethical obligation to protect the planet for future generations. Thus, prioritizing sustainable energy can pave the way for a resilient, equitable, and environmentally sound energy future.

10.10.2. The Future of Energy: Balancing Conventional and Sustainable Sources

The energy landscape is undergoing transformative changes, driven by the urgent need to address climate change and meet the increasing global energy demand.

Traditional energy sources, primarily fossil fuels, have been the backbone of power generation for decades. However, reliance on these sources has led to significant environmental concerns, including greenhouse gas emissions and air pollution. To counteract these effects, the integration of sustainable energy sources such as solar, wind, and hydroelectric power is essential. An optimal energy strategy should thus leverage the strengths of both conventional and renewable sources. By utilizing advanced technologies that enhance efficiency and reduce emissions, it is possible to create a more sustainable energy infrastructure. This shift caters to current energy needs and paves the way for a more eco-friendly future, aligning with global targets like those established by the Kyoto Protocol (Tolis et al., 2010).

As countries adapt to this dual approach, the role of policy and regulatory frameworks becomes crucial. Governments around the world are increasingly recognizing the importance of fostering both technologies and market conditions that encourage clean energy investments. For instance, the European Action Plan for organic food and farming emphasizes the integration of core ethical values into standards and regulations, advocating for more effective decision-making processes in energy production (de Wit et al., 2007). This means that energy policies must promote the use of sustainable resources and support the gradual transition from conventional sources by providing incentives for innovation and infrastructure development. The success of these policies will hinge on a collaborative effort among stakeholders, including governmental bodies, energy producers, and consumers, to strengthen the reliability and efficacy of a mixed energy structure. Looking forward, a balanced energy portfolio that incorporates both conventional and sustainable sources is imperative for achieving energy security and minimizing climate impact. The journey towards this balance will not be without challenges. Stakeholders must navigate the complexities of market dynamics, infrastructure limitations, and consumer demand while addressing the social and economic implications of such transitions. Future electricity generation strategies, as indicated by research, must prioritize a mixed approach to energy production to optimize output while reducing carbon footprints (Tolis et al., 2010). Additionally, ongoing technological advancements in energy storage and smart grid systems will play an essential role in stabilizing energy supplies, making the coexistence of various energy sources a viable option. Ultimately, this balancing act will define the trajectory of global energy and determine our collective ability to achieve sustainability goals in the face of an ever-changing climate.

10.10.3. Call to Action for Energy Transition and Policy Reform

Transitioning to sustainable energy sources is imperative for mitigating the adverse effects of climate change and ensuring long-term environmental health.

Current fossil fuel dependency exacerbates global warming and jeopardizes public well-being through air pollution and habitat destruction. Enacting policy reforms that promote renewable energy investments is essential to facilitate this shift. Governments must implement incentives such as tax breaks, subsidies, and grants for both businesses and households that adopt solar, wind, or geothermal energy technologies. Such measures would stimulate economic growth while prioritizing sustainability, thereby fostering a transition that benefits both the economy and the environment. Energy policy reform needs to encompass a holistic approach that aligns with global climate agreements and local sustainability objectives. This entails robust regulatory frameworks that set ambitious renewable energy targets and facilitate the phasing out of inefficient and polluting energy sources. Policies should also focus on enhancing energy efficiency, promoting innovation in green technologies, and ensuring equitable access to clean energy solutions. Collaboration among government entities, private sector stakeholders, and community organizations is vital to creating a comprehensive strategy that addresses the unique energy challenges faced by diverse populations. Ultimately, a cohesive policy framework will catalyze the systemic change required for a sustainable energy future. The urgency of transitioning to sustainable energy cannot be overstated, as the consequences of inaction are becoming increasingly apparent. There must be a collective call to action involving individuals, businesses, and policymakers to rally around the cause of energy reform. Grassroots movements have proven effective in pushing for change, and public pressure can drive legislative bodies to prioritize renewable energy initiatives. Engaging in advocacy, supporting sustainable energy projects, and holding representatives accountable are critical steps that citizens can take to influence energy policy positively. As we look to the future, our collective commitment to embracing and advocating for an energy transition will shape the resilience of our planet and the sustainability of generations to come.

REFERENCES

1. Ackerman, F., Bernasconi, N., Cabré, M. M., et al. (2016). *Trade in the Balance: Reconciling Trade and Climate Policy: Report of the Working Group on Trade, Investment, and Climate Policy*. Boston University Frederick S. Pardee Center for the Study of the Longer-Range Future. ISBN 978-1-936727-13-1.
2. Adeolu Adenekan, O., Ezeigweneme, C., & Chukwurah, E. G. (2024). *Driving Innovation in Energy and Telecommunications: Next-Generation Energy Storage and 5G Technology for Enhanced Connectivity and Energy Solutions*. Fair East Publishing. https://doi.org/10.51594/ijmer.v6i5.1124.
3. Adger, A., Hope, A., Ballet, A., et al. (2010). *Localism and Energy: Negotiating Approaches to Embedding Resilience in Energy Systems*. Elsevier BV. https://core.ac.uk/download/9304479.pdf (accessed on 5 December 2024).
4. Akhtar, Z. B., Rawol, T., & Ahmed, A. (2024). Unlocking the future for the new data paradigm of DNA data storage: An investigative analysis of advancements, challenges, and future directions. *Journal of Information Sciences, 23*(1). https://doi.org/10.34874/IMIST.PRSM/jis-v23i1.47102 (accessed on 5 December 2024).
5. Alves, H., Latva-aho, M., Alcaraz, O. L., & Martínez, R. (2023). *Sustainable Radio Frequency Wireless Energy Transfer for Massive Internet of Things*. Retrieved from: http://arxiv.org/abs/2306.15051 (accessed on 5 December 2024).
6. Atkinson, R. D., & Hackler, D. (2010). Economic doctrines and approaches to climate change policy. *The Information Technology & Innovation Foundation*. Available at SSRN: https://ssrn.com/abstract=1722821 (accessed on 5 December 2024).
7. Auch, W. E. T., Palmer, R., & Short, D. (2018). *The Human Right to Water and Unconventional Energy*. MDPI AG. https://core.ac.uk/download/161233181.pdf (accessed on 5 December 2024).
8. Ballesteros, A. R., & Nakhooda, S. (2010). *Investing in Sustainable Energy Futures: Multilateral Development Banks' Investments in Energy Policy*. World Resources Institute (WRI). ISBN 978-1-56973-736-1.
9. Bhogal, A., Cormack, B., Costigan, P., Cuttle, S., Pearce, B., Philipps, L., Shepherd, M., & Unwin, R. (2003). *Organic Farming: Technology Transfer (Environmental Impact and Soil Fertility) (OF0405)*. Retrieved from: https://core.ac.uk/download/10923187.pdf (accessed on 5 December 2024).

10. Borras, S. M., Franco, J. C., Isakson, R., et al. (2014). *Towards Understanding the Politics of Flex Crops and Commodities: Implications for Research and Policy Advocacy*. Transnational Institute. https://core.ac.uk/download/20667210.pdf (accessed on 5 December 2024).
11. Borthwick, A., Chatzisymeon, E., Foteinis, S., et al. (2021). *Environmental Impacts of Conventional Versus Organic Eggplant Cultivation Ssystems: Influence of Electricity Mix, Yield, Over-Fertilization, and Transportation*. MDPI AG. https://core.ac.uk/download/425539584.pdf (accessed on 5 December 2024).
12. Braun, G., Byrnes, W. M., & Hellwig, M. K. (2007). *Global Climate Change and Catholic Responsibility: Facts and Faith Response*. https://core.ac.uk/download/163077495.pdf (accessed on 5 December 2024).
13. Buckley, C., Cloete, E., Godfrey, L., et al. (2014). *The State of Green Technologies in South Africa*. Academy of Sciences of South Africa and Ministry of Science and Technology, Republic of South Africa. https://core.ac.uk/download/30610831.pdf (accessed on 5 December 2024).
14. Carsjens, G. J., Farhangi, M. H., Turvani, M. E., & van der Valk, A. (2020). High-tech urban agriculture in Amsterdam: An actor network analysis. *MDPI AG*. Retrieved from: https://core.ac.uk/download/323190791.pdf (accessed on 5 December 2024).
15. Chapman, A., Nguyrgy RevieFarabi-Asl, Đ., Itaoka, K., Hirose, K., & Fujii, Y. (2020). *Hydrogen Penetration and Fuel Cell Vehicle Deployment in The Carbon Constrained Future Energy System*. 409–416. https://doi.org/10.1049/iet-est.2020.0014.
16. Cohen, S. A. (2015). *What is Stopping the Renewable Energy Transformation and What can the U.S. Government Do?* Columbia University Libraries/Information Services. Retrieved from: https://core.ac.uk/download/161454597.pdf (accessed on 5 December 2024).
17. Dam, J. E. G. van (2009). *Options for Sustainability Improvement and Biomass Use in Malaysia: Palm Oil Production Chain and Biorefineries for Non-Food Use of Residues and By-Products Including Other Agricultural Crops*. Agrotechnology & Food Sciences Group. https://core.ac.uk/download/pdf/29249851.pdf (accessed on 5 December 2024).
18. de Wit, J., Alrøe, H. F., Kjeldsen, C., et al. (2007). *Balancing and Integrating Basic Values in the Development of Organic Regulations and Standards: Proposal for a Procedure Using Case Studies of Conflicting Areas*. https://core.ac.uk/download/10925716.pdf (accessed on 5 December 2024).
19. Dessane, D. (2003). *Energy Efficiency and Life Cycle Analysis of Organic and Conventional Olive Groves in the Messara Valley, Crete, Greece*. Thesis, Wageningen University, Ecological Production Systems.

20. Ekbladh, G., Grönlund, E., Ingemarson, F., Karlsson, L., Nilsson, S., & Strid Eriksson, I. (2002). Ecosystem properties and principles of living systems as foundation for sustainable agriculture–Critical reviews of environmental assessment tools, key findings, and questions from a course process. *Ekologiskt lantbruk, No. 32*. Centre for Sustainable Agriculture.
21. Ekins, P., & Watson, J. (2014). *UK Energy in A Global Context: Synthesis Report*. UKERC. https://core.ac.uk/download/20391336.pdf (accessed on 5 December 2024).
22. Elliott, D. (2003). *Sustainable Energy: Choices, Problems, and Opportunities*. Royal Society of Chemistry (RSC). https://core.ac.uk/download/479.pdf (accessed on 5 December 2024).
23. *Energy for Development (2005): The Potential Role of Renewable Energy in Meeting the Millennium Development Goals*. Worldwatch Institute. https://core.ac.uk/download/71338837.pdf (accessed on 5 December 2024).
24. Fawzy, S., Osman, A. I., Doran, J., & Rooney, D. W. (2020). Strategies for mitigation of climate change: A review. *18*, 2069–2094. https://doi.org/10.1007/s10311-020-01059-w.
25. Fekih Hassen, W., & Challouf, M. (2024). *Long Short-Term Renewable Energy Sources Prediction for Grid-Management Systems Based on Stacking Ensemble Model*. Energies, *17*(13), 3145. https://doi.org/10.3390/en17133145.
26. Fulton, M., & Capalino, R. (2014). *Investing in the Clean Trillion: Closing the Clean Energy Investment Gap*. Colegio de Enfermeria de Caceres. https://core.ac.uk/download/71365113.pdf (accessed on 5 December 2024).
27. Gallersdörfer, U., Klaaßen, L., & Stoll, C. (2020). *Energy Consumption of Cryptocurrencies Beyond Bitcoin*. Joule, *4*, 1843–1846. https://doi.org/10.1016/j.joule.2020.07.013.
28. Gupta, A. K., Brandt, A. R., Xu, B., et al. (2017). *A Review of Physical Supply and EROI of Fossil Fuels in China*. Springer Science and Business Media LLC. https://core.ac.uk/download/159617810.pdf (accessed on 5 December 2024).
29. Hartungi, R., & Pye, C. J. (2009). *Life Cycle Costing Applications in Sustainable Building*. Retrieved from: https://core.ac.uk/download/42137540.pdf (accessed on 5 December 2024).
30. Hassan, A., Kouhy, R., & Waziri, B. Z. (2018). The effect of transitioning to renewable energy consumption on the Nigerian oil and gas exports: An ARDL approach. Retrieved from: https://core.ac.uk/download/228178449.pdf (accessed on 5 December 2024).

31. Intergovernmental Panel on Climate Change (2023). *Climate Change 2022–Impacts, Adaptation and Vulnerability*. https://doi.org/10.1017/9781009325844.

32. Jamasb, T., Nuttall, W. J., & Pollitt, M. G. (2005). New electricity technologies for a sustainable future. *Faculty of Economics*. Retrieved from: https://core.ac.uk/download/1248971.pdf (accessed on 5 December 2024).

33. Johnstone, P., & Stirling, A. (2015). *Comparing Nuclear Power Trajectories in Germany and the UK: From 'Regimes' to 'Democracies' in Sociotechnical Transitions and Discontinuities*. SWPS 2015-18. https://ssrn.com/abstract=2744549 or http://dx.doi.org/10.2139/ssrn.2744549 (accessed on 5 December 2024).

34. Keatinge, R. (2000). *A Review of Current European Research on Organic Farming (OF0171)*. ADAS Consulting Ltd, Redesdale. https://core.ac.uk/download/10924020.pdf (accessed on 5 December 2024).

35. Konain, M., Janapati, R., Ahmed, S. M., & Ahmed, M. M. (2023). IoT based solar-powered mushroom farming for sustainable agriculture. *Semantics Scholar*, 944–948. Retrieved from: https://www.semanticscholar.org/paper/77407e89b1e526987be361351d0e0c012e478fca (accessed on 5 December 2024).

36. Kramer, M. R., Mahmud, A., & Makka, S. (2010). *Maximizing Impact: An Integrated Strategy for Grantmaking and Mission Investing in Climate Change*. FSG Communications Ltd. https://core.ac.uk/download/71344782.pdf (accessed on 5 December 2024).

37. Losonc, K. (2012). *Climate Change Impacts on Migration and Labour Market*. Editorial Santa Paula SRL. https://core.ac.uk/download/12355563.pdf (accessed on 5 December 2024).

38. Maghear, D. (2011). Romania's energy potential of renewable energies in the context of sustainable development. *Annals of Faculty of Economics, 1*, 176–180.

39. MeYue, M., Lambert, H., Pahon, E., et al. (2021). *Hydrogen Energy Systems: A Critical Review of Technologies, Applications, Trends, And Challenges*. Renewable and Sustainable Energy Reviews, 146(C).

40. Neuhoff, K., & Sellers, R. (2006). *Mainstreaming New Renewable Energy Technologies*. Faculty of Economics. https://core.ac.uk/download/1248990.pdf (accessed on 5 December 2024).

41. Pati, U., & Mistry, K. (2024). *Cyber-Resilient Trading for Sustainable Energy Management: A Three-Phase Demand-Side Management Solution with Integrated Deep Learning-Based Renewable Energy*

Forecasting. *60*, 2532–2541. https://www.semanticscholar.org/paper/eb0eb681e4e6d60874261c8e3b1150e29fc4fb6c (accessed on 5 December 2024).

42. Pomponi, F., Hart, J., Arehart, J., & D'Amico, B. (2020). *Buildings as a Global Carbon Sink? A Reality Check on Feasibility Limits*. One Earth, *3*, 157–161. https://doi.org/10.1016/j.oneear.2020.07.018.
43. Ramos, A., Guarracino, I., Mellor, A., et al. (2017). *Solar-Thermal and Hybrid Photovoltaic-Thermal Systems for Renewable Heating*. https://doi.org/10.13140/RG.2.2.10473.29280.
44. Ramrao, N., Sridharan, S. A., T. R., & Neelakandan, V. P. (2021). Harnessing renewable energy at Kalasalingam Academy of Research and Education–A role model case. *Institute of Research and Community Services Diponegoro University (LPPM UNDIP)*. Retrieved from: https://core.ac.uk/download/478231103.pdf (accessed on 5 December 2024).
45. Saunois, M., Stavert, A. R., Poulter, B., Bousquet, P., Canadell, J. G., Jackson, R. B., et al., (2020). The global methane budget 2000–2017. *Earth System Science Data, 12*(4), 1561–1623. https://doi.org/10.5194/essd-12-1561-2020.
46. Smith, W. A., Burdyny, T., Vermaas, D. A., & Geerlings, H. (2019). *Pathways to Industrial-Scale Fuel Out of Thin Air from CO_2 Electrolysis*. Joule, *3*, 1822–1834. https://doi.org/10.1016/j.joule.2019.07.009.
47. Tolis, A. I., Rentizelas, A. A., & Tatsiopoulos, I. P. (2010). *Optimisation of Electricity Energy Markets and Assessment of CO_2 Trading on Their Structure: A Stochastic Analysis of the Greek Power Sector*. Renewable and Sustainable Energy Reviews, *14*(9), 2529–2546.
48. World Economic Forum. (2013). *The Green Investment Report: The Ways and Means to Unlock Private Finance for Green Growth*. https://core.ac.uk/download/71361031.pdf (accessed on 5 December 2024).
49. Zira, S. (2022). *Towards Sustainable Livestock Systems: Developing and Applying Methods for Broad Sustainability Assessment of Pig and Cattle Systems*. https://core.ac.uk/download/533952134.pdf (accessed on 5 December 2024).

INDEX

C

D

E

F

M

N

O

P

Q

R

S

FSC
www.fsc.org
MIX
Papier | Fördert
gute Waldnutzung
FSC® C083411

Kolibri 360 GmbH
Ferdinand-Jühlke-Straße 7
99095 Erfurt, Deutschland
produktsicherheit@kolibri360.de